World Water Resources

Volume 9

Series Editor

V. P. Singh, Department of Biological and Agricultural Engineering, Texas A&M University, College Station, TX, USA

This series aims to publish books, monographs and contributed volumes on water resources of the world, with particular focus per volume on water resources of a particular country or region. With the freshwater supplies becoming an increasingly important and scarce commodity, it is important to have under one cover up to date literature published on water resources and their management, e.g. lessons learnt or details from one river basin may be quite useful for other basins. Also, it is important that national and international river basins are managed, keeping each country's interest and environment in mind. The need for dialog is being heightened by climate change and global warming. It is hoped that the Series will make a contribution to this dialog. The volumes in the series ideally would follow a "Three Part" approach as outlined below:

In the chapters in the first Part *Sources of Freshwater* would be covered, like water resources of river basins; water resources of lake basins, including surface water and under river flow; groundwater; desalination and snow cover/ice caps. In the second Part the chapters would include topics like: *Water Use and Consumption*, e.g. irrigation, industrial, domestic, recreational etc. In the third Part in different chapters more miscellaneous items can be covered like impacts of anthropogenic effects on water resources; impact of global warning and climate change on water resources; river basin management; river compacts and treaties; lake basin management; national development and water resources management; peace and water resources; economics of water resources development; water resources and civilization; politics and water resources; water-energy-food nexus; water security and sustainability; large water resources projects; ancient water works; and challenges for the future. Authored and edited volumes are welcomed to the series. Editor or co-editors would solicit colleagues to write chapters that make up the edited book. For an edited book, it is anticipated that there would be about 12-15 chapters in a book of about 300 pages. Books in the Series could also be authored by one person or several co-authors without inviting others to prepare separate chapters. The volumes in the Series would tend to follow the "Three Part" approach as outlined above. Topics that are of current interest can be added as well.

Readership

Readers would be university researchers, governmental agencies, NGOs, research institutes, and industry. It is also envisaged that conservation groups and those interested in water resources management would find some of the books of great interest. Comments or suggestions for future volumes are welcomed.

Series Editor:

V. P. Singh, Department of Biological and Agricultural Engineering, Texas A&M University, College Station, Texas, USA.
Email: vsingh@tamu.edu

More information about this series at http://www.springer.com/series/15410

Muhammad Arif Watto • Michael Mitchell
Safdar Bashir
Editors

Water Resources of Pakistan

Issues and Impacts

Springer

Editors
Muhammad Arif Watto
University of Agriculture Faisalabad,
Sub-campus Depalpur
Okara, Pakistan

Michael Mitchell
Institute for Land, Water and Society
Charles Sturt University
Albury, NSW, Australia

Safdar Bashir
Department of Soil and
Environmental Sciences
Ghazi University, D.G Khan
Punjab, Pakistan

ISSN 2509-7385 ISSN 2509-7393 (electronic)
World Water Resources
ISBN 978-3-030-65678-2 ISBN 978-3-030-65679-9 (eBook)
https://doi.org/10.1007/978-3-030-65679-9

This Springer imprint is published by the registered company Springer Nature Switzerland AG
The registered company address is: Gewerbestrasse 11, 6330 Cham, Switzerland

Foreword 1

I am delighted to introduce *Water Resources of Pakistan: Issues and Impacts*, a detailed and contemporary outlook of Pakistan's water resource potential and exploration through the socio-hydrologic and political economic intricacies of how Pakistan's water is managed, which offers ways forward for Pakistan, especially given climate change impacts on water availability, and the need to manage competing demands of water for food, energy and human development.

The seriousness of any country seeking to make progress towards the United Nations Sustainable Development Goals (SDGs) can be assessed through its approach to water resources management. The relevance of the water sector to the SDGs is not only confined to SDG6 (clean water and sanitation) but also for almost all other goals. For instance, water is a major source of livelihood (SDG1). It is essential for food production and food absorption (SDG2). Implementation of health-sector goals (SDG3) depends on clean drinking water and the availability of water for basic hygiene to counter spread of diseases like COVID-19. Innovations and infrastructure (SDG9) directed at maximising access to water for all can reduce the disproportionate burden on women for ensuring household water security (SDG5), as well as the goal of sustainable cities and communities (SDG11), which in turn partly depends on sustainable production and consumption (SDG 12), especially for countries like Pakistan where agriculture consumes around 96% of available water. Water is also a source of renewable energy (SDG7), contributing to decent jobs and economic growth (SDG8) as well as helping to reduce inequality (SDG10) and poverty (SDG1). In the absence of sustainable water resources management, a country cannot have a successful climate action strategy (SDG13), and life below water (SDG14) and on land (SDG15) would continue to be threatened. Issues of water distribution within and across national boundaries can threaten peace and justice (SDG16), and achieving all the aforementioned goals requires strong "water-partnerships" and transdisciplinary collaboration (SDG17).

It is extremely difficult to capture the centrality of water for SDGs, the multifunctionality of water in our day-to-day lives and the fact that Pakistan is bestowed with some of the world's largest glacial reserves on one hand while rapidly turning into a water-scarce country on the other. The beauty of this book is that it covers a threadbare discussion on different aspects of Pakistan's water resources management in a single volume. Each chapter has been written as a stand-alone contribution, making it very reader friendly. Being edited and authored by international and Pakistani academics and experts of Pakistan's water sector ensures its objectivity and currency.

Pakistan is amongst the most water-stressed countries in the world. The available water resources are under huge pressure from a growing population, a wide range of sectoral water demands, climate change, and the continuous degradation of ecosystem services. Inadequate planning and management further question water security in Pakistan.

Pakistan's water challenge is also driven by the extent its water is used unsustainably, especially across the Indus River Basin. This ultimate challenge underpins the overarching goal for water management in the country, as highlighted in the National Water Policy (NWP) of 2018. Yet, implementation of such a policy needs improved understanding of the complex socio-hydrology of Pakistan. This complexity is due to the high level of human–water interactions, extreme cultural and political diversity, complicated transboundary water dynamics, and elevated climate change with associated increased risks from extreme events. Unfortunately, limited efforts have been made to understand Pakistan's complex socio-hydrology and, more importantly, address its evolving and changing dynamics.

Ensuring water security in Pakistan requires an overall assessment of its water availability and quality, and an understanding of human–water interactions through multiple perspectives at multiple scales. Pakistan's water vulnerabilities from climate change manifested as glacial melt and increase in temperatures, and changes in precipitation patterns add further impetus for the country to invest in improving its water resources management. All these climate change vulnerabilities affect upstream hydrology and groundwater recharge, which in turn affect water demands, cropping patterns and food security.

All of the above are covered by this volume. The concept of "crop per drop" is highlighted with a focus on sustaining irrigated agriculture and unlocking economic growth through reform in policies and practices to improve water productivity. Spatio-temporal variability in groundwater storage and the water-energy-food nexus are discussed in detail. The chapters on the historical geography of the Indus Basin and transboundary water conflicts are of special interest for those keen to understand the political sensitivity of water issues in our region.

Finally, the book provides a roadmap for a comprehensive water forecast system, the role research can play in improving water management and pathways for taking the national water policy forward.

I would like to pay personal gratitude to all authors and editors of this volume for their dedicated efforts in bringing out this much-needed volume. The book is an essential read for water professionals, policy makers, media fraternity, students and the general public who would like to better understand Pakistan's water sector issues, challenges and potential solutions.

Abid Qaiyum Suleri
Executive Director
Sustainable Development Policy Institute
Islamabad, Pakistan
Member, Planning Minister's National Advisory Committee
Islamabad, Pakistan
Member, Prime Minister's Economic Advisory Council
Islamabad, Pakistan

Foreword 2

Indus is Pakistan's lifeline. It casts a shadow on Pakistan's relations with her water neighbours (Afghanistan, China and India) and on the inter-provincial relations between the provinces tied together as custodians and beneficiaries of the Indus civilisation, basin and irrigation system. As population and per capita usage grows, competition between different users of water can lead to conflicts unless, of course, foundations are laid for rational and scientific discourse. This volume is a long-awaited endeavour in this regard.

Several challenges of Indus water usage, management and controls are deeply rooted in history, with many of the underlying policies, practices and approaches a continuation of the colonial period. This, in particular, applies to the issues of water rights, where the country's policies for equitable water distribution and universal access to water are based on those developed under colonial rule. Perhaps the most important colonial legacy has been supply-driven water resources management. From historically well-entrenched irrigation departments in the provinces to the recently created federal institution, the Indus River System Authority, institutions in Pakistan have always focused on water distribution and allocation, without paying much attention to changing cropping patterns, precipitation trends and weather patterns, or managing the steadily growing demand. Direct and indirect subsidies have enabled the country's water managers to disproportionately serve the landed elite by ignoring the needs of those at the tail-end of the canal water distributional system, as well as the growing number of industrial, commercial or even domestic users in rural or urban areas. About 5% of Pakistan's water is budgeted for these non-agricultural users. These distortions have retarded Pakistan's economic growth and strained food security efforts.

Left with no option, these new water users were expected to find their own solutions for their growing water needs, with most resorting to groundwater extraction. Even the municipalities started pumping groundwater to supply their growing populations. Groundwater abstraction for agricultural and other uses has also been rooted in colonial period laws that laid the foundation for extractive, arbitrary and abusive relationships with all natural resources, including groundwater, forests, mining and the use of communal lands or *shamlaats* for individual and commercial purposes.

Using the environmental lexicon, this has been an unfolding tragedy of the commons that has accelerated since the Green Revolution.

The supply-side management of Indus resources has not only continued unabated since independence in 1947, it has skewed the country's development in several ways, particularly with numerous direct and indirect grants and support from the state. The long list has included subsidies on seeds, fertilisers and pesticides in addition to the price support system for several crops, subsidised agricultural loans, reduced duties on the import of agricultural technologies and even import-export policies to support the sector. The beneficiaries have predictably been farmers with larger holdings and, very often, those with political access and influence. Agricultural income is not only exempted from income tax, *abiana* is also determined at nominal levels and its collection has been half-hearted to the extent that the gap between management and maintenance costs of the irrigation system and recovery rates has been widening. With water availability virtually free, the interest in effective water management, water conservation and drive for innovation for new technologies has been weak, resulting in high levels of system inefficiencies. The country has become one of the lowest on several indices that measure crop-water ratios, usually known as crop per drop.

An artificial sense of water *insecurity* has often inadvertently been misconstrued as water *scarcity*. It is stunning to reveal that with almost $300 billion invested in the system, only four major crops use 80% of irrigation withdrawals, while only contributing less than 5% to the nation's GDP. Instead of investing in imbibing water security through efficiency and demand management, the same institutions that had adhered to the *grundnorm* of supply side management have promptly committed mammoth resources to build new reservoirs to ensure greater quantities of water supply for agricultural users. Seldom mentioned in the discourse are non-agricultural users, showing a chronic disinterest with equitable distribution, or universal access to water, or conceding additional water allocation to under-served districts in, for example, southern Punjab, southern Sindh, the delta area or the urban users in all five districts of Sindh.

Finally, the supply–driven institutions have become inward looking and nonresponsive to innovation, entrepreneurship, scientific research, new technologies and best practice. Several attempts to test or introduce new technologies failed or had very limited success as the system worked in a closed loop that did not value scientific approaches or research. It is against this background that Pakistan's water sector began to stagnate and failed to keep pace with global trends and new knowledge. The institutions were slow in their uptake of new techniques, technologies or scientific breakthroughs. The authors have underlined that new management methods and tools are ushering changes in science and technology, with potential to improve water productivity, environmental quality and interstate cooperation. Examples abound and many are illustrated in this book. No wonder Pakistan lacks a comprehensive flood prediction and water resource assessment modelling system; has become one of the highest wastewater producers, missing opportunities to adapt water-saving technologies, recycling and reuse of water; and has failed to adopt drainage provisions to avoid land degradation or curtail ground- and surface water contamination. It has thus far ignored regulating groundwater extraction and allowed dependence on groundwater to become so entrenched. Incentives have rarely been designed to work as effective pull factors among institutions and farmers for operational efficiencies and system innovations. It is argued in this volume that existing institutions are not designed but can be equipped to address these sustainability and socio-hydrologic challenges. Three important institutional strengthening requirements are identified in particular: (i) development of a multi-stakeholder institutional framework to focus on both national- and local-scale sustainable water sector planning; (ii) enhancement of existing water management capacities for simultaneously incorporating surface water and groundwater planning and management; and (iii) ensuring urban water management authorities develop robust infrastructure and practices for climate-resilient, sustainable and water-secure management of cities. Given Pakistan's uncertain water future, it is imperative that the country's institutional water framework is revamped with a climate-adaptive vision and capacity for ensuring water sustainability.

Climate Change and Water Specialist
Member, National Climate Change Council
Ministry of Climate Change
Islamabad, Pakistan

Ali Tauqeer Sheikh

Acknowledgements

First, we would like to offer our special thanks to all chapter authors for their wonderful contributions. Thank you everyone! You responded to all the frequent follow-up emails with good humour in spite of your busy schedules. As lead editor, Arif Watto also extends his sincere gratitude to his co-editors Michael Mitchell and Safdar Bashir who devoted valuable time in support of his dream: it's been a very long journey from the inception of the idea to the receipt of final drafts, and Michael Mitchell's enduring empathy for the vision is especially appreciated.

We owe special thanks to James Wescoat, Stephen Davies, William Young and Taimoor Akhtar for their contributions, keen interest and strong encouragement. Their support throughout the compilation of this work also helped make the dream come true.

We are highly indebted to Thomas E. Adams, Fazilda Nabeel, Kamal Ahmed, Ghulam Hussain Dars, Waqas Ahmed, Muhammad Abid, Muhammad Naveed Anjum, Muhammad Jehanzeb Masud Cheema, Daanish Mustafa and Abubakr Muhammad for their kind support and excellent contributions, and to Tariq Aziz for his support and encouragement from time to time during the book's compilation.

We also thank Margaret Deignan, Marielle Klijn, Jill Ritchie and Neelofar Yasmeen for their help in reshaping this book and ensuring an optimal outcome.

Muhammad Arif Watto

Michael Mitchell

Safdar Bashir

Contents

Chapter 1
Pakistan's Water Resources: Overview and Challenges

Muhammad Arif Watto, Michael Mitchell, and Taimoor Akhtar

Abstract Pakistan has a vision to become one of the top ten global economies by the middle of this century, but has to achieve that transition despite being one of the most water-stressed and arid countries in the world. Its water availability goes through extremes from too much to too little water, and climate change is projected to exacerbate these extremes. For decades, the monumental Indus Basin Irrigation System has been a lifeline, allowing Pakistan's agricultural economy to boom. While the system continues to grow, intensification of agriculture has meant surface water supply is being rapidly replaced with groundwater, and Pakistan has now become the fourth largest groundwater withdrawing country in the world. Yet Pakistan is also among the top five wastewater producing countries, with only 1.2% of that wastewater being treated. This chapter introduces the challenges Pakistan faces in achieving a more sustainable use of its water resources, emphasising that many of these challenges require social and institutional change. It then provides an overview of the chapters, showing how each chapter contributes to a deeper understanding of these challenges, as well as offering practical suggestions for how Pakistan's future challenges can be addressed.

Keywords Indus basin · Water outlook · Water security · Water vision · Water challenge

M. A. Watto (✉)
University of Agriculture Faisalabad, Sub-campus Depalpur, Okara, Pakistan
e-mail: arif.watto@uaf.edu.pk

M. Mitchell
Institute for Land, Water and Society, Charles Sturt University, Albury, NSW, Australia
e-mail: mimitchell@csu.edu.au

T. Akhtar
College of Engineering and Physical Sciences, University of Guelph, Guelph, ON, Canada
e-mail: akhtart@uoguelph.ca

M. A. Watto et al. (eds.), *Water Resources of Pakistan*, World Water Resources 9,
https://doi.org/10.1007/978-3-030-65679-9_1

1.1 Overview of Pakistan's Water Resources

Pakistan is a country bestowed with a well-resourced but complex water landscape that is central to the country's social, economic and environmental health. The Indus Basin is the largest in Pakistan, and is riddled with human interventions. These interventions are a consequence of the monumental Indus Basin Irrigation System (IBIS), which continues to grow in stature as water requirements of Pakistan become increasingly critical and intricate. An added layer of complexity is the transboundary nature of Pakistan's water resources, since the Indus Basin is shared between Pakistan (47%), India (39%), China (8%), and Afghanistan (6%).

With a total area of about 1.12 million km^2 the Indus Basin is one of the largest transboundary basins in the world (FAO 2016). In Pakistan, the Indus Basin covers around 0.52 million km^2, or about 65% of the total territory. It comprises the whole of Punjab and Khyber Pakhtunkhwa (KPK) provinces, most of Sindh province and a part of eastern Balochistan. For Pakistan, the Indus River is the lifeline of roughly 180 million people. The hydrology of the Indus Basin is characterised by glaciated mountain valleys, monsoon plains and a deltaic coastline, each of which has a distinct and extensive water management regime. It is therefore crucial to understand the inextricably intertwined water resources for human development in Pakistan (Khan and Adams 2019). Yet understanding, measurement and assessment of Pakistan's total water resources is inadequate due to scarcity in data availability related to climate, surface water flows, groundwater levels, flood prediction, transboundary hydrological information, and an associated lack of a comprehensive water resource modelling system (Young et al. 2019; Khan and Adams 2019).

Pakistan Vision 2025 sets out targets to include Pakistan in the list of top 10 global economies by 2047, the 100th year of its independence. This aspiring milestone cannot be achieved without having sustainable water resource management as the top agenda item for the country's development (Young et al. 2019). Pakistan's scarce water resources are beset by many management challenges that can undermine the country's development targets, including low water productivity, overdrafting and degradation of water quality. Currently, Pakistan has 2.83% of the global population (Worldometer 2020) but has only 0.45% of the global renewable water resources; i.e. 0.037% of per capita renewable water resources (FAO 2016). Pakistan is not only one of the world's most populous countries but also one of the most arid countries, with an average rainfall less than 240 mm a year (Briscoe et al. 2005).

Pakistan's water supplies mainly come from the Indus River and its tributaries (the Indus, Chenab, Jhelum, Ravi, Beas and Sutlej Rivers). In a recent report Young et al. (2019) has made a careful assessment of Pakistan's total water resources based on extant studies and a range of data. They indicate that Pakistan's current total average annual renewable water resources are 226.4 billion cubic metres (BCM). The Indus Basin contributes 96.3% to the total system, the Kharan desert 1.3% and the Makran coastal drainage 2.4%. Based on data from Young et al. (2019) and others we have presented Pakistan's overall water outlook in Fig. 1.1.

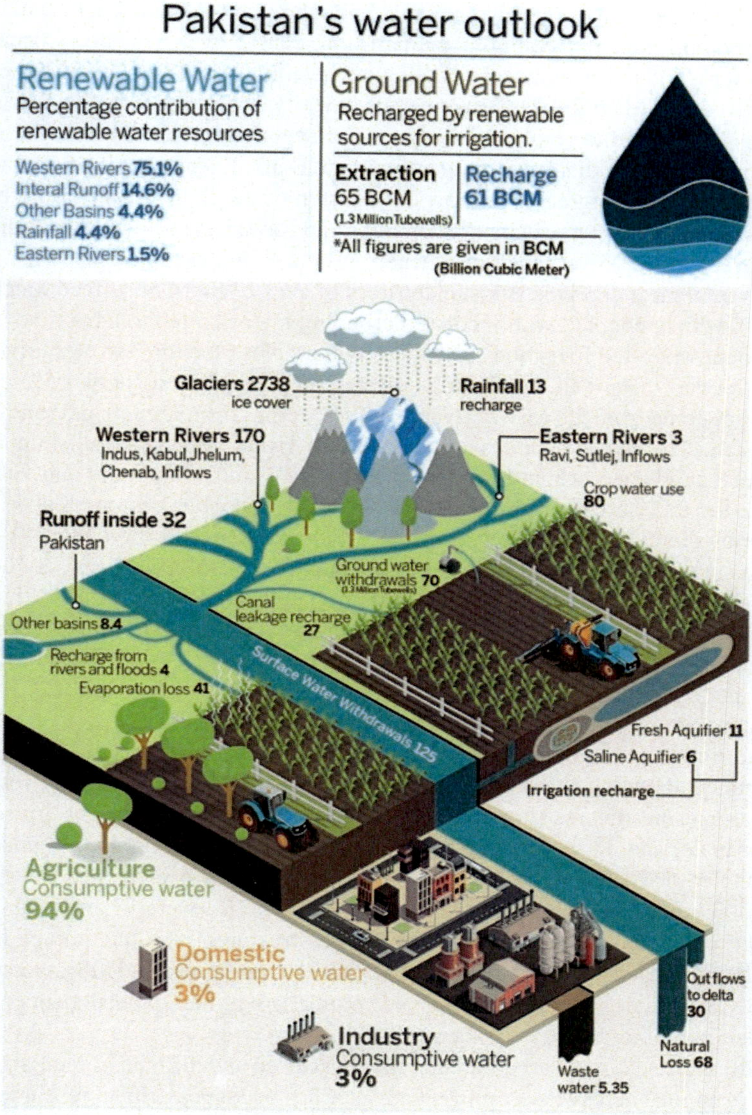

Fig. 1.1 Pakistan water outlook. (The numbers provided in this figure are derived from Young et al. (2019), Qureshi et al. (2010) and Ministry of Finance (2019))

Pakistan has a rich and complex water resource history dating back well before independence in 1947 (Wescoat and Muhammad 2019; Wescoat et al. 2000). The complexity of Pakistan's water resources lies in its dynamic hydrological processes of the upper basin ice melt and snowmelt, and variable monsoon rains and complex groundwater flux in the plains. The partition of the subcontinent between India and

Pakistan in 1947 also created a major division through the Indus Basin, and through its irrigation system. Soon after the partition, claims over river flows became the basis for a wider transboundary conflict between India and Pakistan (Michel 1967; Barrett 1994; Alam 2002). Realising the severity of the conflict, the World Bank offered mediation to resolve the dispute and negotiations started in 1951 between the three parties: India, Pakistan and the World Bank. Eventually, in 1960, India and Pakistan mutually agreed over how the division of the Indus Basin would be managed. The three eastern rivers (Sutlej, Beas and Ravi) were allocated to India while the three western rivers (Jhelum, Chenab and Indus) to Pakistan. The division of the Indus tributaries deprived Pakistan territory of over 19 BCM of surface water (Amir 2005), which, coupled with increased cropping intensity, led to a massive increase in groundwater-fed irrigation across the country. Now, the three western rivers contribute over 75% of the total surface water supplies whereas only 1.5% of water supplies come from the eastern rivers. The rest comes from rainfall and runoff inside Pakistan. The water supplies of the western rivers depend heavily on western Himalayan glaciers which form part of the world's most important and vulnerable water tower reservoirs (Immerzeel et al. 2020). Pakistan has cumulative glacial cover of about 2,738 BCM ice reserves, which is a source of more than 40–50% of downstream surface water demands (Mustafa et al. 2013). Different studies have modelled that snow and ice melt contribute between 50% and 80% of the Indus River's water flow (Bookhagen and Burbank 2010; Shrestha et al. 2015; Koppes et al. 2015).

A large proportion of Pakistan's surface flows (>60%; and around 125 BCM, on average) are diverted for agricultural, domestic and industrial use (see Fig. 1.1). These flows are diverted via the Indus Basin Irrigation System (IBIS). IBIS is a monumental hydraulic asset of Pakistan that includes three major storage reservoirs, 18 barrages/headworks (low-head diversion dams), 12 inter-river links (also called link canals) and 44 major irrigation canal networks (spanning 44,000 kilometres) (Young et al. 2019). While the investment cost of IBIS is estimated to be around US$300 billion, the efficiency of the massive system is severely hampered by inadequate maintenance (especially of hydraulic structures), limited storage capacity and low regulatory control (Young et al. 2019). Consequently, IBIS diversions are supply-driven, i.e., supplies (timing and volume) are primarily dependent on run-off availability rather than system demands.

Given the seasonal variability and highly skewed and difficult to regulate surface water supplies across Pakistan, groundwater has become a critical resource (Watto and Mugera 2016; Young et al. 2019). Currently, Pakistan is the world's fourth-largest groundwater withdrawing country (after China, India, and the USA) – a source of 50–60% of irrigation, 90% of domestic and almost 100% of industrial water supplies (Qureshi and Ashraf 2019). The Indus Basin is a flat unconsolidated region of alluvial deposits, covering 16 million hectares, underlain by an extensive, contiguous and unconfined aquifer (Qureshi et al. 2008; Young et al. 2019). Across Pakistan, natural variations in alluvial deposits, climate, and the pattern of irrigation channels influence the groundwater typology. For instance, in Balochistan, groundwater is mainly confined in consolidated sedimentary landforms (Young et al. 2019).

Pakistan's renewable groundwater resources are replenished in multiple ways, including recharge from rainfall, river flows, floods, seepage from canals, and irrigation return flows. However, exact recharge from these sources and ultimately an overall assessment of the country's renewable groundwater resource is difficult to make, with estimates varying. Briscoe et al. (2005) estimated that Pakistan's annual renewable groundwater resources are 63 BCM, FAO (2016) estimated internally generated groundwater resources as being 55 BCM, while Young et al. (2019) have reported Pakistan's renewable groundwater resources as 74.2 BCM. All three report that groundwater withdrawals have exceeded the corresponding recharge rates with varying estimates. Hassan (2016) reported a net depletion of 7 BCM in the annual groundwater budget of Pakistan. Since 1960, the groundwater share in irrigation water supplies has increased from a mere 8% to over 50% in Punjab province (Qureshi et al. 2010).

Currently, there are over 1.39 million irrigation tube-wells in the country with 1.09 million in Punjab, 0.24 million in Sindh, 0.047 in Balochistan and 0.021 million in Khyber Pakhtunkhwa (Ministry of Finance 2019; Bureau of Statistics 2019, and see Fig. 1.2). These tube-wells irrigate 3.57 million hectares as a single source

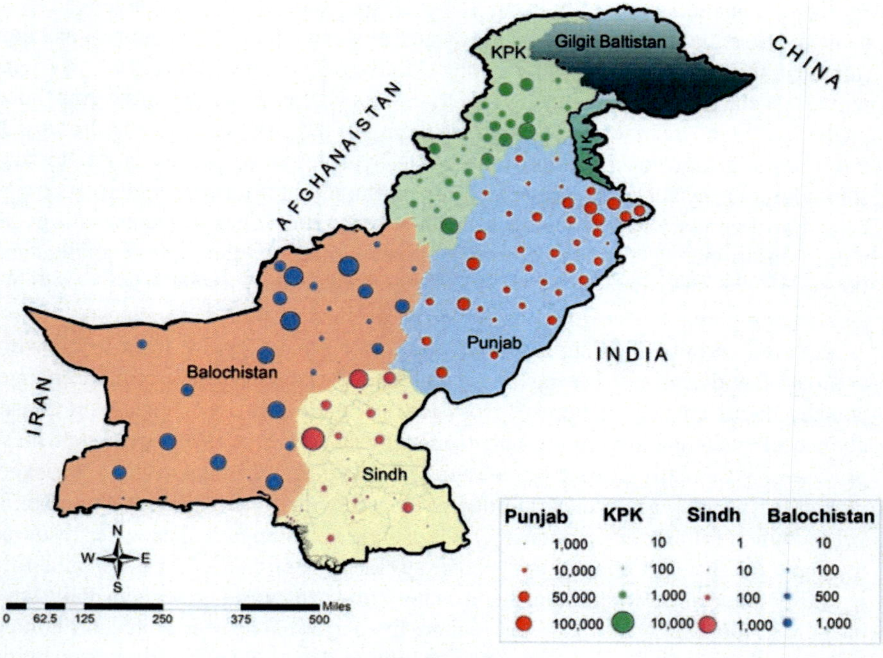

Fig. 1.2 District wise tube-well numbers in different provinces of Pakistan. (Sources: Crop Reporting Services (2018), Bureau of Statistics Punjab (2019), Directorate of Crop Reporting Services (2019) and Sindh Bureau of Statistics (2017). Note: The number of tube-wells is given as a total number in each district of each province except for Sindh province where data is only available on the number of tube-wells added between the years 2003–2004 and 2016–2017)

of irrigation and more than 8 million hectares in conjunction with canal water (Ministry of Finance 2020).

Pakistan does not only have the world's largest contiguous irrigation system and is among the largest groundwater users, it is also among the top five wastewater producing countries. Pakistan treats only 1.2% of its overall wastewater and irrigates about 2.9 million hectares of land with untreated water (Thebo et al. 2017). Based on 2010 estimates, Pakistan produces some 4.5 BCM of wastewater (Qureshi et al. 2010) which is over half of the active storage capacity of the proposed Kalabagh Dam. Based on these 2010 estimates of wastewater production, we extrapolate that under the business as usual scenario the total wastewater production has reached 5.35 BCM across the country in 2020. With only 1.2% of wastewater being treated, 5.28 BCM of wastewater finds its way into freshwater bodies such as rivers and canals, and ultimately into groundwater aquifers.

1.2 Pakistan – Water Challenges

Pakistan's ultimate water challenge is the unsustainable way in which its water is used and managed, especially in the Indus Basin. Addressing this challenge represents the essential overarching goal for water management in the country – as highlighted in the 2018 National Water Policy (Ministry of Water Resources 2018) – with numerous challenges embedded within. A critical challenge among these embedded challenges is to advance existing understanding of the complex socio-hydrology of Pakistan. The high level of Pakistan's socio-hydrology complexity is due to substantial human-water interactions, extreme cultural and political diversity, complicated transboundary water dynamics and elevated climate change and extreme event risks. Unfortunately, limited efforts have been made in the past to understand Pakistan's complex socio-hydrology, and more importantly, to address its evolving and changing dynamics (Siddiqi et al. 2018; Wescoat et al. 2018).

Pakistan's available water resources are under great pressure from its growing population and associated competing inter-sectoral water demands, climate change, and the incessant degradation of ecosystem services being worsened as a consequence of inadequate planning and management. While it is apparent that these pressures will greatly impact water supplies, it is difficult to determine their exact impacts, especially given uncertainties associated with climate change and other as yet unknown future developments. Obviously this challenge makes water resource management difficult in an already water stressed country where water availability is highly skewed between "too much water" and "too little water". In Pakistan's backdrop "too little water" can incapacitate the performance of irrigated agriculture and "too much water" can cause huge economic damage, such as the devastating floods of 2010, which caused an estimated US$10 billion of economic damage (Khan and Adams 2019).

Another critical socio-hydrologic water challenge faced by Pakistan is how to improve its institutional water management framework. The country derives maximum water value from the Indus, with over 90% of its surface flows diverted for agricultural production, as well as serving its domestic and industrial water needs (Hassan 2016). On top of this, approximately 30% of Pakistan's energy generation comes from hydropower dams located in the Upper Indus Basin (Young et al. 2019). Strategies to manage these profound human-water interactions across the Indus Basin of Pakistan are further complicated by extreme variability in surface flows, and thus how to manage extreme drought and flood events (Laghari et al. 2012; Young et al. 2019), especially given that climate change will increase the frequency and intensity of such events (Lutz et al. 2016). Understanding the socio-hydrology of Pakistan is thus paramount to enable its water resources to be managed for sustainability, and to ensure water security of the country's inhabitants now and in the future.

The Indus Basin Irrigation System (IBIS) has historically been the epicentre of Pakistan's water sector, with irrigation and energy demands (and to some extent, flood protection) being the primary forces behind water decision-making. This has driven a managerial focus throughout the evolution of Pakistan's water management framework and authorities on infrastructure to increase irrigation supplies and hydropower generation, and to ensure irrigation delivery to farmers. The most well-established public water authorities of the country are focused on infrastructure development, such as the Water and Power Development Authority (WAPDA), or water delivery, such as the Indus River System Authority (IRSA) and the country's provincial irrigation departments. While the needs for sustainable water management and management of socio-hydrologic dynamics are understood at the policy level in Pakistan (Ministry of Water Resources 2018), existing institutional arrangements are not aligned or equipped to adequately address these sustainability and socio-hydrologic challenges. According to Young et al. (2019), three important institutional strengthening requirements in this regard are:

(i) Developing a multi-stakeholder institutional framework that focuses on both national and local scale sustainable water sector planning.
(ii) Enhancing existing capacities of water management authorities for simultaneously incorporating surface and groundwater planning and management.
(iii) Ensuring urban water management authorities develop robust infrastructure and practices for climate resilient, sustainable and water secure management of cities.

To bolster such strengthening of institutional arrangements, a water sector decision support framework is required that can effectively, rapidly and continuously integrate data, technology and modelling. The importance of integrating data with modelling to achieve environmental sustainability is well documented and well established, both within Pakistan (Ministry of Water Resources 2018; Young et al. 2019) and globally (Liu et al. 2015). While efforts have been made in Pakistan to strengthen hydro-meteorological monitoring and water resources modelling (Young et al. 2019), these have not been sufficiently persistent to ensure continued and successful integration across the water sector. To successfully integrate data collection, modelling and technology-induced knowledge into the water sector in the future, novel strategies, such as public-private technology partnerships and enhancing research-industry synergies, need to be incorporated and persistently pursued.

As noted, all these water management challenges in Pakistan are being further exacerbated from how climate change is impacting the country's water resources. Recent events and future projections demonstrate that climate change is having and will continue to have devastating impacts on Pakistan's water resources in numerous and varied ways (Hussain et al. 2019). Changes in weather patterns (Lutz et al. 2016) and glacial melting is significantly altering upstream hydrology and groundwater recharge with consequent effects on cropping patterns, agricultural and urban water demands and food security. Changes in weather patterns are also projected to increase the frequency and extent of extreme events (Lutz et al. 2019). Given that the country is faced with an extremely uncertain water future, it is even more imperative that Pakistan's institutional water framework is revamped with a climate-adaptive vision and capacity to seek water sustainability, including through the incorporation of inclusive and data-driven management strategies.

1.3 Book Overview and Key Messages

This book provides further detail on the water challenges outlined above and explores a range of strategies to address them. Each chapter has been written as a stand-alone contribution with contributors all holding status as leading international scholars specialising in water-related aspects of Pakistan.

Chapter 2 by **Wescoat et al.** offers a fascinating history into the geography of the Indus Basin, its links with the identity of Pakistan as a nation, and the legacy of past events on how the Indus Basin is currently perceived and managed. The assessment of the past directs readers to the future, and the prospect of a fresh policy approach that sees the basin as a mosaic of gardens, with the gardeners across the complex geopolitical arrangements that make up Pakistan's share of the Indus Basin working together for the common good in a spirit of cooperation, sharing, compassion, and, as and when required, sacrifice.

In Chap. 3, **Nabeel and Cheema** add to this history by exploring how Pakistan as a nation sits alongside the other three riparian nations that share the Indus Basin. Pakistan's position is predominantly dependent and vulnerable, with legal and institutional mechanisms to manage the transboundary water challenges the country faces being inadequate at best. In a similar vein to Wescoat et al., Nabeel and Cheema also call for a more positive approach – one that explores how to share the benefits to, from and beyond the Indus River in a sustainable way. This could replace the current more divisive approach through which shares of Indus Basin water are determined.

In Chap. 4, **Ahktar et al.** provide a helpful overview of the water security situation across Pakistan, with an emphasis on understanding its social implications. They achieve this by drilling down from the nation-wide scale to examine perspectives from a set of sub-national case studies: South Punjab, Karachi and Southern

Sindh. Each case study reveals gross injustices in accessing good quality water, with unsustainable groundwater extraction exacerbating this in South Punjab, inadequate governance and monitoring arrangements magnifying inequality of access among domestic water users in Karachi, and dramatically reduced river flows in the Indus Delta rendering much of Southern Sindh adversely affected by seawater intrusion.

Chapters 5 and 6 examine how climate change is affecting water availability and management in Pakistan. **Anjum et al.** analyse existing historical climate data from 1975 to 2014 to demonstrate the climate has already changed, and is already impacting Pakistan's water availability. They document changes in temperature and precipitation in the Hindu-Kush, Karakoram and Himalayan mountains of Pakistan, part of the afore-mentioned water tower reservoir that is a critical source for Pakistan's surface water supply. Trends towards an increase in temperatures and a decrease in precipitation reduce the storage capacity and will change volume and timing of downstream flow regimes.

In Chap. 6, **Dars et al.** review published studies into climate change trends and analyse future projections based on down-scaled global climate models to reveal how global warming is affecting Pakistan as a whole. They reveal a clear overall warming trend, a trend towards more erratic precipitation patterns, and an associated increase in extreme flood and drought events. They also confirm a significant decrease in glacial coverage over Pakistan's mountainous areas, and the projection that seawater rise will exacerbate the negative impacts of seawater intrusion into the Indus delta. They conclude with an appeal for action, with adaptive actions tailored to the different needs of different places.

Chapter 7 by **Davies and Young** continue this theme by exploring computer-generated models of economic performance according to a range of scenarios for irrigation water use out to 2025. Scenarios varied according to two different rates of economic growth and climate warming, with added scenarios of agricultural reforms, diet shifts and an increase in environmental flows to the delta. Each scenario offers its own conclusions, but the most significant is that policy reforms to reduce water used for irrigation would have the most beneficial long-lasting effect, and that such reforms need to be part of long-term strategic planning: planning that includes how to manage such a transformation for the benefit of Pakistan's economy and environment as a whole.

Chapters 8 and 9 use two different frameworks to analyse the effectiveness of Pakistan's irrigation water management. **Abid et al.** focus on Punjab to analyse its irrigation water management effectiveness in terms of sustainability. The framework they use is based on an established set of indicators and sub-indicators related to risk, environmental integrity and economic equity and acceptability. Their analysis suggests irrigation water management in Punjab is unsustainable and argue that a more comprehensive and inclusive water management approach is needed that can consider the interest of all key stakeholders.

In Chap. 9, **Wescoat et al.** explore how the Water-Energy-Food nexus framework can be used to improve understanding of how these resources are inter-linked and where the gaps may lie. The framework draws on a range of perspectives, levels of analysis and geographic scales such that problems are identified and understood in their historical, theoretical and pragmatic contexts. Pragmatically, a major hurdle to improve coordination along the nexus is that decisions are spread across multiple government departments that rarely connect. Such fragmentation requires a transformational change in approach which incentivises coordination and enables social learning.

Chapters 10 and 11 focus on groundwater. **Nabeel's** chapter challenges the widely held understanding that groundwater over-extraction results from there being no governance arrangements in place to control how groundwater is used. In contrast, Nabeel, using Punjab as a case study, provides a strong case that shows a long history of governance arrangements in place that have encouraged groundwater use, with policies and practices in place driven by a pro-development ethos. Her analysis through colonial, post-colonial and contemporary times demonstrates little focus on nurturing sustainable use, but rather driven by a policy aimed at establishing Punjab's current water intensive export economy.

In Chap. 11, **Ahmed et al.** show how satellite data can be used to offer an analysis of groundwater availability that compensates for a lack of directly derived groundwater data. Their analysis explores changes in groundwater levels over time and across space, and shows massive variability across both. They provide visuals and analysis that clearly show seasonal variations in groundwater levels, as well as broad-scale spatial variations, and a clear overall decline in groundwater levels for much of Pakistan for the period between 2003 and 2016.

Chapter 12 by **Bashir et al.** review the substantial amount of research in Pakistan related to water quality and its impacts on human health. Their extensive review guides the reader through the range of pathogens and toxic metals that characterise key pollutants affecting Pakistan's water quality, where these are sourced from, their indicative impacts, and the extent of monitoring in place. The chapter concludes with a brief history of environmental regulations in Pakistan with recommendations for how these could be more effectively implemented through improved planning, investment and innovative practice.

The remaining three chapters offer suggestions to improve Pakistan's future water management from the perspectives of research, forecasting and policy. In Chap. 13, **Mitchell et al.** evaluate their engagement in a four-year research project to advocate research that engages and builds the capacity of water managers as research users, especially in ways that enables such management agencies to build connections with and engage groundwater users. Their evaluation offers suggestions for how to improve such an interventionist and participatory approach to research. Their recommended use of social-ecological systems analysis links well with the recommendations of several other book chapters in this volume.

Adams addresses a practical matter in Chap. 14: how to establish an effective forecasting system that can assist with flood prediction and management, improve management of water use over time for different purposes of hydropower and irrigation, and establish a comprehensive database and modelling framework to improve decision-making. Adams assesses the advantages and disadvantages of different modelling design elements and strategies, and offers a practical set of recommendations.

Finally, in Chap. 15, **Davies et al.** summarise the challenges for water security in Pakistan, and how these could be addressed by building on the gains recently made through the National Water Policy, and especially through a re-think of how policy delivery can become broader as well as more deep-rooted through a re-evaluation of governance and institutional arrangements. Investments to improve monitoring, modelling and management need to be embedded in governance arrangements that are effective and fit for purpose. Of particular water security concern is how to improve groundwater management and the devastated environment of the Indus Basin below the last barrage.

References

Alam UZ (2002) Questioning the water wars rationale: a case study of the Indus Waters Treaty. Geogr J 168(4):341–353. https://doi.org/10.1111/j.0016-7398.2002.00060.x

Amir P (2005) The role of large dams in the Indus Basin. In: Briscoe J, Qamar U (eds) Pakistan water economy running dry – background papers. Oxford University Press, Oxford

Barrett S (1994) Conflict and cooperation in managing international water resources. Policy Working Paper 1303. World Bank, Washington, DC

Bookhagen B, Burbank DW (2010) Toward a complete Himalayan hydrological budget: spatiotemporal distribution of snowmelt and rainfall and their impact on river discharge. J Geophys Res Earth 115:F03019. https://doi.org/10.1029/2009JF001426

Briscoe J, Qamar U, Contijoch M et al (2005) Pakistan's water economy: running dry. World Bank, Washington, DC

Bureau of Statistics Punjab (2019) Punjab development statistics 2019. Bureau of Statistics, Planning and Development Department, Government of Punjab. http://bos.gop.pk/developmentstat

Crop Reporting Services (2018) Agricultural statistics of Balochistan 2018–2019. Crop Reporting Services, Agricultural and Cooperatives Department, Government of Balochistan

Directorate of Crop Reporting Services (2019) District wise tractor and tube wells data of Khyber Pakhtunkhwa. Directorate of Crop Reporting Services, Government of Khyber Pakhtunkhwa. http://crs.kp.gov.pk/page/district_wise_tractor_and_tube_wells_data_of_khyber_pakhtunkhwa

FAO (2016) AQUASTAT main database. Food and Agriculture Organization of United Nations. http://www.fao.org/nr/water/aquastat/data/query/results.html. Accessed 1 July 2020

Hassan M (2016) Water security in Pakistan: issues and challenges. UN Develop Prog Pak 3(4):1–34

Hussain M, Butt AR, Uzma F et al (2019) A comprehensive review of climate change impacts, adaptation, and mitigation on environmental and natural calamities in Pakistan. Environ Monit Assess 192(1):48. https://doi.org/10.1007/s10661-019-7956-4

Immerzeel WW, Lutz AF, Andrade M et al (2020) Importance and vulnerability of the world's water towers. Nature 577:364–379. https://doi.org/10.1038/s41586-019-1822-y

Khan SI, Adams TE (eds) (2019) Indus River Basin: water security and sustainability. Elsevier, New York

Koppes M, Rupper S, Asay M, Winter-Billington A (2015) Sensitivity of glacier runoff projections to baseline climate data in the Indus River basin. Front Earth Sci 3:59. https://doi.org/10.3389/feart.2015.00059

Laghari AN, Vanham D, Rauch W (2012) The Indus Basin in the framework of current and future water resources management. Hydrol Earth Syst Sci 16(4):1063–1083. https://doi.org/10.5194/hess-16-1063-2012

Liu J, Mooney H, Hull V, Davis SJ et al (2015) Systems integration for global sustainability. Science 347(6225):1258832. https://doi.org/10.1126/science.1258832

Lutz AF, Immerzeel WW, Kraaijenbrink PDA et al (2016) Climate change impacts on the upper Indus hydrology: sources, shifts and extremes. PLoS One 11(11):1–33. https://doi.org/10.1371/journal.pone.0165630

Lutz AF, ter Maat HW, Wijngaard RR et al (2019) South Asian river basins in a 1.5 °C warmer world. Reg Environ Chang 19(3):833–847. https://doi.org/10.1007/s10113-018-1433-4

Michel AA (1967) The Indus waters: a study of the effects of partition. Yale University Press, New Haven

Ministry of Finance (2019) Economic survey of Pakistan 2018–19. Finance Division, Ministry of Finance, Revenue and Economic Affairs, Government of Pakistan. http://www.finance.gov.pk/survey_1819.html

Ministry of Finance (2020) Economic survey of Pakistan 2019–20. Finance Division, Ministry of Finance, Revenue and Economic Affairs, Government of Pakistan. http://www.finance.gov.pk/survey_1920.html

Ministry of Water Resources (2018) National Water Policy. https://mowr.gov.pk/index.php/national-water-policy-2018/

Mustafa D, Akhter M, Nasrallah N (2013) Understanding Pakistan's water-security nexus. US Institute of Peace, Washington, DC

Qureshi R, Ashraf M (2019) Water security issues of agriculture in Pakistan. Pakistan Academy of Sciences (PAS), Islamabad. https://www.paspk.org/wp-content/uploads/2019/06/PAS-Water-Security-Issues.pdf

Qureshi AS, McCornick PG, Qadir M, Aslam Z (2008) Managing salinity and waterlogging in the Indus Basin of Pakistan. Agric Water Manag 95(1):1–10. https://doi.org/10.1016/j.agwat.2007.09.014

Qureshi AS, McCornick PG, Sarwar A, Sharma BR (2010) Challenges and prospects of sustainable groundwater management in the Indus Basin, Pakistan. Water Resour Manag 24(8):1551–1569. https://doi.org/10.1007/s11269-009-9513-3

Shrestha M, Koike T, Hirabayashi Y et al (2015) Integrated simulation of snow and glacier melt in water and energy balance-based, distributed hydrological modeling framework at Hunza River Basin of Pakistan Karakoram region. J Geophys Res-Atmos 120(10):4889–4919. https://doi.org/10.1002/2014JD022666

Siddiqi A, Wescoat J, Muhammad A (2018) Socio-hydrological assessment of water security in canal irrigation systems: a conjoint quantitative analysis of equity and reliability. Water Security 4(5):44–55. https://doi.org/10.1016/j.wasec.2018.11.001

Sindh Bureau of Statistics (2017) Development statistics of Sindh 2017. Bureau of Statistics, Planning and Development Department, Government of Sindh. http://sindhbos.gov.pk/category/publications/

Thebo AL, Drechsel P, Lambin E, Nelson K (2017) A global, spatially-explicit assessment of irrigated croplands influenced by urban wastewater flows. Environ Res Lett 12:074008. https://doi.org/10.1088/1748-9326/aa75d1

Watto MA, Mugera AW (2016) Groundwater depletion in the Indus Plains of Pakistan: imperatives, repercussions and management issues. Int J River Basin Manag 14(4):447–458. https://doi.org/10.1080/15715124.2016.1204154

Wescoat JL Jr, Muhammad A (2019) Irrigated gardens of the Indus River Basin: new models for water resource management. In: De Wolff K, Faletti R, López-Calvo I (eds) Water and the humanities: transforming currents for uncertain futures. University of California Press, Oakland

Wescoat JL, Halvorson SJ, Mustafa D (2000) Water management in the Indus Basin of Pakistan: a half-century perspective. Int J Water Resour Dev 16(3):391–406. https://doi.org/10.1080/713672507

Wescoat JL, Siddiqi A, Muhammad A (2018) Socio-hydrology of channel flows in complex river basins: rivers, canals, and distributaries in Punjab, Pakistan. Water Resour Res 54(1):464–479. https://doi.org/10.1002/2017WR021486

Worldometer (2020) Countries in world by population (2020). https://www.worldometers.info/world-population/population-by-country/. Accessed 15 July 2020

Young W, Anwar A, Bhatti T et al (2019) Pakistan: getting more from water. World Bank, Washington, DC. https://openknowledge.worldbank.org/handle/10986/31160

Chapter 2
Pakistan's Water Resources: From Retrospect to Prospect

James L. Wescoat Jr., Abubakr Muhammad, and Afreen Siddiqi

Abstract This chapter introduces the subject of Pakistan's water resources with a retrospective view, beginning with pre-independence water issues in the Sutlej River valley that brought the governments of India, Punjab, Sindh, and the princely states of Bahawalpur and Bikaner together in regional negotiations. Concurrent nationalist movements placed limited emphasis on water resources and the Indus Basin at that time. After partition, however, protracted water negotiations in the context of wider boundary uncertainties led to the Indus Waters Treaty of 1960. The late-twentieth century witnessed water transfers that redistributed water from Pakistan's western rivers to eastern river canal commands to compensate for reduced flows from India, which was followed by a period of unprecedented private tube-well development and associated problems. The massive 2010 flood marked another turning point in Indus Basin consciousness. New management methods and tools are ushering in changes in science and technology, though their impact on water productivity, environmental quality, and interstate cooperation remains uncertain. We conclude therefore with reflections on emerging challenges and a new vision of the Indus Basin as a garden.

Keywords Socio-hydrology · Canal irrigation · Tube-well irrigation · Indus Basin · Gardens

J. L. Wescoat Jr. (✉)
School of Architecture + Planning, MIT, Cambridge, MA, USA
e-mail: wescoat@mit.edu

A. Muhammad
Center for Water Informatics & Technology, LUMS, Lahore, Pakistan
e-mail: abubakr@lums.edu.pk

A. Siddiqi
Department of Aeronautics and Astronautics, MIT, Cambridge, MA, USA
e-mail: siddiqi@mit.edu

M. A. Watto et al. (eds.), *Water Resources of Pakistan*, World Water Resources 9,
https://doi.org/10.1007/978-3-030-65679-9_2

2.1 Introduction to Pakistan's Water Resources

Pakistan's water resources have a rich and complex history that, depending on how one frames it, dates back well before independence in 1947 to antecedent water disputes between Punjab, Sindh, and the princely states, and to much earlier eras of Indus Basin development (Wescoat 1999; Wescoat and Muhammad 2021). The complexity of Pakistan's water resources arises in part from dynamic processes of upper basin ice melt and snowmelt, and in part from variable monsoon rains and complex groundwater flux on the plains. Socio-economic and political struggles over water, land, and related resources have shaped Indus Basin irrigation systems and settlement patterns (Akhter 2015; Gilmartin 2014; Mustafa 2010). Evolving historical conceptions of Pakistan have further complicated the assessment of its water resources, and for that reason, this chapter is subtitled from retrospect to prospect. Uncertainties abound about the magnitude of both water flows and uses, which make a socio-hydrologic approach relevant for understanding Pakistan's water issues (Siddiqi et al. 2018a, b; Wescoat 2013; Wescoat et al. 2018).

Socio-hydrology treats social and hydrologic processes jointly, and in a historical context (Sivapalan et al. 2012, 2014). To that, we add a geographical dimension, as Pakistan's water resources have evolved as much through space as through time (Haines 2017; Michel 1967). The Greek philosopher Heraclitus wrote that, "Into the same rivers we step and do not step, we are and are not," in part because rivers always change, but also because the person and in this case the country of Pakistan are also always changing (Graham 2015, p. 8, quoting fragment B49a). In this chapter, we consider how Pakistan has changed in social, hydrologic, and socio-hydrologic terms over the past century, in ways that have profound implications for water management in the coming century.

A historical perspective on Pakistan's water resources could focus on an apparent overall shift from surplus to scarcity, but that would be misleading. One can certainly appreciate concerns about scarcity, as Pakistan's population is growing rapidly relative to its hydroclimatic water budget (Khan and Adams 2019; Young et al. 2019). Estimates vary, but most agree that Pakistan is rapidly approaching the gross water stress threshold of 1000 m^3 of water per capita per annum (e.g., Young et al. (2019) provide an estimate of 1100 m^3/cap). However, finer-grain assessments of surplus and scarcity are not so clear (Akhter 2017). Each period in the history of Indus Basin management has recorded acute periods of both water scarcity and intense surpluses, including the massive flood of 2010 (Panhwar 2010). Such surpluses of rainfall, runoff, and waterlogging have proven as hazardous for human settlement as scarcity. It is erroneous to regard discharge of river runoff into the delta as a surplus, and even worse as "waste". These flood flows help build and sustain floodplains and deltaic lobes (Meadows and Meadows 1999). While there are rapidly increasing population pressures on Pakistan's water resources, enough water exists in aggregate for basic human water and environmental needs, if managed wisely and fairly. By contrast, unregulated water extraction, wastewater discharge, and environmental degradation are unsustainable and unjust.

This chapter reviews the socio-hydrologic record of water surpluses, scarcity, development, degradation, and management in Pakistan, and it culminates in a socio-hydrologic vision of the Indus Basin as a "garden". The historical roots of this story have much greater depth than can be considered here, evolving from the Harappan era to those of Sultanate, Mughal, and regional kingdoms. This chapter begins instead in the relatively recent past of the early twentieth century, with competing development proposals in the Sutlej and Indus River basins by the governments of India, Punjab, Sindh, and the princely states of Bahawalpur and Bikaner. In parallel with this complex political geography of water management, there arose calls for "Pakistan" in the 1930s, the boundaries of which were at that point still unclear. The ambiguous relationship between struggles for water and territorial independence continued through Independence in 1947 and continue to shape transboundary disputes.

The 1930s and 1940s section of this chapter addresses formative conceptions of "Pakistan". Many sources have debated the political and territorial dimensions of the Pakistan movement, but few of them gave much attention to gaps and linkages with parallel Indus River water issues (Michel 1967). Jawaharlal Nehru referred to the "story of Ganga as the story of India" (Haines 2017, p. 21). Who spoke for the Indus? How was the "story of the Indus" related to the "story of Pakistan?" Did the Indus and its tributaries have salience in the ideological development of Pakistan? While the answers to these questions deserve greater archival investigation, we offer some preliminary responses here.

The third section of the chapter focuses on water resource traumas and uncertainties from the moment of Independence up to the signing of the Indus Waters Treaty from 1947 to 1960. The story of India's canal closure on April 1, 1948, and the aftermath of distrust and intra-basin transfers have been told so often that we treat it briefly here. In its place, we survey the territorial boundaries of Pakistan as they related to Indus Basin water resources at that time. On each of Pakistan's borders – east, north, and west – political boundaries and water entitlements were indefinite and tenuous in ways that have had persistent implications for water management.

Part four charts out a pair of radical transformations in Pakistan's water system that followed the Indus Waters Treaty of 1960. The first involved dramatic expansion of canal irrigation via transfers of water from the western rivers allocated to Pakistan (Indus, Jhelum, and Chenab) through link canals to canal commands of the rapidly diminishing eastern rivers awarded to India (Ravi, Beas, and Sutlej). Link canals enabled the expansion of the area under irrigated food production. Concurrently, in the 1960s the government of Pakistan installed large public tube-wells to manage waterlogging and salinity. Much more dramatic, however, was the subsequent revolution in private tube-well development (over a million in Punjab alone), which now withdraws more water for irrigation than is delivered by canal flows. Canal irrigation has been transitioning from its primary function of water delivery to one of groundwater recharge through seepage and field infiltration. Nearby tube-wells tap these seepage waters, and in some areas increase canal losses. Failure to regulate groundwater pumping and salinity have been the twin hazards of irrigation agriculture in the twentieth and early twenty-first centuries.

It is worth asking how important irrigation agriculture will be in Pakistan in the coming century. As its overall contribution to the national GDP declines and the urban population grows, the case for irrigation subsidies and crop price supports weakens. As in other irrigated regions of the world, such policies stress food security, along with the cultural value and virtues of agrarian livelihoods, but it is not clear how those values will fare in future visions and plans for the Indus Basin in Pakistan. What difference will these trends in socio-hydrologic processes, policies, and cultural values make for Pakistan's water resources? What difference can new historical insights and management visions make for Pakistan's water future?

We chose 2010 as the timestep for the penultimate section of the paper on new water management technologies and policies, due to the 2010 Indus mega-flood that year and its association with hydroclimatic variability and trends. In subsequent years, drought concerns have risen again. As other chapters in this book address flood and drought issues, we concentrate in this chapter on emerging developments in the science and technology of water management in what Daanish Mustafa has called the "hydro-hazardscape" of climate variability and change. These developments include new technologies for water measurement, modelling, precision cultivation, and decision support. We briefly discuss innovations in measurement and modelling of socio-hydrologic processes in the context of climate change, which will require these innovations and more.

The concluding section of the chapter connects these scientific, technological, and policy advances with a philosophical vision for water management in Pakistan that conceives of the Indus Basin as a garden as well as a complex socio-hydrologic system. This garden philosophy takes its guidance and direction from the cultural heritage of gardens in Islam and the history of Pakistan (Wescoat 2012). This vision is gaining traction on the ground through the invention and adoption of nature-inspired cultivation practices that produce greater crop yields with fewer water and fertiliser inputs – and it is thus an emerging approach for linking retrospect with prospect.

2.2 Pakistan's Water Origins in the Middle Indus Valley

It is interesting to study how modern consciousness of Indus Basin water management has developed, and how that story relates to the formation of the new nation of Pakistan. Several principles help define what we mean by a modern consciousness of Indus Basin water management. First, it relates to an area that extends from the glaciated headwaters to the deltaic shelf. Second, it takes into account the geographic linkages as well as the tensions between upstream and downstream provinces and nation-states. Third, it encompasses the massive canal irrigation network, extensive groundwater resources, and variable climates of the basin as part of a single complex socio-hydrologic system. And finally, it embraces the strong record of innovation in water and related land management in the Indus. The consciousness of these key concepts has developed slowly over time, as will be shown below. Early

Timurid geographies like the *Zafarnama* briefly describe the Indus and its tributaries from Tibet down to Thatta. Even when apprehended in principle, the broader consciousness of Indus Basin water management has proven difficult to realise in practice, though that is the prospect we seek.

One place and time to explore how the consciousness of Indus Basin water management developed in limited but important ways involves the Sutlej River valley of southern Punjab, from the 1920s onwards. The Sutlej is the longest tributary of the Indus, rising in western Tibet and flowing through the Himalayan mountains and the Sivalik foothills in India where it would later be impounded behind Bhakra dam. The Sutlej debouches onto the Punjab plains, flowing along the boundary between Jalandhar and Ludhiana districts toward the former princely state of Kapurthala where it receives the waters of the Beas River on its right bank. It constitutes the boundary between Firozpur and Tarn Taran districts, and between Firozpur and Lahore districts, in what would later become an international boundary between India and Pakistan. Within Pakistan, the river establishes the boundary between Sahiwal and Multan districts on the right bank and the former princely state of Bahawalpur on the left bank. It becomes the Panjnad River when it is joined by the Jhelum and its upstream waters from the Chenab and Ravi rivers. These combined tributaries then enter the Indus River main stem near Mithankot, and about 100 km downstream it flows into the province of Sindh.

The middle and lower Sutlej valley had large inundation canals that irrigated its broad floodplains from antiquity. By one account, proposals for Sutlej irrigation development to establish perennial canals dated back to 1854, and only came to fruition after a 1918 technical agreement in Delhi (Government of Punjab, Public Works Department 1920, p. 1). Bahawalpur and Bikaner states negotiated vigorously on their behalf (Government of India 1935). The Sutlej Valley Project (SVP) transformed the inundation channels into perennial canals, which involved major diversion barrages and headworks – Panjnad, Islam, Suleimanki, and Ferozpur – as well as long-term planning for the Bhakra Dam upstream. These impoundments and diversions generated serious concerns downstream in Khairpur state and Sindh, where the Government of Bombay had its major development plans underway for Sukkur Barrage to support cotton production (Haines 2013). Concerns between Punjab and Bahawalpur, and between them and Sindh, led to the establishment of the Anderson Committee to ascertain the adequacy of river flows and formulation of discharge measurement methods (Government of India 1935).

While these water tensions among diverse governments were growing, so too were nationalist movements in India and ideas about the new nation-state of Pakistan. An early formative document for the creation of Pakistan was Allama Iqbal's presidential speech at the All India Muslim League conference of 1930 in Allahabad (Iqbal 1944a, p. 3). In that well-known speech, Iqbal declared that *"I would like to see the Punjab, North-West Frontier Province, Sindh, and Baluchistan* amalgamated *into a single State. Self-government within the British Empire, or without the British Empire, the formation of a consolidated North-West Indian Muslim State appears to me to be the final destiny of the Muslims, at least of North-West India"* [italics in the original]. A Muslim majority population, along with the political and cultural needs of the Muslim community, provided the rationale for this territorial argument. However, Iqbal did not link his territorial argument with the physical geography of the region or refer to the Indus or its water resources. Indeed, in other places, Iqbal argued for a Pan-Islamic vision of Indian Muslims unbounded by territory (Zaman 2018, p. 36).

In a speech earlier that year, Iqbal (1944b, p. 87) responded to criticism of a report that attributed a fiscal deficit in Punjab to Indus and Jhelum flood damages and crop failures by arguing that the real fiscal problem lay in high costs of colonial administration that should be reduced or eliminated. "There IS no other alternative!" Contrasting increased costs of flood damages and jails he stated, "Well, floods are a natural phenomenon and cannot be prevented, but unless we are complete fatalists, crime is a preventable affair" (ibid., p. 87). This is one of the occasional references to water in the collected *Speeches, Writings, and Statements of Iqbal.*

One gains deeper insights from Iqbal's poetic and philosophical writings. The *Kinar-e-Ravi* ((کنار راوی، بانگ درا (اردو)) poem evokes the flow of human existence and experience along the "Bank of the Ravi" between Lahore Fort on one side and minarets of Jahangir's tomb on the other (Iqbal 2017). In keeping with the traditions of Persian and Urdu poetry, Iqbal used the imagery of the river in his deepest poetic works. In his landmark Persian work *Payam-e-Mashriq* (پیام مشرق) (Message from the East), Iqbal freely renders Goethe's poem *Mahomet's Gesang,* in praise of the Prophet of Islam,[1] as *Ju-e-Aab*[2] (Stream of Water), invoking the imagery of the prophetic mission as a stream as it is joined by other streams (representing other prophets) and eventually falling into the ocean of divinity (Iqbal 2010). The other prominent reference is in Iqbal's *Javed Nama*[3] (Book of Eternity), arguably the finest poetical work by Iqbal, in which his entire philosophy is presented as he journeys to the heavens (Iqbal 2007). Most interestingly for this chapter, he adopts the pseudonym *Zinda Rud*[4] (Sentient River) for himself in this journey. We may ask, how might the Indus be, or become, likened to a sentient river?

Finally, in *The Reconstruction of Religious Thought in Islam,* which is Iqbal's magnum opus in prose (Iqbal 1989, p. 42), he wrote that:

> "Even our acts of perception are determined by our immediate interests and purposes. The Persian poet 'Urfī has given beautiful expression to this aspect of human perception. He says[5]:
>
> 'If your heart is not deceived by the mirage, be not proud of the sharpness of your understanding; for your freedom from this optical illusion is due to your imperfect thirst.'
>
> The poet means to say that if you had a vehement desire for drink, the sands of the desert would have given you the impression of a lake. Your freedom from the illusion is due to the

---

[1] محمد صلی اللہ علیہ وسلم

[2] جوئے آب۔

[3] جاوید نامہ (فارسی)

[4] رود -

[5] ز نقص تشنہ لبی دان بعقل خویش مناز
دلت فریب گر از جلوۂ سراب نخورد

absence of a keen desire for water. You have perceived the thing as it is because you were not interested in perceiving it as it is not."

Throughout this paper, we ask when in history water managers in Pakistan appear to have suffered from illusion, when they appear to have suffered from a lack of thirst, and when inspired visions of a living river basin appear to have been crafted by those with an utmost desire for the well-being of its people.

The next touchstone in the creation of Pakistan was Choudhary Rahmat Ali's (1933) *Now or Never: Are we to Live or Perish Forever?* which called for "Pakistan" to be comprised of Punjab, Sindh, Kashmir, Baluchistan, and the North West Frontier Province. It stressed the identity and needs of Muslim communities in different regions of India, each of which had its name (e.g., Bangistan and Osmanistan), in addition to Pakistan (Fig. 2.1). It does not mention the Indus or any aspect of its land or water resources (though Ali (1947, 224) later criticised the word "Indus" as an Hellenic concept). As discussed above, there was no lack of regional water development projects and tensions in the lands that would become Pakistan well before independence. From the early 1920s, disputes arose on the Sutlej between

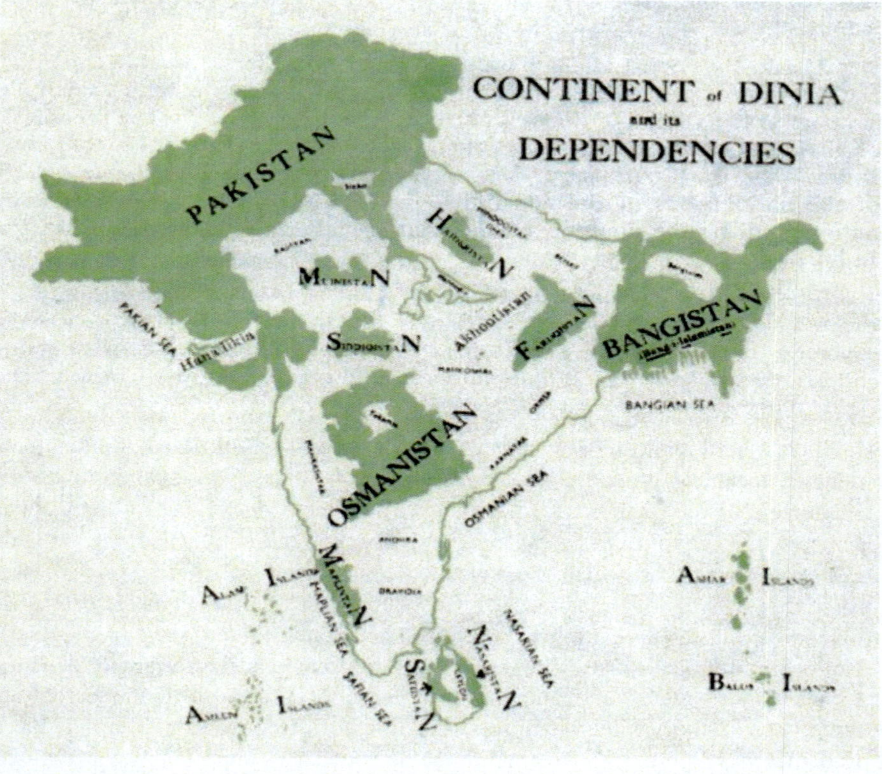

Fig. 2.1 A pre-Independence map of Choudhary Rahmat Ali from pamphlets that coin the name "Pakistan". (Source: https://en.wikipedia.org/wiki/File:MAPOFRAHMATPLAN.jpg)

Bahawalpur state, Bikaner state (a non-riparian), and Punjab state that led to a Tripartite Agreement in 1921. In the same year, the Government of Bombay objected to the proposed upstream developments on behalf of Sindh, which led to the formation of the Indus Discharge Committee that initiated continuous flow measurement from 1922 onwards. A decade later, construction of Sutlej Valley Project canals and barrages, and approval of Bhakra Dam upstream, threatened to exceed annual discharge, which led to the creation of the SVP Committee in 1932, followed by the Anderson Committee in 1935, the same year that Sindh became a province separate from Bombay. These events drew close attention from administrators at the time, and have figured prominently in subsequent histories of Indus Basin development as well, so much so that we will not recite them here (e.g., Alam 1998; Ali 1988; Haines 2017; Gilmartin 2015; Michel 1967). It is worth noting that 1930 was also the publication year of the first doctoral dissertation on canal irrigation in Punjab in the United States, underscoring the perceived international significance of irrigation in the region (Paustian 1930). The key point here is that they do not appear to have figured prominently in the most pivotal nation-building efforts of leading political and religious figures in the 1930s.

It is important to consider how long this apparent gap between water development and emergent nationalism persisted in the 1940s. Muhammad Ali Jinnah (1940) made his Presidential Speech in the Lahore Resolution calling for Pakistan and stating, "…that geographically contiguous units are demarcated into regions which should be constituted, with such territorial readjustments as may be necessary that the areas in which the Muslims are numerically in a majority as in the North Western and Eastern Zones of (British) India should be grouped to constitute 'independent States' in which the constituent units should be autonomous and sovereign." While the Lahore Resolution did not mention water issues, subsequent work by the Muslim League Planning Committee authorised by Jinnah did so, along with other substantive economic development considerations (Zaidi 2004, pp. 708–746).

On the ground, water disputes between Sindh and Punjab escalated during the 1940s. The Rau Commission created in 1941 produced a report the following year on proposed water allocations that Punjab and Sindh both rejected. Reports and rejoinders between those governments continued on legal, financial, political, and detailed technical grounds (e.g., Government of Sindh 1944). Gilmartin (2015) sheds light on the socio-political and territorial dimensions of these water disputes between the province of Sindh, on one hand, and the entrenched yet parochial Punjab Unionist Party (that the Muslim League later challenged), on the other. Michel (1967, p. 132) reports that while Punjab and Sindh engineers drafted an agreement in 1945, it was eclipsed by political and territorial negotiations. Nationalist political and religious leaders, on the other hand, seem somewhat distant from concrete water issues in the run-up to partition, perhaps due to the secrecy and rapidity of territorial delimitation by the Radcliffe Commission (Chester 2009). Radcliffe was reportedly concerned about the boundary's impact on water resources (e.g., to ensure that the Suleimanki headworks lay in Pakistan) (Michel 1967, p. 177).

2.3 The Indefinite Indus Basin of Pakistan – August 1947 to September 1960

It is important to reflect upon the geographical associations between the Indus Basin and Pakistan. A large proportion of the basin has rough conformity with the territory of Pakistan, and a massive program of Indus Basin development took shape after the Indus Waters Treaty of 1960. Events between 1947 and 1960 have pivotal importance in constructing these relationships between the water resources and territory of Pakistan.

The historiography of Indus Waters Treaty antecedents and negotiations is rich in documentary detail and interpretation. It includes Michel's (1967) monograph on *The Indus Rivers: A Study of the Effects of Partition,* Gulhati's (1973) perspective as a negotiator in the process, a doctoral dissertation with World Bank primary sources by Undala Alam (1998), a volume with chapters on the political history of the Indus by Gilmartin (2015), and scores of other books and scholarly journal articles (e.g., Haines 2017; Hussain 2017). The literature on the partition is even more vast.

With this large body of research in mind, we briefly outline some key points before opening up the discussion in a broader way. First, the implications of partition for irrigation management were broadly anticipated in 1947 but not examined in depth at that time. Engineers assumed the "standstill" arrangement would maintain the status quo of canal deliveries pending more detailed negotiations. When cooperative management was suggested, Nehru and Jinnah both took strong territorial positions on the claims of their future states, perhaps because the process was so rushed (with only roughly a month allocated to delineate the boundary) (Chester 2009; Talbot and Singh 2009, p. 45). Earlier debates over equitable apportionment between Punjab and Sindh in the Anderson and Rau reports were far more rigorous, as would be the subsequent Indus Waters Treaty negotiations.

India's closure of flows into Pakistan from three major canals on April 1, 1948, the day after the completion of the Arbitral Tribunal, came as a shock (Michel 1967, pp. 195 ff.). Although flows resumed a month later in May, along with an Inter-Dominion Agreement, the unilateral cut-off triggered expressions of existential anxiety in Pakistan. Pakistan responded in part by digging a new alignment of the Main Branch Lower (MBL) along its new boundary with India. A subsequent canal shut-off in 1951 by the Government of East Punjab, acting on its own without approval from Prime Minister Nehru, underscored the two levels of federal and state vulnerability, and the inability of the former to control the latter.

Former TVA director David Lilienthal's visit to the countries and his influential *Colliers* magazine article in 1951 titled "Another Korea in the Making?" renewed proposals for cooperative water management. This, in turn, helped spark the commitment of the World Bank to mediate international negotiations between India and Pakistan that became the basis for the Indus Waters Treaty (IWT) of 1960 (Michel 1967). Although the IWT partitioned, rather than integrated, the Indus and its tributaries, it set new standards for water treaty-making and new legal and institutional foundations for multilateral investment in developing the "Indus Basin of Pakistan" as an emerging concept. Detailed scholarly histories of these events are widely available even if they are not widely known.

With this brief overview of the extensive literature covering IWT negotiations, we now take a step back to reflect on the historical geographic relationships between Indus watershed boundaries and the territorial boundaries of Pakistan. During the formative period of 1947–1960, the borders of the new nation of Pakistan were uncertain on all fronts. We begin with the western border with Afghanistan, which was delineated in principle by the Durand Line in 1893. It represented an enormously wide borderland as a simple line that cut through Pushto and Baluch speaking regions of what is now Afghanistan and Pakistan. It comes as little surprise that this border has remained highly porous and contested for more than a century. Afghanistan formally renounced the Durand Line boundary in 1949. Most important for our purposes, the Durand Line cut across the middle of the Kabul Basin, leaving the upper basin in Afghanistan and the lower basin in what would later become Pakistan. Aside from a superficial one-page and long-obsolete agreement between the King of Afghanistan and the British Crown in 1924, no water treaty addresses the waters of the Kabul River tributary of the Indus. In the absence of upstream water development, Pakistan developed Warsak Dam and canals (including civil canals) to meet water demand in the North West Frontier Province downstream. Recently, however, Afghanistan and international donors have proposed a cascade of dams on the upper Kabul (e.g., World Bank 2010). Fortunately, there have been even more recent bilateral meetings on Kabul River management. But overall, the legacy of the Durand Line created a large indefinite water and territorial border on the western side of the country.

On the eastern side, the partition on August 14, 1947, established a hurried boundary line between India and Pakistan that split East from West Punjab and separated Sindh from Gujarat. While more definite than the Durand Line, the Radcliffe Award was not announced until days after Independence, which contributed to the chaos. Operation and inspection of canal headworks remained unclear for engineers, security forces, and irrigators. Unilateral canal closure by India on April 1, 1948, made it clear that relations between East and West Punjab would be at least as fraught with uncertainties as the decades of colonial conflict between Punjab and Sindh.

Even less clear are the northern basin and territorial boundaries that involve the upper Chenab, Jhelum, and Indus watersheds. In October 1947, chiefs in areas of Muzaffarabad, Mirpur, and Poonch districts declared control over the territory of Azad Jammu and Kashmir (AJK), which borders on Jammu and Kashmir District controlled by India, all of which remain disputed. AJK is the territory in which Pakistan built its first major storage reservoir behind Mangla Dam on the Jhelum River in the 1960s.

The Karachi Agreement of 1949 acknowledged the AJK territory, and it also acknowledged the territories of Gilgit and Baltistan as the Federally Administered Northern Areas, which drain the Upper Indus and its tributaries and that border on the Kashmir and Ladakh districts controlled by India. The Simla Agreement that ended the 1972 war between India and Pakistan reinforced a ceasefire line as a "line of control" that endures to this day notwithstanding frequent violations. Where the Line of Control ends at the Siachen Glacier, whose accelerating melting drains into a tributary of the Indus, territorial conflict and hydrologic uncertainties have also escalated. And lastly, in the upper basin, India and China fought a war in 1962 in part over the disputed territories of Aksai Chin and the Shaksgam Tract that drain into the Upper Indus main stem.

This survey of boundary water issues, sparked by the more closely studied implications of the partition of Punjab, is completed in the lower basin by the dispute between India and Pakistan over Sir Creek, which flows east of the Indus Delta, and which also has ramifications for the marine boundary between the two countries. This dispute dates back to a treaty provision between the Government of Bombay and the Princely State of Kutch in 1914. Negotiations over Sir Creek date from the 1960s to the present.

Thus, the water resources of Pakistan, from the time of its founding as a new nation in 1947 to the present have been ambiguous on *every* border. This situation amplifies uncertainties in Indus Basin water management. Territorial conflicts and ambiguities diminish the prospects for collaboration on hydrologic measurement let alone water resources management. Transboundary collaboration is challenging in the best of circumstances. At the same time, it may be precisely these uncertainties on all borders that enable one to imagine Pakistan as having extensions into, as well as interests in, the extended Indus Basin that shapes the imagined space known as the "Indus Basin of Pakistan."

2.4 From State Canals to Private Tube-Wells – 1960 to 2010

The Indus Basin of Pakistan is shaped not only by its land surface but also by its subsurface aquifers, tributaries, and canals. The extensive alluvial aquifers provide natural storage that human activity and agricultural enterprise increasingly tap for irrigation and drinking water needs. Groundwater use in the region dates back to the Harappan civilisation. Some historians suggest that "brick-lined wells were a Harappan invention," with "vertical water-supply systems being virtually unknown in contemporary Egyptian and Mesopotamian cities" (Fagan 2011, p. 207). The bricks supported earthen walls, prevented erosion, and filtered the water while keeping out silt and other contaminants. Mohenjodaro was served by an estimated 700 wells, with each well serving an average service area of 1326 m^2 (about one-third of an acre). In subsequent eras, wells remained the primary source of water for domestic use, animal use, and small-scale drinking water for villages, animals, and irrigation with Persian wheels on the *doab* and *bar* lands. The technology of well-digging recorded in the nineteenth century involved laying courses of brick upon a wooden ring. The digger excavated below the ring, which sank into the ground by virtue of its weight (Crooke 1989).

Massive canal irrigation eclipsed traditional dug wells and Persian wheels in the late-19th and early-20th centuries, and helped establish the Indus as the largest contiguously irrigated region in the world. The equity and efficiency of canal deliveries have historically been low. While collective action has been documented in some tail-end and tank irrigation commands (Wade 2007), this phenomenon appears to have been limited in the former canal colonies (Mustafa 2010).

The development of low-cost pumps dramatically increased groundwater withdrawals from the mid-twentieth century onwards. This major transition to accelerated groundwater withdrawals occurred during the 1960s with the Salinity Control and Reclamation Projects (SCARP) in the plains of Punjab and Sindh. Up to 20,000 of these government-run tube-wells were installed to alleviate waterlogging and salinity problems. In areas with good groundwater quality, pumped water initially served as a supplemental supply for irrigation. The results and impacts of SCARP public wells were mixed, however, with some areas recovering productivity while other areas suffered from salinisation of soils, groundwater, and well maintenance problems. However, the government-run tube-well program was eclipsed by a rising tide of private tube-wells starting in the 1970s. Agricultural expansion since the mid-1970s is also credited with the increased use of tube-wells (van Steenbergen and Gohar 2005). As the number of tube-wells increased over time, in some districts more than 18 tube-wells per acre, some non-tube-well owners purchased well water from their neighbours. Data on tube-well installations from the 1960s to the present shows a consistently rapid rise, along with some notable jumps in years of drought (Fig. 2.2).

In the early years of tube-well development, the adoption of electric pumps was encouraged and incentivised by the government. In Punjab and Sindh, electricity charges were subsidised up to 40% for private tube-well use (van Steenbergen and Gohar 2005).

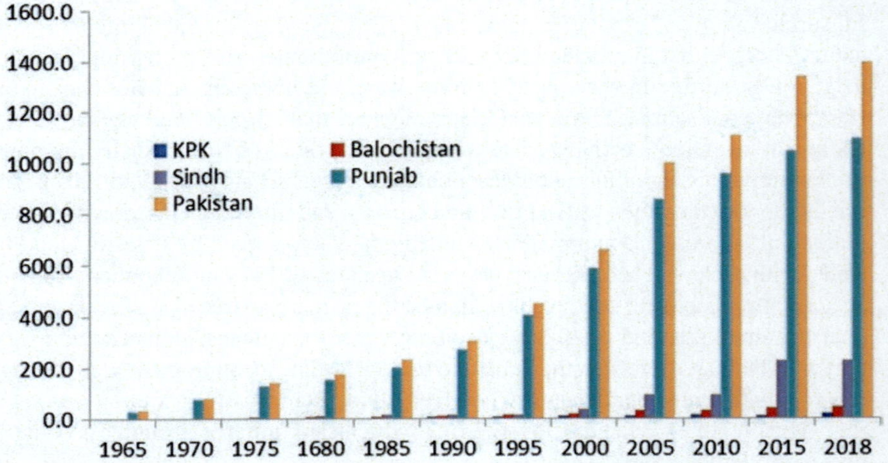

Fig. 2.2 Number of tube-wells in provinces of Pakistan, 1965–2018. (Source: Pakistan Bureau of Statistics 2018)

However, the availability of low cost locally manufactured diesel pump-sets accelerated the adoption of diesel-powered systems. The current pumping system in Punjab, with reportedly over a million tube-wells, is primarily powered by diesel fuel (Fig. 2.2).

Groundwater pumping places significant energy demands on the system that was historically not a concern in the gravity-fed canal-based irrigation system in the Indus plains (Siddiqi and Wescoat 2013). However, the pumping system provides water on-demand to farmers who no longer have to rely on semi-fixed timings and volumes of canal deliveries. Recent studies have shown that the major drivers of groundwater use are the variability and uncertainty of surface irrigation water deliveries (Mekkonen et al. 2016). The increased reliability of supply obtained through groundwater pumping consumes up to 20% of total primary energy use in Punjab (Siddiqi and Wescoat 2013).

This partially conjunctive irrigation system in Pakistan emerged from early efforts for salinity control and land reclamation that over time have transitioned through private tube-wells to become a key enabler of agricultural expansion and productivity enhancement. However, in this latter phase, the system is only partially conjunctive, as there are no regulations on the quantity or timing of withdrawals, or common problems like well interference, water table decline, subsidence, and source water protection. Groundwater economist Tushaar Shah (2008) has called this situation that has spread across South Asia "anarchic". Unregulated tube-well pumping follows a primitive "rule of capture" derived from wildlife law in which the principle was: if you can get it, you can keep it. It leads to declining water table levels, accelerated subsidence, and profound water injustice between those who have and do not have ever-deepening tube-wells. Whether these trends and problems of groundwater withdrawal can be sustained into the future remains a vitally important question. In the present age, as climate change and upstream control of river inflows alter surface water availability in the plains, preliminary studies indicate that tube-well installations are likely to continue growing in Punjab (Siddiqi et al. 2018a, b). So far, the importance of groundwater in Pakistan's Indus Basin has been acknowledged, accounted, and documented – but not managed. What is needed is a transition to a new phase where smart and strategic management is operationalised for a fully conjunctive water use system, i.e., in ways that are socially just as well as efficient. Towards this end, a socio-hydrological framing of the system may provide useful insights.

Socio-hydrological analysis requires that one, "…understand which way the water is flowing and why this is so" (Sivapalan et al. 2012). In hydrology, flows follow the gradient of potential energy, but in socio-hydrology flows follow human actions that include pumping water against the pull of gravity. Groundwater pumping is thus especially well-suited for socio-hydrological analysis, and the Indus Basin can serve as a particularly useful case for examining fundamental theory building, as well as practical application. Up until now, groundwater use in Pakistan has been examined within the disciplinary approaches of hydrogeology, applied economics, and institutional analysis (Qureshi et al. 2003; Mekonnen et al. 2016). While there have been efforts in integrated groundwater modelling and agro-economic modelling in the Indus (Khan et al. 2017), new studies that focus on joint consideration of social and hydrological drivers can point to new questions and yield new insights. For instance, the analysis of surface water resource flows across boundaries was expounded in the previous section. A related question concerns the flux of regional *subsurface* resources. The boundaries and connectivity of regional aquifers are not well understood. Data availability, integration, and modelling have been limited. As pumping activities continue to rise on both sides of the international political boundary, it is crucial to understand subsurface hydrology and to devise social policies for just and sustainable use. Related questions include: (i) investigating the effects of groundwater quality (including arsenic) on crop yields and human health; (ii) understanding crop yield trade-offs between surface and groundwater based irrigation; (iii) identifying canal operations that improve groundwater recharge while also supplying surface water for irrigation; and (iv) using high-efficiency pumps with metered pumping and tier-based water tariffs to balance recharge and withdrawals.

In summary, the late twentieth century marked a massive transition from canal irrigation to tube-well pumping for agricultural and municipal and industrial supplies. By the end of the century, the majority of water use came from groundwater. While initially beneficial in waterlogged areas, and for improved timing of water deliveries in less reliably served areas of deficit irrigation, unregulated tube-well withdrawals have posed new problems for irrigation agriculture in the basin. As municipal and industrial water withdrawals increase, coordination between sectors will become more pressing. Municipal tube-wells outcompete agricultural ones, but municipal and industrial wastewater discharge is also contaminating riparian and urban aquifers for future use. These groundwater problems are not unique to the Indus plains of Pakistan. One finds them in many, if not most, large irrigated regions worldwide.

An important dynamic in the Indus plains involves the massive delivery of canal water along with groundwater irrigation. In a period when many water managers in Pakistan have called for increased dam and reservoir construction for water storage upstream, with slow and limited results, it began to be realised that the snow and ice fields are the great Himalayan water towers of the Upper Indus Basin while the surface water distribution system functions as a source of groundwater recharge and water spreading, as well as surface water delivery. Had this not occurred, groundwater depletion would be even faster and deeper. These dynamics lead one to realise that the great storage process and indirect storage investments in Pakistan during the late-twentieth century have been through canal water seepage, rather than through dams. Managing that stored seepage water requires expenditure for pumping, and more rigorous water quality analysis (e.g., of total dissolved solids). Focusing on balancing groundwater recharge, pumping, and salt budgets could go a long way toward addressing Pakistan's storage needs.

2.5 The Great Flood and Back to the Future (2010–2020)

The great flood of 2010 was a watershed moment in Pakistan's history of water resources management (e.g., Deen 2015; Fair et al. 2017). No other recent event has shaken the public consciousness of water, and attitudes of Pakistan's water managers, more than this catastrophe. Unrelenting monsoon rains flooded almost one-fifth of Pakistan's area, affecting close to 20 million lives directly. Many parts of Pakistan remained underwater for months even after the rains subsided, and the economic losses were estimated to be more than $10 billion or 6% of that year's GDP (Hashmi et al. 2012). In this section, we assess how perspectives on water management in the three decades leading up to the flood took new directions in the years afterward, discovering new strengths, weaknesses, and opportunities.

The period following completion of Indus Waters Treaty infrastructure and leading up to the great flood (1976–2010) is widely recognised as one of decline and neglect in water resources management (Briscoe et al. 2005), though there were important planning efforts (e.g., the Water Sector Investment Planning Study of 1990). Not only did institutions and infrastructure related to water services including flood control *bunds* deteriorate, calls for fresh perspectives on water management (e.g., Faruqui 2004) failed to generate public awareness or political consensus. World Bank-funded irrigation reforms of the 1990s did not succeed in catalysing the desired participatory approaches towards water management (Ul Hassan 2009). The Water Accord of 1991 among the provinces was not able to stop political opposition to new proposals for irrigation diversions or remove mistrust between upstream and downstream users (Anwar and Bhatti 2017). Unexpected dividends from private groundwater development in the irrigated parts of the basin, described in the previous section, only delayed the crisis by a couple of decades. The deterioration of surface water irrigation services reached a point of crisis in the 2000s. Unease in the minds of some keen water practitioners was crystallised in gloomy reports, the most influential of which was titled *Pakistan's Water Economy: Running Dry* (Briscoe et al. 2005), which included predictions that Pakistan would run out of the water by a date in the not too distant future. Climate change science began to yield contradictory yet worrisome scenarios for the sources of water in the Upper Indus Basin (Akhtar et al. 2008) and agricultural production on the irrigated plains (Iqbal and Arif 2010). Drought-like conditions during the 1990s and 2000s seemed to align with these assessments. The slow dynamics of diminishing supplies and increasing demands had limited salience in the year-to-year planning of government ministries. Amidst political turmoil within the country, post-9/11 events in Pakistan's neighbourhood, and public attention focused on a crippling energy crisis, water issues went into the backdrop with other crucial development challenges such as healthcare and education. Life for marginalised communities in the coastal areas, arid regions, mountainous areas, and tail-ends of canals on the irrigated plains, for the poor and women, became more challenging with every year (Mustafa 2010). The fight for finite and diminishing water resources often led to elite-capture at every level and compromised operations for water managers (Rinaudo 2002).

During this time, successive governments occasionally proclaimed their resolve to solve the crisis by building more infrastructure (Faruqui 2004). But the singular focus on reservoirs failed to gain political consensus and left other practicable measures out of the debate. Attempts to reduce mistrust by scientific efforts also did not succeed as advanced technologies such as an expensive river telemetry system were brought to their knees by political conflict and mismanagement (Yousafzai 2017). The water crisis was real but could perhaps be left for the next government to address, or postponed until some imagined political consensus was reached. It was in this numbed background that the 2010 mega-flood caught the country and shook it deeply.

During the post-flood decade, the strongest realisation amongst water practitioners and other stakeholders of the Indus Basin has been that old ways have not worked and there have to be new out-of-the-box approaches (Mustafa 2013). A critical analysis of the flood and its preceding years has led many to question the capture of water discourse in Pakistan by the obsession with conventional infrastructure construction. On one level, recognition of the Water-Energy-Food nexus rose among thought leaders and leading water practitioners of Pakistan (e.g., Yang et al. 2016). It asserts that water issues are inseparable from issues related to energy, land use, climate change, food security, and healthcare. On another level, it is also increasingly recognised that, while vital, agriculture is not the only sector with future water needs. With the rapid urbanisation of an industrialising society, Pakistan's future water profile will become more diverse and more complex (Amir and Habib 2015; Young et al. 2019). It will require increases in agricultural water efficiency and transfers of water from agriculture to other sectors. In this context, water quality will also be at least as important as surface and groundwater scarcity. Some industries have started to see water insecurity as a business, brand, and supply chain risk and have begun engagement with communities-at-large to mitigate their risks, promoting ideas of stewardship, conservation, and partnerships (Qureshi and Sayed 2014). Others continue to discharge untreated wastewater directly into urban horticultural fields and water bodies.

The previously murky understanding of climate science has become clearer with the realisation that climate change is already here, as evidenced by an increase in the frequency of extreme hydrological events. Not only is the per capita *average* availability diminishing with each passing moment, but climate change is also causing *unpredictable variations* in supplies and hazards (Mukhopadhyay and Khan 2015; Young et al. 2019). Thus, to the established threat of water scarcity, new problems of climate change, environmental degradation, and population explosion have been added.

Revocation of the Indus Waters Treaty seemed a remote theoretical possibility until Indian Prime Minister Modi threatened to stop the flow of rivers into Pakistan from Kashmir during recent escalations of hostilities between the two countries (PTI and IANS 2016). At the same time, Track II transboundary discussions of joint management of the basin suggest transformational opportunities for the four countries that occupy the basin: Pakistan, India, China, and Afghanistan (Adeel and Wirsing 2016).

The post-flood years have reinforced the idea that water and land are the biggest endowments given by nature to build a prosperous Pakistan. In years leading up to the great flood, some have questioned why irrigated agriculture should remain the main focus of the country's water policy. An overwhelming majority of policy-makers, practitioners and researchers continued to think that despite agriculture's mixed performance, water's full agro-economic potential is yet to be realised.

The response to these old, new, and emerging opportunities has come in multiple forms. There have been calls to train a new type of workforce with a holistic understanding of water issues that extend beyond engineering hydrology (Mustafa et al. 2013). New centres of interdisciplinary research have emerged, e.g., at Mehran University of Engineering and Technology (MUET) and the Lahore University of Management Sciences Centre for Water Informatics and Technology. MUET's (2020) USAID-funded program has developed flagship projects on Indus Basin Modelling, Clean Water, and Water Sustainable Development Goals. LUMS (2020) Centre for Water Informatics and Technology is developing new water sensing, measurement, information and communications technology, and precision agricultural innovations in close engagement with policy debates.

The government has produced policy papers for water, agriculture, and environment which, despite their shortcomings, are more holistic and cross-cutting. The approved National Water Policy (Ministry of Water Resources 2018) is one such document. The provinces are developing similarly expansive water policies. In multi-stakeholder forums, future visions have been imagined for the Indus Basin in which economy, society, and environment have each been given importance. The possibility of regional cooperation between the basin member countries has been examined, cross-sectoral optimisations of the Water-Energy-Food nexus have been investigated, and the importance of integrated planning has been underlined. These visions remind us that there are multiple promising possibilities for the future, and they are at the same time explicit about the uncertainties of climate change, technological transitions, regional rivalry, and political struggles (Langan 2018).

The quest for new types of solutions has sought new saviours, the most prominent among them being technology. Despite the setbacks of the Indus River telemetry system, the unenthusiastic adoption of high-efficiency irrigation technologies, weak extension services, and failure by local research institutions to produce new and better crop varieties, water managers of Pakistan remember that the world's largest contiguous canal irrigation system and wide-scale adoption of tube-well pumps were themselves technological revolutions. What could be the new technological revolutions of the twenty-first century? Information-related technologies (*water informatics*) will play a transformative role in coping with emerging water management challenges and possibilities. Issues ranging from reducing water demand (e.g., when, how much, and who should irrigate?) to conserving available supplies (e.g., through accounting and auditing surface flows and groundwater levels), and estimating uncertainties (e.g., forecasting demand and supply fluctuations) are fundamentally related to information collection and decision-making. This is why the establishment of water information systems, national and transboundary hydro-meteorology networks, decision support systems, and irrigation advisory services are featured among the least complex yet most urgent interventions in recent policy documents (PCRWR 2016; Ministry of Water Resources 2018; Young et al. 2019). The prospect of using more futuristic technologies such as robotics, artificial intelligence, and Industry 4.0 has also been raised, opening up new ways to imagine water resources and agricultural practices in the Indus Basin (Muhammad 2016).

Public awareness of water issues took another major turn in 2018 when the Chief Justice of Pakistan (later joined by the Prime Minister of Pakistan) led a widely publicised fund-raising campaign for building dams. This was a step forward and a step backward in many ways. On the positive side, water has not featured so prominently in public discourse except in times of major disasters. However, the singular focus on dam building indicates that there is still a long way to go in imagining creative alternatives and transforming them into action. One hopes that this newly emerged public awareness can be transformed into a deeper cultural and moral as well as technological awareness around water issues.

Recalling Iqbal's commentary on 'Urfī's verse in this chapter, we have seen that water managers in the period leading up to the 2010 flood suffered from a *lack of thirst* in imagining novel approaches. Additionally, progress in the post-flood period has at times proven illusionary, as when complex issues have found champions who revert to overly simple narratives and options. On the other hand, their *keen desire* contrasts with the lack of thirst in the earlier pre-flood period. The sincerity of their cause has found resonance with people who have contributed generously to the dam fund, and with groups that have committed themselves to improve water management in Pakistan. As we will see below, there are still others who have imagined visions of the river basin as a mosaic of irrigated gardens, and of achieving greater river sentience both through the increasingly ubiquitous availability of information and an expanding sense of purpose and historical consciousness. It remains to be seen whether this too will prove fruitful, illusory, or still lacking in desire and imagination sufficient to address Pakistan's water challenges.

2.6 Prospect

With this perspective on the past, present, and emerging future of Pakistan's water resources, we can better reflect on what is at stake, and how one might think about it in fresh ways. We have shown that the role of water in Pakistan's origins was less central than one might imagine – at least for those of us in the water resources field. General histories of Pakistan give the Indus River limited attention (e.g., Long 2015). Although the Indus Basin is one of the world's great water resources laboratories, we found that it has only occasionally been at the forefront of national consciousness, concern, and vision. The 1948 canal closures, 1960 Indus Waters Treaty, and extreme events like the 2010 flood have focused attention on water resources (Mustafa 2013). Climate change and environmental impacts, whether on vegetation and soil ecology or the charismatic Indus dolphin, have also drawn increasing public concern (Mustafa 2013; Yu et al. 2013).

At other times, the news in Pakistan revolves around domestic and international politics, personalities, and sports – like the news in most countries of the world. These preoccupations are a reflection of societal interest in water relative to other topics, but they also remind us that water issues are often contingent upon events outside of the water sector. Domestic and international political issues may draw attention to or deflect it away from current and long-term water issues.

This assessment warrants a fresh policy approach, as Indus Basin water management has little prospect in the absence of much greater public concern and consciousness of the sort called for earlier in the chapter. In this final section, we link the challenges discussed above with an emerging vision of the Indus Basin as a garden cultivated with multiple levels of innovation and care (Wescoat 2012; Wescoat and Muhammad 2021).

We began the chapter with the pivotal role of the lower Sutlej River valley development in the early twentieth century. The parties negotiated, in some respects effectively, and ushered in a period of scientific water measurement, negotiation, and allocation that has led to a process of continuous adjustment. While those early negotiations by no means resolved interstate competition, they established some principles, methods, and precedents for large-scale water management. A century later, the proposed new province in southern Punjab may again give this region a transformative role in Indus Basin water management. The proposed new province, whose boundaries are once again indefinite, could establish a new type of linkage between Sindh and Punjab, and thus again become a creative nexus in the basin. How its water claims and uses unfold will shape the next chapter of water use in the central plains of the Indus Basin.

We then showed how the revolutionary expansion of canal irrigation from the nineteenth century onwards, and tube-well irrigation in the twentieth century, produced both the greatness and great challenges in the basin. For the past 150 years, scholars have noted how the "protective irrigation" approach in colonial Punjab spread water as far as possible across the land, as compared with "productive irrigation" that seeks an optimal level of water allocation. Deficit irrigation produced low yields and induced large scale groundwater withdrawals. The conventional solution increases water and fertiliser application to increase crop yields. However, new models of crop intensification may revolutionise irrigation agriculture in Punjab. For example, the System of Rice Intensification (SRI) has 30 years of history that originated in Madagascar and is diffusing rapidly across South Asia (Uphoff 2016). Mechanisation of raised bed preparation in Pakistan has enabled lower seed and fertiliser application rates and reduced water delivery through furrows in between the raised beds (Sharif 2020, pers. comm.; and Pedaver Facebook page). These new systems of crop intensification turn production functions on their heads: the lower the inputs, the higher the yields. If field plot experiments are verified, one can imagine a situation where deficit irrigation allocation rates become optimal, and where yields per unit of water and land increase to much higher levels. Punjab and Sindh could once again become the garden of South Asia, but in altogether different ways from the Green Revolution.

Upstream, the geopolitics of water today are arguably more complicated than a century ago. The most tractable issue may involve negotiations with Afghanistan over upstream and downstream rights on the Kabul River, a river along which the first Mughal ruler Babur (d. 1526 CE) constructed some of the earliest gardens in the region. The Kabul River has only one long-lapsed letter of agreement between the King of Afghanistan and the British government which was negotiated in 1924. Over the past decade, however, several studies have generated post-conflict water development scenarios that involved only Afghanistan (World Bank 2010). Current international approaches spark hope for a more scientifically informed and negotiated agreement. Another promising precedent for scientific modelling has focused on Upper Indus snow and ice hydrology under the auspices of the International Centre for Integrated Mountain Development (ICIMOD).

A much less tractable challenge involves the Jammu and Kashmir region, which is ironically a region of some of the finest gardens and waterworks in the subcontinent. Kashmir did not receive detailed consideration as to its long-term needs and claims during the Indus Waters Treaty negotiations. In recent years, disputes have arisen over the design of the run-of-river Baglihar dam on the Chenab River. It was decided by a neutral expert that supported India's position on most points. A more mixed result was rendered by the International Court of Arbitration on India's Kishanganga Dam in the upper Jhelum Basin, which will divert water from Pakistan's Neelum-Jhelum project. Most recently, a stalemate occurred between the two parties as to which clause of the Indus Waters Treaty should prevail when one party requests a neutral expert under one clause of the treaty while the other party demands the court of arbitration under a different treaty clause. At the time of writing, that dispute had not resolved and was indeed eclipsed by broader geopolitical processes involving conflict between India and Pakistan over non-water related territorial, human rights, and terrorism issues. It is also possible that Kashmiri advocates will put forward new water claims and arguments to negotiate on their behalf. We do not know what geopolitical scenarios could or should unfold in the Upper Indus Basin. But the examples above and the history recounted in this chapter suggest that it will probably require in Iqbal's terms *thirst* on a level that is yet to be imagined.

These political tensions raise a second set of issues regarding Pakistan's water future, which involves ethical values associated with water resources management. A dominant theoretical paradigm among water professionals is integrated water resources management (IWRM) (Briscoe et al. 2005). Although some progress has been made toward incorporating social and environmental factors in IWRM, ethical issues receive brief treatment. Instead, complex river basins like the Indus are shaped more by political forces. Among political scientists, a realist approach suggests that water users, managers, and polities are all well-served if each acts in their self-interest (see critiques by Mustafa 2013; and Naqvi 2013). In that way of thinking, existing states would rarely make space for new claims, or needs, of other polities like South Punjab, Kabul, or Kashmir, or new needs like environmental protection or climate adaptation.

Counterarguments to realist political economic perspectives on water management include ethical positions that underscore the role of cooperation, sharing, compassion, and even sacrifice. Under these paradigms, scarcity and pain are shared, just as prosperity and profit are shared. For some, this paradigm involves the sacrifice of part of one's share or turn (*wari*) so that others can also live and thrive, which reminds one of the Punjabi Sufi poem *Sammi Meri Waar*. For others, water ethics have a pragmatic rationale, preparing all for suffering, coping, adjustment, and adaptation so that all may survive.

What are the models of water resource management that can advance such ethical aspirations? Elsewhere we have introduced the vision of the Indus Basin as a mosaic of gardens (Wescoat 2012; Wescoat and Muhammad 2021). This vision of the garden evokes deep cultural and historical aspects of water management associated with the care, cultivation, and meaning of the land. The gardens of the Indus Basin have taken many forms, from kitchen gardens to medicinal gardens, flower gardens, fields, woodlands, pastures, and parks. Some are rain-fed, others irrigated, all depend upon careful water management. While they are by no means ubiquitous, there are uncounted millions of them in Pakistan, the prospect for millions more, which collectively offer hope and inspiration for the prospect of Indus Basin management.

Acknowledgments We would like to thank Dr. Watto for encouraging us to prepare this chapter, along with extraordinarily probing reviewers, and our respective universities for supporting our collaboration over the years. We thank Rachel Hirsch for her help in copyediting.

References

Adeel Z, Wirsing RG (2016) Imagining Industan: overcoming water security in a new world. Springer, Cham

Akhtar M, Ahmad N, Booij MJ (2008) The impact of climate change on the water resources of Hindukush–Karakorum–Himalaya region under different glacier coverage scenarios. J Hydrol 355(1–4):148–163

Akhter M (2015) The hydropolitical cold war: the Indus waters treaty and state formation in Pakistan. Polit Geogr 46(3):65–75

Akhter M (2017) The political ecology of the water scarcity/security nexus in the Indus Basin: decentering per capita water supply. In: Adeel Z, Wirsing R (eds) Imagining Industan: overcoming water security in a new world. Springer, Cham, pp 21–33

Alam UZ (1998) Water rationality: mediating the Indus Waters Treaty. Unpublished dissertation, Durham University

Ali CR (1933) Now or never: are we to live or perish forever? Pakistan National Movement, Cambridge

Ali CR (1947) Pakistan: the Fatherland of the Pak Nation, 3rd edn. Pakistan National Movement, Cambridge

Ali I (1988) The Punjab under imperialism, 1885–1947. Princeton University Press, Princeton

Amir P, Habib Z (2015) Estimating the impacts of climate change on sectoral water demand in Pakistan. https://cyphynets.lums.edu.pk/images/Readings_concluding.pdf. Accessed 25 Mar 2020
Anwar AA, Bhatti MT (2017) Pakistan's water apportionment accord of 1991: 25 years and beyond. J Water Resour Plan Manag 144(1):05017015
Briscoe J, Qamar U, Contijoch M et al (2005) Pakistan's water economy: running dry. World Bank, Washington, DC
Chester LP (2009) Borders and conflict in South Asia: the Radcliffe Boundary Commission and the partition of Punjab. Manchester University Press, Manchester
Crooke W (1989) A glossary of north Indian peasant life. Oxford University Press, New Delhi
Deen S (2015) Pakistan 2010 floods. Policy gaps in disaster preparedness and response. Int J Disaster Risk Reduct 12:341–349
Fagan B (2011) Elixir: a history of water and humankind. Bloomsbury Press, New York
Fair CC, Kuhn P, Malhotra NA, Shapiro J (2017) Natural disasters and political engagement: evidence from the 2010–11 Pakistani floods. Stanford University Graduate School of Business Research Paper No. 17–42
Faruqui NI (2004) Responding to the water crisis in Pakistan. Int J Water Resour Dev 20(2):177–192
Gilmartin D (2014) Civilization and modernity: narrating the creation of Pakistan. Yoda Press, Delhi
Gilmartin D (2015) Blood and water: the Indus River Basin in modern history. University of California Press, Berkeley
Government of India (1935) Report of a committee of the Central Board of Irrigation on distribution of the waters of the Indus and its tributaries. Superintendent, Government Printing, Lahore
Government of Punjab Public Works Department (1920) Sutlej Valley project. Superintendent, Government Printing, Lahore
Government of Sindh (1944) Rejoinder of the Government of Sindh to the representations of Punjab and other units on the report to the Indus Commission. Government Press, Karachi
Graham D (2015) Heraclitus. Stanford Encyclopedia of Philosophy online. https://leibniz.stanford.edu/friends/members/view/heraclitus/. Accessed 25 Mar 2020
Gulhati ND (1973) Indus Waters Treaty: an exercise in international mediation. Allied, New Delhi
Haines D (2013) Building the empire, building the nation: development, legitimacy, and hydropolitics in Sindh, 1919–1969. Oxford University Press, Karachi
Haines D (2017) Rivers divided: Indus Basin waters in the making of India and Pakistan. Oxford University Press, Karachi
Hashmi HN, Siddiqui QTM, Ghumman AR, Kamal MA (2012) A critical analysis of 2010 floods in Pakistan. Afr J Agric Res 7(7):1054–1067
Hussain I (2017) Indus Waters Treaty. Oxford University Press, Karachi
Iqbal M (1944a) Sir Muhammad Iqbal's 1930 Presidential Address to the 25th Session of the All-India Muslim League, Allahabad, 29 December 1930. In: Sherwani LA (ed) Speeches, writings, and statements of Iqbal. Iqbal Academy, Lahore, pp 3–26
Iqbal M (1944b) Speech on the budget, 1930-31, delivered in the Punjab Legislative Council on the 7th March 1930. In: Sherwani LA (ed) Speeches, writings, and statements of Iqbal. Iqbal Academy, Lahore, p 86
Iqbal M (1989) The reconstruction of religious thought in Islam. Iqbal Academy, Lahore
Iqbal M (2007) Javid nama (trans: Arberry AJ). Kazi Publishers, Chicago
Iqbal M (2010) A message from the east. Iqbal Academy, Lahore
Iqbal M (2017) Iqbal: a selection of his Urdu and Persian verse (trans: Mathews D). Oxford University Press, New York
Iqbal MM, Arif M (2010) Climate-change aspersions on food security of Pakistan. J Sci Dev 15(1):15–23
Jinnah MA (1940) Address by Quaid-i-Azam Mohammad Ali Jinnah at Lahore Session of Muslim League, March, 1940. Directorate of Films and Publishing, Ministry of Information and Broadcasting, Government of Pakistan, Islamabad

Khan SI, Adams TE III (2019) Indus River Basin: water security and sustainability. Elsevier, Amsterdam
Khan HF, Yang YCE, Ringler C et al (2017) Guiding groundwater policy in the Indus Basin of Pakistan using a physically based groundwater model. J Water Resour Plan Manag 143(3):05016014
Langan S (2018) Integrated solutions for water, energy, and land (ISWEL). IBKF Workshop, Laxenburg. http://pure.iiasa.ac.at/id/eprint/15442/1/180602_Simon_Langan_ISWEL_Project.pdf. Accessed 25 Mar 2020
Long RD (ed) (2015) A history of Pakistan. Oxford University Press, Karachi
Meadows A, Meadows P (eds) (1999) The Indus River: biodiversity, resources humankind. Oxford University Press, Karachi
Mekonnen DK, Siddiqi A, Ringler C (2016) Drivers of groundwater use and technical efficiency of groundwater, canal water, and conjunctive use in Pakistan's Indus Basin Irrigation System. Int J Water Resour Dev 32(3):459–476
Michel AA (1967) The Indus waters: a study of the effects of partition. Yale University Press, New Haven
Ministry of Water Resources (2018) National Water Policy and Water Charter. Ministry of Water Resources, Government of Pakistan. https://mowr.gov.pk/index.php/national-water-policy-2018/. Accessed 25 Mar 2020
MUET (2020) Flagship projects. Mehran University of Engineering and Technology. http://water.muet.edu.pk/research/flagship-projects/. Accessed 23 Mar 2020
Muhammad A (2016) Managing river basins with thinking machines. In: 2016 IEEE conference on Norbert Wiener in the 21st century (21CW), pp 1–6
Mukhopadhyay B, Khan A (2015) A reevaluation of the snowmelt and glacial melt in river flows within Upper Indus Basin and its significance in a changing climate. J Hydrol 527:119–132
Mustafa D (2010) Hydropolitics in Pakistan's Indus Basin. US Institute of Peace, Washington, DC
Mustafa D (2013) Water resource management in a vulnerable world: the hydro-hazardscape of climate change. I.B. Tauris, London
Mustafa D, Akhter M, Nasrallah N (2013) Understanding Pakistan's water-security nexus. US Institute of Peace, Washington, DC
Naqvi SA (2013) Indus waters and social change: the evolution and transition of agrarian society in transition. Oxford University Press, Karachi
Pakistan Bureau of Statistics (2018) Pakistan statistical year book 2018 (Provisional). Pakistan Bureau of Statistics, Ministry of Planning Development and Special Initiatives, Government of Pakistan. http://www.pbs.gov.pk/content/pakistan-statistical-year-book-2018-provisional
Panhwar NA (2010) The Indus flood, 2010: perspectives issues strategies. Centre for Environment and Development, Karachi
Paustian P (1930) Canal irrigation in the Punjab: an economic inquiry relating to certain aspects of the development of canal irrigation by the British in the Punjab. Columbia University Press, New York
PCRWR (2016) National research agenda on water 2016–25. Pakistan Council of Research in Water Resources, Islamabad
PTI, IANS (2016) Blood and water cannot flow together, says Modi. [online] Khaleej Times. https://www.khaleejtimes.com/international/india/blood-and-water-cannot-flow-together-says-modi. Accessed 25 Mar 2020
Qureshi AS, Sayed AH (2014) Situation analysis of the water resources of Lahore establishing a case for water stewardship. WWF-Pakistan and Cleaner Production Institute (CPI), Lahore
Qureshi AS, Shah T, Akhtar M (2003) The groundwater economy of Pakistan. International Water Management Institute Working Paper 64
Rinaudo JD (2002) Corruption and allocation of water: the case of public irrigation in Pakistan. Water Policy 4(5):405–422
Shah T (2008) Taming the anarchy: groundwater governance in South Asia. Routledge, New Delhi

Siddiqi A, Wescoat JL Jr (2013) Energy use in large-scale irrigated agriculture in the Punjab Province of Pakistan. Water Int Spec Issue Water Food Secur Chall Pak 38(5):571–586. https://doi.org/10.1016/j.wasec.2018.11.001
Siddiqi A, Tran J, Wescoat J (2018a) Modeling growth and diffusion of groundwater pumping at multiple sub-provincial scales. American Geophysical Union Fall Meeting Abstracts
Siddiqi A, Wescoat J, Muhammad A (2018b) Socio-hydrological assessment of water security in canal irrigation systems: a conjoint quantitative analysis of equity and reliability. Water Secur 4(5):44–55. https://doi.org/10.1016/j.wasec.2018.11.001
Sivapalan M, Savenije HHG, Bloschl G (2012) Socio-hydrology: a new science of people and water. Hydrol Process 26:1270–1276
Sivapalan M, Konar M, Srinivasan V et al (2014) Socio-hydrology: use-inspired water sustainability science. Earth's Future 2(4):225–230
Talbot I, Gurharpal S (2009) The partition of India. Cambridge University Press, Cambridge
Ul Hassan MM (2009) The conception, design and implementation of IMT in Pakistan's Punjab: a public policy reflection. S Asian Water Stud 1(2):132–142
Uphoff N (2016) The system of rice intensification: frequently asked questions in black and white, 2nd edn. CreateSpace Publishing, Ithaca
van Steenbergen F, Gohar S (2005) Groundwater development and management. In: Briscoe J, Qamar U (eds) Pakistan water economy running dry – background papers. Oxford University Press, Oxford
Wade R (2007) Village republics: economic conditions for collective action in South India. Cambridge University Press, Cambridge
Wescoat JL Jr (1999) The historical geography of Indus Basin management: a long-term perspective, 1500-2000. In: Meadows A, Meadows P (eds) The Indus River: biodiversity, resources, humankind. Oxford University Press, Karachi, pp 416–428
Wescoat JL Jr (2012) The Indus River Basin as garden. Die Gartenkunst 24(1):33–41
Wescoat JL Jr (2013) Reconstructing the duty of water: a study of emergent norms in socio-hydrology. Hydrol Earth Syst Sci 17:4759–4768. http://www.hydrol-earth-syst-sci.net/17/4759/2013/hess-17-4759-2013.pdf
Wescoat JL Jr, Muhammad A (2021) Irrigated gardens of the Indus River Basin: new models for water resource management. In: De Wolff K, Faletti R, López-Calvo I (eds) Water and the humanities: transforming currents for uncertain futures. University of California Press, Oakland
Wescoat JL, Siddiqi A, Muhammad A (2018) Socio-hydrology of channel flows in complex river basins: rivers, canals, and distributaries in Punjab, Pakistan. Water Resour Res 54(1):464–479. https://doi.org/10.1002/2017WR021486
World Bank (2010) Scoping strategic options for development of the Kabul River Basin: a multi-sectoral decision support system approach. World Bank, Washington, DC
Yang YE, Ringler C, Brown C, Mondal MAH (2016) Modeling the agricultural water–energy–food nexus in the Indus River Basin, Pakistan. J Water Resour Plan Manag 142(12):04016062
Young W, Anwar A, Bhatti T et al (2019) Pakistan: getting more from water. World Bank, Washington, DC. https://openknowledge.worldbank.org/handle/10986/31160
Yousafzai F (2017) 'Govt scraps Indus telemetry system'. The Nation, 24 March [Online]. https://nation.com.pk/24-Mar-2017/govt-scraps-indus-telemetry-system. Accessed 25 Mar 2020
Yu W, Yang YC, Savitsky A et al (2013) The Indus Basin of Pakistan: the impacts of climate risks on water and agriculture. World Bank Publications, Washington, DC
Zaidi ZH (2004) Quaid e Azam Muhammad Ali Jinnah Papers. Series 2, Volume 10. Quest for political settlement in India. 1 October 1943 to 31 July 1944. Printing Corporation of Pakistan Press, Islamabad, pp 708–746
Zaman MQ (2018) Islam in Pakistan: a history. Princeton University Press, Princeton

Chapter 3
Pakistan's Transboundary Water Challenge

Fazilda Nabeel and Muhammad Jehanzeb Masud Cheema

Abstract Pakistan's water economy is overwhelmingly reliant on the Indus River – a river that flows across international political boundaries, and is thus transboundary in nature. Over the last two decades, the waters of the Indus have become subject to contentious hydro-politics between Pakistan and its neighbouring riparians – India, Afghanistan and China – due to population and development pressures on one hand, and climate change induced water variability on the other. This chapter aims to analyse the underlying complexity and political economy of riparian relations on the Indus. It starts by outlining Pakistan's dependency and vulnerability with respect to its position in the Indus Basin, and summarises the most pressing dimensions of Pakistan's transboundary water challenge related to each of its riparian neighbours. The chapter then examines the existing legal and institutional apparatus for managing transboundary water issues in the Indus Basin with a view to revealing the gaps and weaknesses that must be addressed for more effective transboundary water and benefit sharing. As the way forward for greater water security on Pakistan's transboundary front, the chapter advocates small steps in hydro-diplomacy to move from a "water sharing" perspective to a "benefit sharing" approach that binds all riparian countries to work for a sustainable water future for the Indus.

Keywords Transboundary water resources · Water challenge · Water sharing · Water inflows · Indus river basin

---

F. Nabeel (✉)
Department for International Development, University of Sussex, Brighton, UK
e-mail: f.nabeel@sussex.ac.uk

M. J. M. Cheema
Faculty of Agricultural Engineering and Technology, PMAS-Arid Agriculture University, Rawalpindi, Pakistan
e-mail: mjm.cheema@uaar.edu.pk

M. A. Watto et al. (eds.), *Water Resources of Pakistan*, World Water Resources 9,
https://doi.org/10.1007/978-3-030-65679-9_3

3.1 Pakistan's Transboundary Water Profile: Dependencies and Development Pressures in the Indus Basin

Pakistan is overwhelmingly reliant on the Indus Basin waters yet most of this water originates outside the country's political boundaries. Out of the six main tributaries of the Indus, the Sutlej originates in China's Tibetan Plateau, while the Ravi, Beas, Jhelum and Chenab originate in India. The Kabul River originating in Afghanistan, is the sixth tributary of the Indus and is shared between Pakistan and Afghanistan. Pakistan thus shares the Indus Basin with India, Afghanistan and China (see Fig. 3.1). Of the basin's drainage area, 47% lies in Pakistan, while the remaining is split between India (39%), Afghanistan (6%) and China (8%) (FAO 2011). The Indus Basin covers 65% of Pakistan's territory, comprising parts of all provinces Punjab, Sindh, Khyber Pakhtunkhwa as well as a small part of Balochistan (FAO 2011).

The most challenging aspect of Pakistan's transboundary water governance is its riparian relationship with India. The only water sharing agreement that Pakistan has on sharing the Indus is the Indus Waters Treaty (IWT) of 1960. This treaty allocated the three eastern rivers (Ravi, Beas and Sutlej) to India and the three western rivers (Indus, Jhelum and Chenab) to Pakistan. Although India is a huge country with numerous river basins, the Indus Basin remains of critical significance for irrigation in northwest India, which serves as the country's breadbasket (Haines 2017). India is the upper riparian in the transboundary Indus Basin, thus all major tributaries flow through India into Pakistan. In recent years India has planned to divert all flows from the east flowing rivers and construct hydropower structures on western rivers to fulfil its growing irrigation and hydropower requirements. For this purpose, the Madhupur Beas Link and Beas Sutlej Link Canals were constructed diverting water from Ravi to Beas, and Beas to Sutlej, respectively (Cheema and Pawar 2015). On the western rivers allocated to Pakistan, India has initiated numerous hydropower projects in recent years. These include the Kishanganga and Uri II on Jhelum River, while Salal, Baglihar, Dul Hasti, Bursar Dam, Pakal Dul, Ratle, Swalkote, Bursar, Kirthai-Naunatu, Kirthai-II, Kiru and Kwar are located on the Chenab, and Chutak and Nimoo Bazgo are located on the Indus itself (Ali 2020). The cumulative effect of water diversion projects as well as hydropower projects on shared rivers has exacerbated pressure on water availability in Pakistan. Pakistan also shares the Indus Basin with Afghanistan and draws up to 12% of its flows from the Kabul River (Akhtar and Iqbal 2017). In Afghanistan, post war development has increased the demand for water resources from the Kabul, thus destabilising the sustainability of flows to Pakistan. China does not directly depend on the Indus to meet its water resource demands but the basin has strategic importance for China as it is home to Tibetan glaciers that feed the freshwaters of the Indus River.

Mean annual flows in the transboundary Indus Basin are estimated to be more than 200 km^3 (Cheema 2012). Of these, the inflow into Pakistan through the rivers of the Indus Basin is estimated at 184 km^3, out of which 21.6 km^3 is through the Kabul and other Indus tributaries flowing from Afghanistan, 1.9 km^3 through the

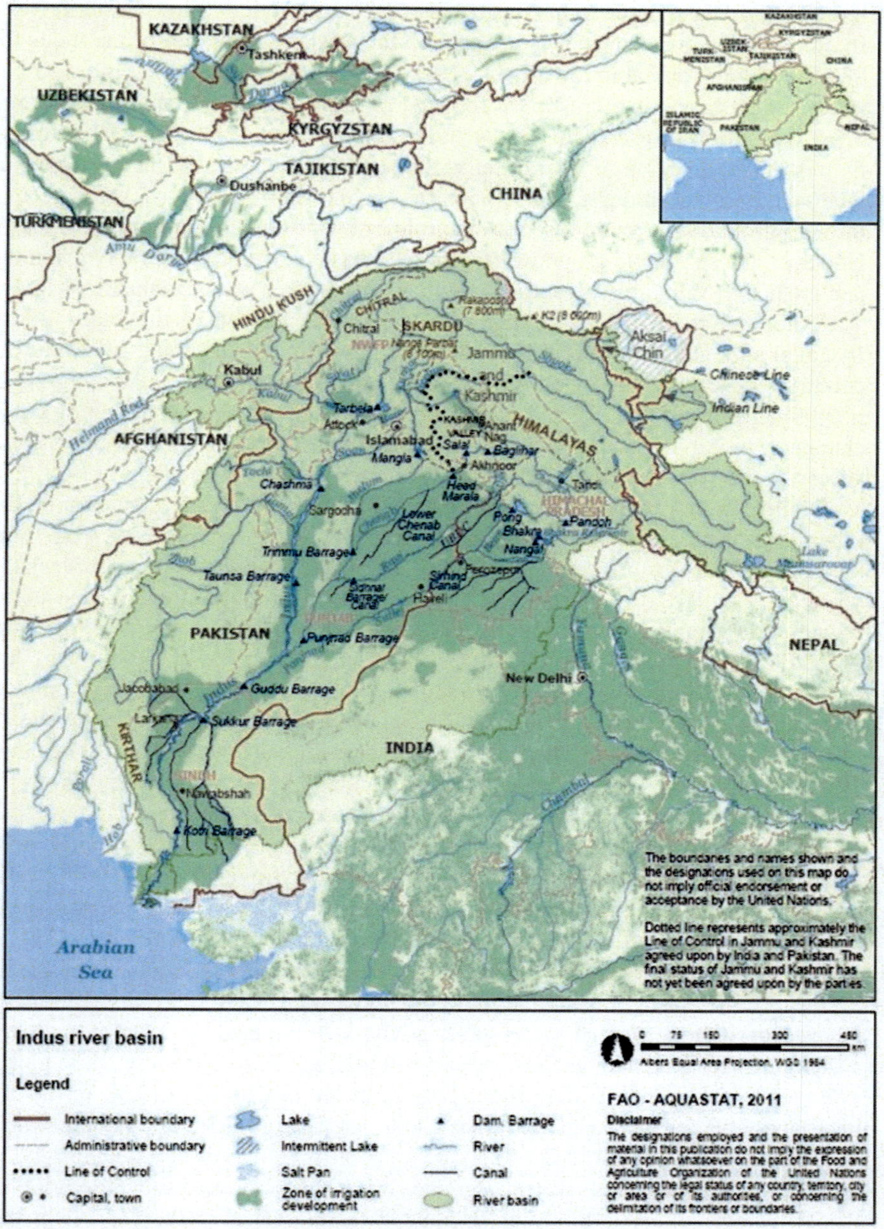

Fig. 3.1 Main river system of transboundary Indus Basin covering Pakistan, India, China and Afghanistan. (Source: FAO 2011)

three eastern rivers of the Indus Basin and 161 km^3 through the three western rivers. In other words, the country is dependent on the western tributaries of the Indus for 88% of its total annual inflow (FAO 2011). Of the total inflows in Pakistan, 91.6% is consumed by agriculture, 3.3% for environment, 2.6% for domestic uses and 2.5% for industrial use (Ahmad 2016). According to the FAO Aquastat database figures of 2011, 66.4% of total water withdrawals in the Indus Basin are from surface waters of the Indus and its tributaries, while the remaining 33.6% are met through groundwater pumping from the transboundary Indus Aquifer (FAO 2011). However, the surface to groundwater use ratio has changed considerably given the decline in the availability of, and variability in, the surface waters in the Indus Basin.

Almost all (95%) of Pakistan's irrigated area is located in the Indus Basin. Irrigated agriculture is critical for Pakistan's economy, contributing 21.4% to the country's GDP and employing 45% of its labour force (Adeel and Wirsing 2016, p. 9). Since most of the irrigated agriculture in the basin is based on these flows, any change in availability can have severe effects on food security in the basin. Pakistan's access to the waters of the Indus is becoming precarious in the face of declining availability in its tributaries on one hand, and rapid groundwater depletion on the other hand. In recent years, the average annual flows in the eastern rivers from India has reduced from historic levels by 92%, when compared to the average between the years 2007 and 2010 (Cheema 2012). Even more worrisome for Pakistan is the reduction in average flows of the western tributaries of the Indus by 17% after the year 2007 in comparison to historic years (Cheema and Pawar 2015). This has mostly been attributed to the completion of significant Indian hydroelectric projects upstream including the Baglihar I, II and Dul Hasti dams. The impact of upstream interventions through hydropower development in India as well as climate change associated water variability have been documented as key factors responsible for reduction in available river flows in Pakistan (Ahmad 2009; Ahmad and Iqbal 2016).

Despite experiencing a demonstrated decline in surface flows as well as groundwater overdraft, Pakistan's water use is anything but frugal. The country's water productivity is lower compared to the global average as well as in comparison to other countries in South Asia. For example, water productivity of rice, which is the second largest export crop of Pakistan is 0.45–0.69 kg m^{-3} is less than the global average of 1.09 kg m^{-3}, while the figure for India stands at 1.18 kg m^{-3}. Similarly, in case of wheat, average water productivity is estimated at 0.76 kg m^{-3} and 1.08 kg m^{-3} for Pakistani and Indian part of Indus Basin, respectively while global average is 1.09 kg m^{-3}. Water productivity of maize is also lower (0.58 kg m^{-3} for local and 1.59 for hybrid maize) than global averages (1.8 kg m^{-3}) (WaterWatch 2003; Cai et al. 2010; Zwart 2010).

While inefficiency of water use and high comparative water footprint of Pakistan's agriculture is likely unsustainable (as discussed in the chapter in this volume by Abid et al.), this chapter focuses on aspects of Pakistan's international transboundary governance, which are constraining the country's access to a sustainable supply of water from the Indus to meet the needs of its ever thirsty water economy. The next section highlights the breadth and depth of the transboundary water challenge in the Indus Basin vis-à-vis Pakistan's riparian relationship with each of riparian countries on the Indus – India, Afghanistan and China – and its political economy and water security implications for water resource availability in the country. The discussion ends with the challenge posed by climate change on a basin-wide scale and how it exacerbates Pakistan 's transboundary water challenge.

3.2 Nature of the Transboundary Water Challenge

3.2.1 Pakistan–India

Pakistan shares six rivers with India – Indus, Chenab, Jhelum, Sutlej, Beas and Ravi – that flow through northern India into Pakistan. When Pakistan and India became independent countries in 1947, the boundary was drawn right through the Indus Basin, making Pakistan the lower riparian state and giving India control over Indus headwaters in general and the Chenab in particular.

The IWT of 1960 that guides water sharing arrangements between Pakistan and India was crafted around an approach of "dividing" rather than "sharing" the rivers. According to the treaty, the three western rivers including Indus, Chenab and Jhelum, were allocated to Pakistan, while India was given rights to the three eastern rivers Sutlej, Bias and Ravi. In addition, the treaty gave India certain rights of "non consumptive" use of the three western rivers to Pakistan. It is this part of the treaty, and the extent of India's use of western tributaries of the Indus before they enter the Pakistani territory, that has become increasingly contentious.

While the IWT of 1960 is generally held as an example of a successful water sharing treaty, having survived two wars as well as a range of other political hostilities between India and Pakistan (Biswas and Iwra 1992), the treaty's ability to provide answers to contentious water issues between the two riparian neighbours is increasingly coming under stress. Increased demand for water in both countries coupled with its inefficient and wasteful use and the growing need for hydropower development for economic growth further pressurise already dwindling surface water resources. Increasing water stress in the two countries, which is further reinforced and exacerbated by climate change, is also contributed to escalating tensions on the issue of water (Mahbub ul Haq Human Development Centre 2013). Some observers hold that the IWT is a static treaty, which is increasingly proving to be inflexible for dealing with some of the grave challenges for transboundary water sharing between India and Pakistan (Biswas 2011). The Treaty is silent on the issue of climate induced water variability, speaks minimally to the issue of environmental degradation, and does not address the equitable and sustainable sharing of transboundary aquifers between India and Pakistan (United Nations 1962). Further, the Treaty only guides transboundary water sharing on the tributaries of the Indus shared between Pakistan and India, without including other riparians (Afghanistan and China) in the Basin.

Within the Indus Basin, the Chenab sub-basin has been the hotpot for contentious hydropower projects on shared western rivers. The Chenab is seen as a critical water resource for Pakistan, as it combines the waters of the four rivers, Jhelum, Sutlej, Beas and Ravi to form a single water system, which then joins the Indus in Pakistan. Indian control over the Chenab as the upper riparian along with the fact that all the three western rivers assigned to Pakistan under the IWT originate or flow through the volatile disputed region of Jammu and Kashmir are key drivers of trans-boundary water issues between India and Pakistan (Mahbub ul Haq Human Development Centre 2013). Pakistan's objections to Indian projects on western rivers have centred on India's ability to store water, which goes against Pakistan's rights to these western rivers as provided by the IWT. Almost all Indian projects on the western rivers are run-of-the-river projects as allowed under the IWT. However, the cumulative live storage of these projects could give India a critical degree of control over the western rivers, as acknowledged in the US Senate Foreign Relations Committee Report (2011).

Major upstream Indian projects that have become controversial from time to time and involved issues around the compliance of IWT include Salal, Wullar Barrage/ Tulbul Navigation project, Baglihar, Kishanganga, Dul Hasti, Bursar Dam, Uri II and Nimoo Bazgo. Indian plans for a cascade of hydropower projects on the Chenab river presents a serious threat to Pakistan as a lower riparian. There are 12 major sub-tributaries of the Chenab River above Marala barrage, and a total of 15 small hydroelectric projects and three dams (Salal, Baglihar and Dul Hasti) have already been completed by India. An additional five dams are under construction, i.e.: Pakal Dul, Ratle, Miyar, Lower Kalnai and Khari-I (a trench weir with 2 MW power generation capacity). Although no data has been shared by India, it is reported that a total of 55 projects are planned for western rivers, including major dams like Swalkote, Bursar, Kirthai-Naunatu, Kirthai-II, Kiru and Kwar (Ali 2020). The enhanced control over the waters of the western tributaries of the Indus due to upstream Indian Projects will further escalate water security fears in downstream Pakistan.

Pakistan has raised objections on several Indian hydropower projects on shared rivers under the IWT in the last decade, yet the treaty's weak dispute resolution mechanism meant that cases had to be referred to international arbitration. These include the construction of the Baglihar dam on the Chenab and the Wullar barrage/ Tulbul hydropower projects on the Jhelum, and the Kishanganga hydropower project on the Kishan Ganga River, a tributary of Jhelum (Zawahri 2009). The planned 850 MW Ratle Hydroelectric Project on the Chenab is the latest in the series of contentious projects on the Chenab that has been subjected to international arbitration between India and Pakistan. India has unilaterally begun construction on the Ratle Dam while the two countries decided between the appointments of a neutral expert as preferred by India as opposed to the appointment of the seven-member international court of arbitration as preferred by Pakistan. Pakistan believes Ratle's design would reduce Chenab flows by 40% at Marala barrage causing considerable loss to crops as experienced after the construction of the Baglihar Dam. This belief is further strengthened by a recent study that reveals an increase in peak flows in the Chenab River at Marala during summer (monsoon) after commencement of the Baglihar Dam, while winter flows are reduced. The result is artificial flood and drought conditions in summer and winter respectively (Ali 2020). The Ratle dam is believed to be three times larger than the Baglihar dam (Hill 2016).

Apart from water scarcity, transboundary water quality issues between India and Pakistan pose an equally grave risk to the livelihoods of populations dependent on the Indus. About 55 km^3 of waste water is dumped into the Indus every year, which is daunting when compared to the average annual flows of 180 km^3 (Iqbal 2013). Agriculture, industry and municipal waste are the primary sources of pollution in the transboundary basin, contaminating it with chemicals, fertilisers, pesticides, heavy metals and pathogens that are detrimental to health (Michel and Sticklor 2013). India and Pakistan have been narrowly focused on volumetric allocations of water and the race to building hydroelectric dams, without focusing on the issue of transboundary water pollution. The IWT also does not provide conclusive solutions for the issue of water pollution. In Article IV, Clause 10 of the treaty, it does refer to the intent of each riparian to conserve quality of waters of the Indus Basin, but does not provide for appropriate monitoring and surveillance mechanisms to ensure this.

While competition over visible surface water resources between India and Pakistan in the Indus Basin is evident, the invisible nature of groundwater and its subsurface location has meant that shared groundwater resources in the Indus Basin have escaped any policy attention. Groundwater use in the Indus Basin lends itself to somewhat similar challenges as transboundary water sharing as surface water. India and Pakistan share the vast transboundary Indus Aquifer that provides groundwater as an important resource used for irrigation in the basin, with abstraction ranging between 40% and 60% of irrigation water requirements depending on land usage (Cheema et al. 2014). However, the aquifer is rapidly deteriorating, and currently stands as the second most overstressed aquifer in the world (Richey et al. 2015). The Indus Aquifer is unconfined, thus excessive abstraction of water on one side of the border can decrease the water table on the other side of the border. It can create a water gradient where water begins flowing to the region with lower water table, which has implications for water availability, quality and equity. Recent studies have reported how over abstraction of groundwater in the Indian Punjab is lowering the water table in bordering regions of Pakistan (IUCN 2010a). Depletion of groundwater forces Pakistani farmers to dig deeper wells, at increasing costs to them, and the water abstracted is more saline, thus exacerbating soil salinisation. This impacts local livelihoods and makes Pakistani farmers vulnerable to the adverse effects of groundwater abstraction in India without having a share in the benefits of the water abstracted.

Despite the existence of a water sharing treaty, the transboundary relationship between Pakistan and India is mired by years of political mistrust. Both riparians regard the issue of control over transboundary waters as important for national security. This has undermined the extent to which hydrological data is generated and shared between the two countries, thus further fuelling mistrust. The securitisation of water within India and Pakistan means they blame each other to cover for internal water resource inefficiency. In addition, water issues are deeply embroiled with political issues, and particularly the Kashmir dispute. In the recent past, nationalist Indian governments have threatened to renounce the IWT and "cut off" Pakistan's supply of the Indus Rivers in the aftermath of the Uri attacks of 2016 saying "blood and water cannot flow together"(The Indian Express 2016).

3.2.2 *Pakistan–Afghanistan*

Notwithstanding the increased frequency of transboundary disputes between Pakistan and India in recent years, a water sharing treaty, no matter how inadequate, does exist to guide transboundary water sharing between them. Pakistan's relationship with Afghanistan over the sharing of water resources of the Kabul River presents a more ominous challenge as there are no agreed upon rules for water and/or benefit sharing.

The Kabul River rises in the Hindu Kush, receives substantial flows from the Kunar and several minor rivers originating in Pakistan's Chitral region, flows east past Kabul and Jalalabad before entering back to Pakistan (Fig. 3.2). In Pakistan, the Kabul River is augmented by the Swat river and its tributaries before it drains into the Indus at Attock. Thus, Pakistan is in a unique position of being both the upper and lower riparian in case of the Kabul River.

Both Afghanistan and Pakistan are significantly dependent on the waters of the Kabul River. Overall, 12% of Pakistan's total water supply comes from the Kabul River and its tributaries, especially in the winter months when Indus flows decline

Fig. 3.2 Kabul and Kunar basins, sub basins of Indus Basin. (Source: Khan et al. 2020)

(Akhtar and Iqbal 2017). Within Pakistan, the province of Khyber Pakhtunkhwa (including FATA) is particularly dependent on four specific rivers flowing from Afghanistan – Kabul, Kurram, Kaitu and Gomal.

Afghanistan's renewable water resources are estimated at 57 km^3; however these are distributed unevenly across five transboundary river basins – Northern, Haridud Murghab, Kabul, Helmand and Panj Amu. Out of Afghanistan's transboundary water resources, there is only an international agreement with Iran on the use of the Helmand River. In Afghanistan, the Kabul Basin, including the important tributary Kunar River, is the most important river basin, representing approximately 26% of the available water resources in Afghanistan and containing almost half of the country's urban population. It is crucial to the livelihoods of the millions of people sharing its water resources for drinking water, sanitation, agriculture, power generation, and industry (IUCN 2010b).

Population and development pressures on both sides of the Kabul River have compelled riparian countries to consider a water sharing agreement from time to time. Processes of climate change further complicate the future of water security; yet efforts at hydro-diplomacy have been thwarted by a long history of political issues. Similar to Pakistan's relationship with India, the relationship between Pakistan and Afghanistan is hindered by political mistrust and a range of other political economic issues including territorial disputes along the Durand line. In addition, the asymmetry of power between the two countries is a significant hurdle in getting an agreement on the Kabul Basin. What further complicates hydro-diplomatic issues between Pakistan and Afghanistan is increased prominence of India's role in post-war Afghanistan. India remains Afghanistan's largest development donor focusing on hydropower development in the region. According to a Pakistani newspaper's report, India plans to assist in building 12 water-control structures on the Kabul River in Afghanistan, increasing the impression of Pakistan's vulnerability to upstream state developments (Kugelman et al. 2011; The News 2011). Most notable among Indian-funded projects is the 42 MW India-Afghanistan Friendship and Salma Dams that supply water to irrigate 75,000 hectares, and the new Afghan parliament building. India is also in a tripartite agreement with Iran and Afghanistan to develop Iran's Chabahar port.

Pakistan has engaged discussions with Afghanistan for a water-sharing agreement on the Kabul River and announced plans for a joint hydropower plant on the Kunar tributary (Vick 2014). In August 2013, the two countries made a joint public announcement to engage in the development of a 1500 MW hydropower dam cascade on the Kunar River and work towards a bilateral formula of cooperation (Sadeqinazhad et al. 2018), but no significant progress has yet been made.

3.2.3 Pakistan–China

While Pakistan and China do not share transboundary rivers in the same manner as Pakistan does with India and Afghanistan, in many ways China is the most critical relationship that needs to be managed for sustainable transboundary water flows to Pakistan. China is the largest source of transboundary flows to much of the Indus Basin. The Indus River originates in Tibet – also known as the "water tower of Asia" and the world's "third pole" – and China's plans to harness the massive Tibetan glaciers means that much of the source of river flows to South Asia are at risk. The development of water infrastructure projects on Tibet's transboundary rivers has already infuriated many downstream countries and triggered international criticism. For example, China's construction of hydroelectric dams along the Brahmaputra River has become a source of friction between China and India. China has also dammed the upper Mekong River, which has become a major source of conflict between China and Southeast Asian countries. There are no legal safeguards or formal agreements between China and downstream countries over the use of shared river systems. While China has the fourth largest freshwater reserves in the world, it also has a formidable population, which results in water scarcity compelling investment in massive infrastructure projects to secure its water future. China recently launched several water diversion initiatives and programmes such as the South-to-North Water Diversion Project (SNWDP), which could have grave implications for water availability for lower riparians in the Indus and Ganges Basins (Xie et al. 2018). India and China also do not currently have a water sharing treaty, which allows China to construct dams and reservoirs on shared rivers thus reducing share of water coming from China. This has implications for water availability in India, and further repercussions on Indian plans to harness water as an upper riparian in the Indus Basin.

The critical importance of the Tibetan glaciers to the region's water security, and the effect of climate change on this resource, has been documented by the International Centre for Integrated Mountain Development (ICIMOD)'s glaciers study (Maharjan et al. 2018). The ICIMOD assessment report is one of many reports confirming the melting of glaciers on the Tibetan plateau, which could cause significant disruptions to future water scarcity (Williams 2018) (See Sect. 3.2.4).

The political economic significance of China's role in the Indus is perhaps as critical as its control over the source of major water flows to the river basin. China's increasing influence in Pakistan under the China-Pakistan Economic Corridor (CPEC) entails a planned emphasis on the construction of hydroelectric power stations. According to reports, more than half of the planned 46 billion USD is to be spent on energy projects, specifically hydroelectric power stations (Nabeel 2020). China's hydropower investments have been focusing on the Pakistani controlled side of Kashmir. These include the Neelam Jhehlum, Dasu, Phandar, Bashu, Harpo and Yulb hydroelectric projects. It is worth noting that the Pakistani government pulled out of the Diamir Basha Dam agreement with China because of the stringent conditions against the country's interest for financing the dam, leading to Pakistan's decision to finance the 14 billion USD project on its own. The Belt and Road

Initiative, and the investment in hydropower that come with CPEC, carry important implications for politics as well for water and environment in Pakistan (Moore 2019).

3.2.4 *Effects of Climate Change on a Basin-Wide Scale*

While Pakistan is faces peculiar challenges in its riparian relationship with respect to India, Afghanistan and China, the problem of climate change and associated water variability entails disastrous consequences for all basin countries. Climate change results in water variability including floods, water shortages and disrupted monsoon cycles. A recent study by ICIMOD estimates that even an average of global warming of 1.5 degrees Celsius would actually result in a warming of 2.1 degrees Celsius in the Hindukush Himalayan Region, which is the source of river flows to the Indus Basin. The same study estimates that if current emissions continue unabated, it will lead to a loss of 64% of the ice cover in the Hindukush region by the year 2100 (Maharjan et al. 2018; Henderson et al. 2019). Such a change would be catastrophic for livelihoods of the Hindukush region, as well as for downstream countries like India and Pakistan that depend on river systems originating in the Hindukush.

The impact of climate change on the nine million people who depend on the Kabul Basin for livelihoods in Afghanistan and Pakistan can be catastrophic. Recent studies estimate that the effect of climate change may induce up to 50% decline in precipitation, especially in the western parts of the Kabul Basin (Henderson et al. 2019). The rise in mean annual temperatures due to global warming will accelerate snow and glacier melt, thus leading to an increased frequency of flash floods in the basin. Climate change threatens food security and livelihoods on both sides of the Afghanistan-Pakistan border, thus there is a critical and urgent need for adaptation and mitigation measures.

The incentive to build hydroelectric dams and reservoirs is heightened in the current discourse, as water flows may increase until 2050 because of the rapid glacial melt. Of the Indus Basin countries, only China's hydropower development is at 86% of its total potential. Developed hydropower as a percentage of the total potential is at a mere 29% for India, 14% for Pakistan and 4% for Afghanistan according to the Hindu Kush Himalayan Monitoring and Assessment Programme (HIMAP) Report 2019 (Wester et al. 2019). Many large hydropower projects have been planned in the Indus, with a large number in the disputed territory of Jammu and Kashmir. China's role in the construction of these dams in Pakistan under CPEC is particularly critical because of the combined effect of these dams on the Indus Basin. It has been reported that the combined effect of these dams that form the "Indus Cascade" can result in a reduced silt load downstream, which will affect agricultural productivity negatively. In addition, the dams may also reduce the water flows to Punjab and Sindh during the non-monsoon months of October to June (Gupta 2017).

While the construction of dams might help riparian countries deal with medium to long-term storage dilemmas, they have significant socio-economic and environmental impacts, as well as an impact on global warming. Recent research suggests that the emphasis on the construction of dams and reservoirs globally can further accentuate the production of greenhouse gases (Deemer et al. 2016). Part of the methane produced at the bottom of reservoirs, where oxygen is low and bacteria decompose, escapes through the surface of the water as bubbles. Over the course of 20 years, methane produces three times more to global warming than carbon dioxide, hence making hydroelectric dams more of a threat to climate change than previously thought.

3.3 Legal and Institutional Architecture for Transboundary Water Sharing: Voids, Omissions and Spaces for Action

The legal architecture and institutional arrangements for transboundary water sharing between Pakistan and its neighbours are inadequate to say the least. In the case of Pakistan's relationship with India, the 60-year-old IWT is falling short in its ability to settle water sharing issues. In the case of Pakistan's access to water from Kabul River shared with Afghanistan, there is no legal or institutional framework that has yet come to fruition, despite some diplomatic efforts on both sides. Perhaps the most critical and worrying aspect of Pakistan's transboundary relationship concern the role of China. As detailed in Sect. 3.2.3, China not only has control over the source of the origin of Indus flows, but is also systematically and gradually increasing its investments in hydropower in Pakistan-controlled Kashmir. This section looks at the efficacy and adequacy of the existing legal and institutional architecture for transboundary water sharing between the riparians in the Indus Basin, to highlight the need for strong basin-wide institutions and tactful trust building in order to take hydro-diplomacy on the Indus forward.

The IWT has created a legal framework for governing transboundary water resources between India and Pakistan and is largely regarded as a successful framework for cooperation on shared water resources, having survived three wars and other hostilities between the two neighbours. The Treaty has served to moderate "the worst impulses of India and Pakistan toward each other" and had guided water sharing through political turmoils (Mustafa 2010). However, several important areas of concern fall outside the ambit of the treaty and are increasingly becoming a source of hydro-conflict between the two riparian nations. As discussed earlier, the Treaty focuses narrowly on "water sharing" as opposed to benefit sharing and cooperation on a basin-wide scale. The treaty is also quiet on the ominous challenges of climate change induced water variability, environmental degradation and the (over)use of transboundary groundwater aquifers.

Apart from the glaring omissions in the IWT, the agreement is also increasingly inflexible in addressing the concerns of India and Pakistan for issues that fall within its ambit. The treaty allows run-of-the-river projects on western rivers allocated to Pakistan, but does not place any limits on the number of hydropower projects planned by India on western tributaries of the Indus, fuelling Pakistan's apprehensions about the technical specifications and control potential of these. Similarly, while the treaty has mechanisms for data sharing and exchange of information (under Annexure D of the Treaty), in practice both countries are reluctant to share hydrological data especially on new hydropower projects planned on shared rivers (Dawn Newspaper 2019).

The treaty's dispute resolution mechanism has also come under scrutiny in recent years due to its inability to resolve disputes on hydropower projects constructed on shared waters. Under the IWT Article VIII, Pakistan and India have established the Permanent Indus Water Commission (PIWC) to supervise the sharing of waters between the two countries. The commissioners from both countries are scheduled to meet at least once a year to exchange hydrological data and information regarding planned water development projects, alternately in Pakistan and India. The functions of the PIWC are to establish and maintain cooperative agreements for IWT implementation, provide a report at the end of each year, inspect the rivers once every 5 years, and settle disputes. The commission is also responsible for sharing data on agricultural use, hydro-electric power generation, water storage, and flows in the rivers. Under Article VI, both countries are liable to share daily gauge and discharge data, reservoir extractions, canal withdrawals, and escapes, at the meeting of the Indus Water Commissioners. However, there is increasingly less trust in the quality of data shared between the two countries partly because it cannot be verified. In addition, there is no provision to share data on water quality of shared rivers, as well as data for the shared transboundary Indus Aquifer (Cheema and Pawar 2015).

In case of a dispute between the two countries, the Permanent Indus Water Commission presents the first point of referral where the issue can be solved bilaterally. In case the issue cannot be solved bilaterally, it can be referred to a neutral expert appointed by the two members of the commission or by the World Bank. Both parties also have the option of resolution of the dispute through the seven-member International Court of Arbitration. Over the last decade, on several occasions both India and Pakistan have had to go beyond the treaty's dispute resolution mechanism and resort to expensive arbitration in the case of recent hydropower projects such as Baglihar, Wullar, Neelam Jhelum and Ratle.

While Pakistan's transboundary relationship with India is precarious despite the existence of a treaty, there is no agreement on water sharing with Afghanistan on the use of Kabul River. Past efforts at establishing a Kabul Basin treaty have been stymied by political issues. In 2003, a nine-member technical committee headed by the Chairman of the Federal Flood Commission Pakistan visited Kabul to begin drafting a treaty on the use of shared waters, but their efforts were forestalled reportedly due to a lack of sufficient river flow data from the Afghan side (King and Sturtewagen 2010). There have been several efforts by international organisations such as the World Bank and USAID to provide platforms for renewed negotiation and science diplomacy since then, but these have not been able to get any political traction. Perhaps the strongest statement by the two countries was the Islamabad Declaration of 2009, adopted after the Regional Economic Cooperation Conference on

Afghanistan, but no concrete steps regarding water cooperation on Kabul River have materialised so far.

Both Pakistan and Afghanistan need to take into account the principles of earlier international conventions, accords and agreements and become a signatory to these, where possible. In this regard, the 1997 Convention on the Law of the Non-Navigational Uses of International Watercourses (the Convention) needs to be ratified by both countries. The countries have so far been hesitant in ratifying the Convention because of its open-ended terms and phrases such as "equitable use" which do not have an agreed upon legal definition and hence may be interpreted differently by riparians. For example, a less socio-economically developed riparian like Afghanistan may argue that it needs a larger share of water. In addition to the Convention, the 1911 Madrid Declaration (the Declaration) should also be looked at for inspiring a water sharing agreement on the Kabul River. The Declaration was a landmark international agreement on joint watercourses. It laid down several principles, such as discouraging unilateral alterations in the use of transboundary basins and the creation of joint water commissions to settle disputes between two or more riparian states.

As far as China's position in the Basin is concerned, the institutional and legal architecture for the protection of Tibetan glaciers under China's control are absent. While China has made significant investments in hydropower in Pakistan, much of this is without any basin-wide institutional or legal arrangement for protection of downstream countries in South Asia. The Indus Basin has not witnessed basin-wide multilateral cooperation among India, Pakistan, Afghanistan and China. There have been half cooked bilateral efforts at hydro-diplomacy, sometimes on the India-Pakistan, and sometimes on the Pakistan-Afghanistan front. Pakistan and China have had a long history of "friendship" ties and recently have been strengthening their economic partnership under the CPEC initiative, much to the chagrin of India. India and China have had a history of tumultuous transboundary relationship convoluted by territorial disputes and balance of power issues. Lonstanding political issues between basin countries have hindered progress on a basin-wide legal arrangement for sustainable transboundary water use.

3.4 Transboundary Water Governance for Pakistan's Future

This chapter has delved into the recent challenges that complicate sustainable management of Pakistan's transboundary water resources from the Indus Basin. It discusses the nature and underlying political economy of Pakistan's riparian relationship with each country that shares the Indus, highlighting the critical areas that need to be addressed to ensure Pakistan's water security. These challenges open a lot of questions that need to be explored, issues to be thought through, and decisions to be made for Pakistan's approach to sustainable transboundary water governance in the Indus Basin.

As discussed in the chapter, the first and most ominous of the challenges concerns the inadequacy of the only water sharing agreement in the basin – the IWT of 1960. Increasingly contentious hydro-politics between India and Pakistan, and the need to frequently resort to expensive international arbitration, suggests the treaty is falling short in its ability to address contemporary water sharing issues. Despite having a provision for incorporating additional areas for 'future cooperation' under Article VII of the treaty, it has not been fruitful in promoting any cooperation that goes beyond what was initially agreed in 1960. Some experts have argued for a renegotiation of the treaty to bring it more in line with contemporary international watercourse law, the Helsinki rules, and emerging concerns with water quality, environmental sustainability, climate change, and principles of equitable sharing (Mustafa 2010). While principles of international law such as the UN Watercourses Convention can serve as a useful starting point on which to base water sharing rules in the Basin, in reality Indus riparian countries have not supported these global principles. There are also differences between the extents to which each riparian country upholds principles of international law for transboundary water governance. For example, in case of the 1997 Convention on the Law of the Non-Navigational Uses of International Watercourses, both Afghanistan and Pakistan have been hesitant in ratifying the convention because of open-ended phrases such as "equitable use", the perspective on which will differ according to the country in question. In addition, both India and China have also not ratified the Convention.

Another key issue is that transboundary cooperation on the Indus has so far been restricted to bilateral efforts. Even if the scope and terms of the IWT end up being expanded to include issues of water quality, environmental sustainability and climate change, the treaty does not include Afghanistan and China. While basin countries periodically engage in bilateral efforts for transboundary water diplomacy, there is no multilateral institution that engages all riparians of the Indus on a basin-wide scale. Strengthening existing multilateral institutions to serve as platforms for cooperative exchange between countries could be useful in making the transition to basin-wide governance that would enable a more systematic approach to basin-wide issues. A region-wide institution that enrols all the four riparians of the Indus Basin would foster mechanisms for exchange of data and information to improve the current trust deficit between countries as well as steer the future of basin-wide management with regards to climate change and hydrological variability. The regional institution should be able to address issues of pollution and degradation, promote better flood management, and be able to emphasise benefit sharing for contentious hydroelectric projects on the shared watercourses. The potential for South Asian Association for Regional Cooperation (SAARC) to serve as a basin-wide organisation for transboundary water governance in the Indus cannot be overstated, as it could allow for a deepening of relations on a range of issues fostering economic interdependence among the countries (Magsig 2015). SAARC could bring together the two hegemons India and China to counter the current power asymmetries in the bilateral relationship between Pakistan and India on one hand, and Pakistan and Afghanistan on the other. It could also be effective in voicing lower riparian concerns for the massive diversion and river interlinking projects planned in India and

China. China currently has observer status in SAARC and has expressed interest in becoming a member. However, the history of China's transboundary relationship with basin countries suggests that the former has adopted a largely bilateral approach in the management of transboundary waters. China's lack of support for international legal principles governing the use of transboundary waters is further testified by its refusal to attend the World Commission for Dams 2000 meeting, thus allowing it to conduct dam construction activities in Tibet unhindered (Adeel and Wirsing 2016, p. 169).

While there is considerable potential for SAARC as a basin-wide institution for water cooperation, in reality its efficacy as a platform for South Asian cooperation has been limited. In the face of long standing political issues as well as disputed territorial claims between member countries in the Basin, SAARC remains ineffective in taking forward the discourse on the Basin's future. In addition, trust deficit between riparians stemming from concealment of hydrological information and planned hydropower construction on shared rivers, as well as securitisation of the water problem within each country, means that basin countries are mostly unwilling to come together on a shared basin-wide platform.

Sustainable transboundary water governance for Pakistan's water future would necessitate basin countries moving from a "water sharing" approach to a "benefit sharing" approach. The riparians of the Indus Basin have been thus far focused on sharing water leading to contentious hydro-politics between basin countries. Instead, moving to a benefit sharing approach can enable better management of ecosystems (benefits to the river), an increase in food and energy production (benefits from the river), as well as increased economic integration among riparian countries (generate benefits beyond the river) (Sadoff and Grey 2002). An integrated basin-wide and cooperative benefit sharing approach focusing on the water food and energy nexus can yield benefits for basin countries (Wada et al. 2019). Recent studies on the Kabul River Basin have shown the benefits of joint development and operation of water infrastructure for both Afghanistan and Pakistan as opposed to unilateral action by riparian countries (Khan et al. 2020). In this regard, basin countries can learn from international best practice case studies. The Indus can draw on the institutional design and efficacy of the Colorado and Murray-Darling Basin as an initiative to evolve and improve its institutions (Ahmad 2019; Sattar et al. 2018). For reaching an agreement on the Kabul River, Pakistan and Afghanistan can look towards the Columbia River Treaty as a useful blueprint to initiate the particulars of benefit sharing in the Kabul River Basin. Basin countries can also look at how the Danube has emerged as an important global example in transboundary river management, particularly in enhancing the water quality and environmental standards in the basin (Zavadsky 2019).

Notwithstanding the proposed benefits from multilateral coordination and benefit sharing on transboundary waters of the Indus, political issues have so far thwarted basin countries' experiences with hydro-diplomacy. The importance of track 2 and track 3 initiatives when official initiatives reach a political deadlock is critical for effective hydro-diplomacy on the Indus (Romsho 2012). In this regard, the Indus Basin can learn from examples of conflict-ridden basins where science diplomacy was effective, such as the Jordan River Basin and Lake Champlain. Pakistan and its neighbours must engage in science and environmental diplomacy in order to overcome the political deadlock over transboundary waters.

References

Adeel Z, Wirsing RG (2016) Imagining Industan: overcoming water security in a new world. Springer, Dordrecht

Ahmad S (2009) Water availability in Pakistan. Paper presented at National Seminar on Water Conservation, Present Situation and Future Strategy. Project Management and Policy Implementation Unit, Ministry of Water and Power, Islamabad, Pakistan

Ahmad S (2016) Water sector of Pakistan: a situational analysis. Develop Advocate Pak 3(4):2–9. United Nations Development Programme (UNDP), Islamabad, Pakistan. https://issuu.com/undppakistan/docs/dap_volume3__issue4_english

Ahmad B (2019) Is the Murray-Darling Basin success relevant and beneficial to Indus Basin? Lead Pakistan Talk Series No. 18. http://www.lead.org.pk/talks/talk18.htm

Ahmad S, Iqbal J (2016) Transboundary impact assessment of Indian dams: a case study of Chenab River Basin in perspective of Indus Water Treaty. Water Policy 18(3):545–564. https://doi.org/10.2166/wp.2015.137

Akhtar SM, Iqbal J (2017) Assessment of emerging hydrological, water quality issues and policy discussion on water sharing of transboundary Kabul River. Water Policy 19(4):650–672. https://doi.org/10.2166/wp.2017.119

Ali S (2020) Integrated modelling of Chenab River basin for flood forecasting. PhD dissertation, University of Agriculture, Faisalabad

Biswas AK (2011) Cooperation or conflict in transboundary water management: case study of South Asia. Hydrol Sci J 56(4):662–670. https://doi.org/10.1080/02626667.2011.572886

Biswas AK, Iwra F (1992) Indus Water Treaty: the negotiating process. Water Int 17(4):201–209. https://doi.org/10.1080/02508069208686140

Cai X, Sharma BR, Matin MA, Sharma D, Gunasinghe S (2010) An assessment of crop water productivity in the Indus and Ganges river basins: current status and scope for improvement (IWMI research report 140). International Water Management Institute (IWMI), Colombo. https://doi.org/10.5337/2010.232

Cheema MJM (2012) Understanding water resources conditions in data scarce river basins using intelligent pixel information – case: transboundary Indus Basin. Unpublished PhD Dissertation, Delft University of Technology, Delft, The Netherlands

Cheema MJM, Pawar P (2015) Bridging the divide: transboundary science and policy interaction in the Indus Basin. VF report. Environmental Security Program, Stimson Center, Washington, DC

Cheema MJM, Immerzeel WW, Bastiaanssen WGM (2014) Spatial quantification of groundwater abstraction in the irrigated Indus Basin. Groundwater 52(1):25–36. https://doi.org/10.1111/gwat.12027

Dawn Newspaper (2019) India shares with Pakistan data of three hydropower projects, February 16. https://www.dawn.com/news/1464148/india-shares-with-pakistan-data-of-three-hydropower-projects

Deemer BR, Harrison JA, Li S et al (2016) Greenhouse gas emissions from reservoir water surfaces: a new global synthesis. Bioscience 66(11):949–964. https://doi.org/10.1093/biosci/biw117

FAO (2011) AQUASTAT – Country profile – Pakistan. http://www.fao.org/3/ca0403en/CA0403EN.pdf

Gupta J (2017) Indus cascade a Himalayan blunder. The Third Pole, May 22. https://www.thethirdpole.net/2017/05/22/indus-cascade-a-himalayan-blunder/

Haines D (2017) Rivers divided: Indus Basin waters in the making of India and Pakistan. Oxford University Press, Karachi

Henderson G, Fogarty J, Gibb S (2019) Warming could decimate high-altitude South Asia water reserves. Bloomberg Law. https://news.bloomberglaw.com/environment-and-energy/warming-could-decimate-high-altitude-south-asia-water-reserves

Hill D (2016) The Indus Basin: the potential for basin-wide management between India and Pakistan. In: Adeel Z, Wirsing RG (eds) Imagining Industan: overcoming water security in a new world. Springer, Cham, pp 141–158

Iqbal AR (2013) Environmental issues of Indus River basin: an analysis. ISSRA Papers 5(1):89–112. https://ndu.edu.pk/issra/ISSRA_paper_2013_IssueI.php

IUCN (2010a) Beyond Indus Water Treaty: ground water and environmental management – policy issues and options. IUCN Pakistan, Karachi, 10 p

IUCN (2010b) Indus Water Treaty and managing shared water resources for the benefit of basin states: policy issues and options. IUCN, Karachi

Khan HF, Yang YCE, Wi S (2020) Case study on hydropolitics in Afghanistan and Pakistan: energy and water impacts of Kunar River development. J Water Resour Plan Manag 146(9):05020015. https://doi.org/10.1061/(ASCE)WR.1943-5452.0001262

King M, Sturtewagen B (2010) Making the most of Afghanistan's river basins: opportunities for regional cooperation. East West Institute, New York. https://reliefweb.int/sites/reliefweb.int/files/resources/57B4DF724C74B9DF492576D90010276F-Full_Report.pdf

Kugelman M, Alam AR, Bakhshi G (2011) Peace through water. The South Asia Channel, Foreign Policy, December 2. https://foreignpolicy.com/2011/12/02/peace-through-water/

Magsig BO (2015) Water security in Himalayan Asia: first stirrings of regional cooperation? Water Int 40(2):342–353. https://doi.org/10.1080/02508060.2015.1011463

Maharjan SB, Mool PK, Lizong W et al. (2018) The status of glacial lakes in the Hindu Kush Himalaya (ICIMOD research report 2018/1). International Centre for Integrated Mountain Development (ICIMOD), Kathmandu. https://lib.icimod.org/record/33736

Mahbub ul Haq Human Development Centre (2013) Water for human development. Human development in South Asia annual report 2013, Mahbub ul Haq Human Development Centre, Lahore. https://mhrc.lums.edu.pk/reports

Michel D, Sticklor R (2013) Toxic brew: water quality, public health, and human welfare in the Indus Basin. The Stimson Center, Washington, DC. https://www.stimson.org/2013/toxic-brew-water-quality-public-health-and-human-welfare-indus-basin-0/

Moore S (2019) The Blue Belt and road: China's belt and road and the world's water resources. Reconnecting Asia, February 6. https://reconnectingasia.csis.org/analysis/entries/blue-belt-and-road/

Mustafa D (2010) Hydropolitics in Pakistan's Indus Basin. US Institute of Peace, Washington, DC

Nabeel F (2020) Pakistan meets climate SDG, but that is not enough. The Third Pole, July 29. https://www.thethirdpole.net/2020/07/29/opinion-pakistan-meets-climate-sdg-but-that-is-not-enough/

Richey AS, Thomas BF, Lo MH et al (2015) Quantifying renewable groundwater stress with GRACE. Water Resour Res 51:5217–5238. https://doi.org/10.1002/2015WR017349

Romsho SA (2012) Indus River Basin common concerns and the roadmap to resolution. Centre for Dialogue and Reconciliation, University of Kashmir, Srinagar

Sadeqinazhad F, Atef SS, Amatya DM (2018) Benefit-sharing framework in transboundary river basins: the case of the Eastern Kabul River Basin-Afghanistan. Cent Asian J Water Res 4(1):1–18

Sadoff CW, Grey D (2002) Beyond the river: the benefits of cooperation on international rivers. Water Policy 4(5):389–403. https://doi.org/10.1016/S1366-7017(02)00035-1

Sattar E, Robison J, McCool D (2018) Evolution of water institutions in the Indus River Basin: reflections from the law of the Colorado River. Univ Mich J Law Reform 51(4):715–775. https://repository.law.umich.edu/mjlr/vol51/iss4/3

The News (2011) India to help Afghanistan build 12 dams on Kabul River, May 12. https://www.the-news.com.pk/archive/print/613757-india-to-help-afghanistan-build-12-dams-on-kabul-river

The Indian Express (2016, September 27) https://indianexpress.com/article/india/india-news-india/indus-water-treaty-blood-and-water-cant-flow-together-pm-modipakistan-uri-attack/

United Nations (1962) The Indus Waters Treaty 1960 between the Government of India, the Government of Pakistan and the international Bank for Reconstruction and Development. Signed at Karachi, on 19 September 1960. United Nations Treaty Series 6032(419), pp 126–292

US Senate Foreign Relations Committee Report (2011) Avoiding water wars: water scarcity and central Asia's growing importance for stability in Afghanistan and Pakistan. A Majority Staff Report. Committee of Foreign Relations United State Senate, 112 Congress 1st Session, February 22. https://www.foreign.senate.gov/imo/media/doc/Senate%20Print%20112-10%20Avoiding%20Water%20Wars%20Water%20Scarcity%20and%20Central%20Asia%20Afgahnistan%20and%20Pakistan.pdf

Vick MJ (2014) Steps towards an Afghanistan–Pakistan water-sharing agreement. Int J Water Resour Dev 30(2):224–229. https://doi.org/10.1080/07900627.2014.886471

Wada Y, Vinca A, Parkinson S et al (2019) Co-designing Indus water-energy-land futures. One Earth 1(2):185–194. https://doi.org/10.1016/j.oneear.2019.10.006

WaterWatch (2003) Crop and water productivity of the Pakistan wheat and rice systems. A Water Watch Study for the World Bank. http://www.waterwatch.nl/fileadmin/bestanden/Project/Asia/0053_PK_2002_CropWaterProductivity.pdf

Wester P, Mishra A, Mukherji A, Shrestha AB (eds) (2019) The Hindu Kush Himalaya assessment: mountains, climate change, sustainability and people. International Centre for Integrated Mountain Development (ICIMOD), Kathmandu. https://lib.icimod.org/record/34383

Williams JM (2018) Stagnant rivers: transboundary water security in South and Southeast Asia. Water 10(12):1819. https://doi.org/10.3390/w10121819

Xie L, Zhang Y, Panda JP (2018) Mismatched diplomacy: China–India water relations over the Ganges–Brahmaputra–Meghna River Basin. J Contemp China 27(109):32–46. https://doi.org/10.1080/10670564.2017.1363014

Zavadsky I (2019) Transboundary cooperation on Danube River: some lessons for Indus Basin in pollution management. Lead Pakistan Talk Series No. 33. http://www.lead.org.pk/talks/talk33.htm

Zawahri NA (2009) India, Pakistan and cooperation along the Indus River system. Water Policy, 11(1):1–20. doi:https://doi.org/10.2166/wp.2009.010, 1

Zwart SJ (2010) Benchmarking water productivity in agriculture and the scope for improvement – remote sensing modelling from field to global scale. https://ris.utwente.nl/ws/files/18476489/PhD_thesis_SanderZwart.pdf

Chapter 4
Water Security in Pakistan: Availability, Accessibility and Utilisation

Taimoor Akhtar, Hassaan F. Khan, and Daanish Mustafa

Abstract Water security is the assessment of a region's capacity to provide safe and adequate water for all (including humans and the environment). In Pakistan, water security has predominantly been studied and assessed at the national scale, and without due understanding of societal implications. However, given Pakistan's cultural, economic, and environmental diversity, assessment of water security in the country requires an understanding of human-water interactions at multiple scales, including national, regional, and local scales. This chapter is an attempt at analysing Pakistan's water security both at the national scale and the sub-national scales and with emphasis on societal implications. The sub-national scales analysed are: (a) South Punjab; (b) Karachi; and (c) Southern Sindh. Key outcomes of the local scale water security assessment are: (i) a lack of water quality data and high-resolution water quantity data is a major hurdle to adequate water security assessment, especially in urban settings; (ii) existing and antiquated energy policies are promoting unsustainable groundwater extraction; and (iii) the 'water for all' notion is non-inclusive in practice and should, in future, include water rights of the neglected, e.g., farmers of South Punjab who do not have access to groundwater, or the deltaic communities that have been severely impacted by the lack of flows to the Indus Delta. More local perspectives need to be understood to truly assess and alleviate Pakistan's water security challenges.

Keywords Water security · Water challenge · Water demand · Agricultural water use · Pakistan

---

T. Akhtar (✉)
College of Engineering and Physical Sciences, University of Guelph, Guelph, ON, Canada
e-mail: akhtart@uoguelph.ca

H. F. Khan
Dhanani School of Science and Engineering, Habib University, Karachi, Sindh, Pakistan
e-mail: hassaan.khan@sse.habib.edu.pk

D. Mustafa
Department of Geography, King's College, London, UK
e-mail: daanish.mustafa@kcl.ac.uk

M. A. Watto et al. (eds.), *Water Resources of Pakistan*, World Water Resources 9,
https://doi.org/10.1007/978-3-030-65679-9_4

4.1 Water Security in Pakistan: A Broad Perspective

4.1.1 Defining Water Security

'Water security' is a term with definitions that vary across academic disciplines, policy forums, governing institutions, and stakeholders (Cook and Bakker 2012). The definitions also vary across spatial scales, with broad definitions and metrics focusing on understanding water security at national and regional scales, such as the Falkenmark Water Stress Indicator (Falkenmark et al. 1989), and narrow definitions focusing on local scales (Vörösmarty et al. 2010). However, a comprehensive qualitative definition, that covers the essence of the word 'water security' is given by the United Nations:

> Water security is defined as the capacity of a population to safeguard sustainable access to adequate quantities of acceptable quality water for sustaining livelihoods, human well-being, and socio-economic development, for ensuring protection against water-borne pollution and water-related disasters, and for preserving ecosystems in a climate of peace and political stability. (UN WATER 2013, p. vi)

According to the above definition, water security of a society is embedded in the balance between three potentially contrasting components, i.e.: (i) water access to sustain human needs; (ii) protection against water-borne pollution and water-related hazards; and (iii) preservation and sustainable management of the coupled human-nature ecosystem. Within the context of Pakistan, this definition has extremely complex connotations, as Pakistan is a country with diverse livelihoods, ecosystems and socio-economic conditions, and like many other countries, also exhibits a wide range of spatial and temporal variability in its water resources and water uses (see chapter by Ahmed et al. in this volume). However, much of the discourse related to Pakistan's water security revolves around agriculture, hydropower, and related structural development requirements (including canals, mega dams etc.), and that too, when viewed from the lens of a bureaucracy, is disconnected from society (Mustafa 2002). This disconnect has manifested into a framework in which the varied water security challenges across the country are overshadowed and not addressed adequately by the country's water resource management systems. A case in point is the Pakistan Water Charter of 2018 that emphasises structural developments, but fails to emphasise engagement between water managers and society (Ministry of Water Resources 2018).

This chapter aims to highlight the variety of water security challenges in Pakistan through an agent-based perspective. Over the past two decades, the use of agent-based modelling and bottom-up water management approaches have emerged as alternatives to top-down centralised management (Khan et al. 2017; Yang et al. 2009; Zechman 2011). Among other things, an agent-based approach provides a more realistic framing of water challenges from end users' perspectives (Akhbari and Grigg 2013). Understanding the dynamics at the agent level and studying the interactions among agents can help reveal the emergent properties of a system, allowing for a more proactive water management policy (Berglund 2015).

We present three agent-based and local perspectives to provide a varied discussion of water security in Pakistan that goes beyond the broad techno-centric perspective dominating most of the water security discourse. The perspectives have been selected to highlight challenges related to groundwater management, urban water systems, and environmental flows in the context of societal justice. We recognise that these perspectives alone are not representative of the entire country, and there are other equally, if not more critical, water security perspectives across Pakistan.

We also understand that any robust water security analysis within an agent-based perspective requires sufficient data availability at local scales. However, there are severe gaps in the availability of such data in Pakistan at local scales. Hence, our analysis will also attempt to identify such data gaps. Moreover, we will look at three key sub-components of water security in this analysis, namely: (i) availability, defined as the renewable water available; (ii) access, which includes both physical and socio-economic access to water; and (iii) use, to explore how humans and the environment need water, and how it is being used.

4.1.2 Water, Population and Uncertainty

While water security is multi-dimensional and complex to assess, a first step in assessment of water security of a society is to comprehensively account for its water supply and consumptive water demand (Vörösmarty et al. 2010). According to numerous 'estimations' (FAO 2011; Laghari et al. 2012; van Steenbergen et al. 2015), Pakistan's average annual renewal water availability is around 220 billion cubic metres. However, as Young et al. (2019) pointed out, due to lack of adequate water monitoring and accounting (especially of water resources in Balochistan and of groundwater-surface water interactions), these estimations are to be treated with caution.

Due to lack of publicly available water consumption data, comprehensive estimations of water demand in Pakistan have rarely been attempted. Instead, the assessment of water security is typically tied to the water supply-demand debate by correlating water demand with population (Falkenmark et al. 1989). For instance, a recent notion that is very popular in Pakistan is that the country is becoming water-scarce (Nabi et al. 2019) because annual per capita water availability has gone below 1000 m^3 (Schewe et al. 2014).

Simply attributing the country's water security and stress to population is a dangerously simplistic approach to understanding the multi-dimensional nature of water security. Even to adequately assess water stress and scarcity at the national scale, a more comprehensive effort is required to analyse and assess the spatio-temporal variations in water supply and demand. Firstly, freshwater supply/availability cannot be assumed to be fixed, neither annually, nor intra annually. Figure 4.1 provides an illustration of the uncertainty of annual and intra annual renewable freshwater availability in Pakistan (estimated). The dotted black line in Fig. 4.1a is the average annual water availability (estimated from data of years 1961–2020).

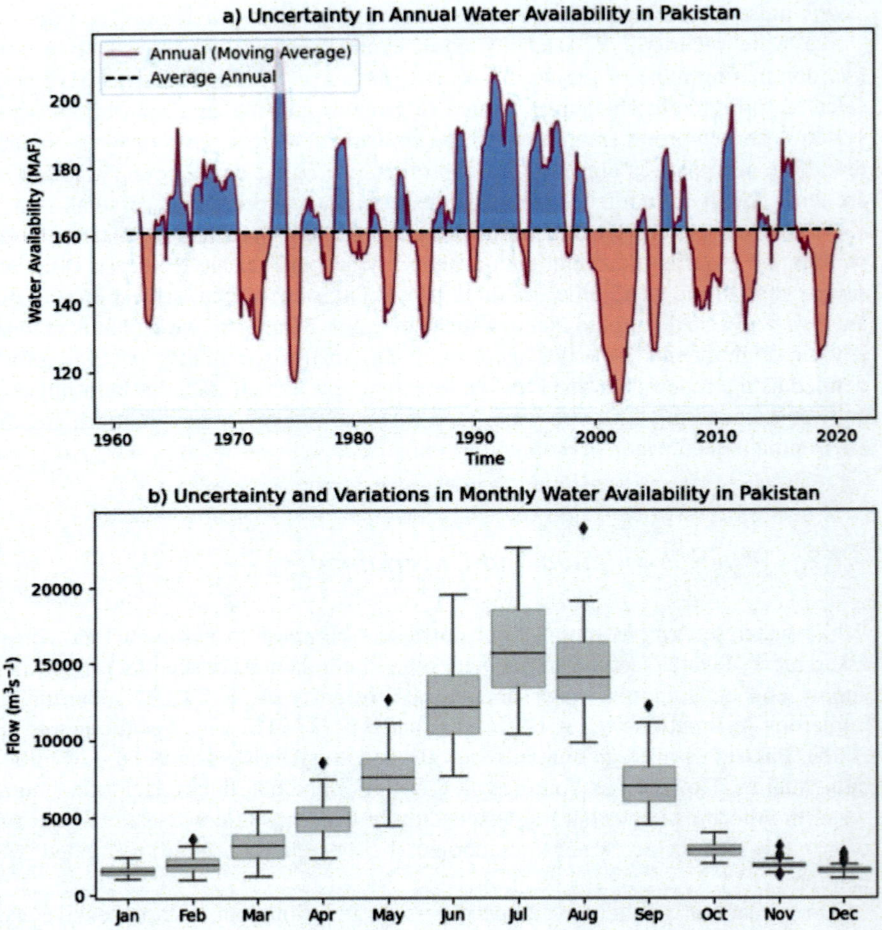

Fig. 4.1 Uncertainty in (**a**) annual (Million Acre Feet [MAF]) and (**b**) intra annual (monthly) water availability (m^3/s) in Pakistan. Panel **a**) includes a 12-month moving average of annual renewable water availability in Pakistan and panel **b**) shows the distribution of monthly renewable freshwater availability within the country. Estimates of both panels are derived from observed inflows from Indus (at Tarbela), Jhelum (at Mangla), Chenab (at Marala Barrage) and Kabul (at Nowshera) rivers (also called western rivers), and the authors' estimates of internal runoff in the Indus Basin and runoffs generated in Kharan and Makran basins. Based on prior studies, we have assumed that freshwater availability from precipitation, internal system run-off, eastern rivers and Kharan, and Makran basins is 25% of total annual water availability

The years shaded in blue report higher than average water availability and the years shaded in pink report lower than average water availability.

As is depicted in Fig. 4.1a, freshwater availability at the national level varies significantly across years. For instance, in the mid-1990s, annual freshwater availability in the country was, approximately, 25% higher than average, whereas in the early 2000s it was 25% lower than average. Moreover, the last two decades have seen a prevalence of lower than average water availability (drought conditions). A plausible explanation of this trend may be that the Indus Basin is experiencing a drought cycle since there is no clear evidence that these reductions in overall annual flows is due to the variable impacts on precipitation from climate change (Young et al. 2019). Indeed the graph in Fig. 4.1a points to the futility of focusing on average flows. Like all water environments, Pakistan's too is a variable one and the management systems have to respond to that uncertainty instead of trying to impose the monochromatic lens of average flows upon that variegated reality.

While annual renewable water availability in Pakistan can vary significantly, seasonal and monthly availability variations are even starker (see Fig. 4.1b for illustration), and spatial variations in water availability are also high. Hence, attributing water demand to Pakistan's population and national-scale annual renewable water availability only is an over-simplification. Water demand in Pakistan is extremely variant both temporally and spatially (due to local-scale consumptive demands and population densities). However, even an aggregate estimation of annual water demand in the country is not trivial. In the next sub-section, we will attempt to understand water supply and demand in Pakistan through the lens of water flows through the socio-hydrologic cycle.

4.1.3 Understanding the Resource at the National Scale

Figure 4.2 provides an illustrative overview of the human-water interface (also called socio-hydrologic cycle) of Pakistan at the national scale, including sources of renewable water availability, physical infrastructure that allows human access to water (canal system and groundwater) and the dominant water demand sectors.

Inflows from the transboundary upper catchments of Indus, Jhelum, Chenab and Kabul Rivers (also called western rivers) are the primary source of renewable water availability in the country (75% of total renewable water comes from western rivers). A large magnitude of this water is diverted through a vast network of surface canals, primarily for use in agriculture. Since a large proportion of the diverted water goes into the ground as recharge, it can be and is accessed for re-use via groundwater extraction. Spatio-temporal equity in surface/canal water distribution to provinces, to canal systems and within canal systems is ensured via an allocation and water-rights based distribution mechanism. At the provincial level, the water allocation mechanism follows the allocation framework prescribed in the 1991 Provincial Water Apportionment Accord (Anwar and Bhatti 2018), whereas within canal systems, water distribution is based on an equity-based water rationing and scheduling mechanism called *Warabandi* (Wescoat et al. 2018).

As is depicted in Fig. 4.2, Pakistan's water demand (dominated by agriculture) is heavily reliant on groundwater, with around 45–60% of agricultural water demand satisfied by groundwater (Hassan 2016; Watto and Mugera 2016). This high demand has put immense pressure on the country's groundwater resources over the last two

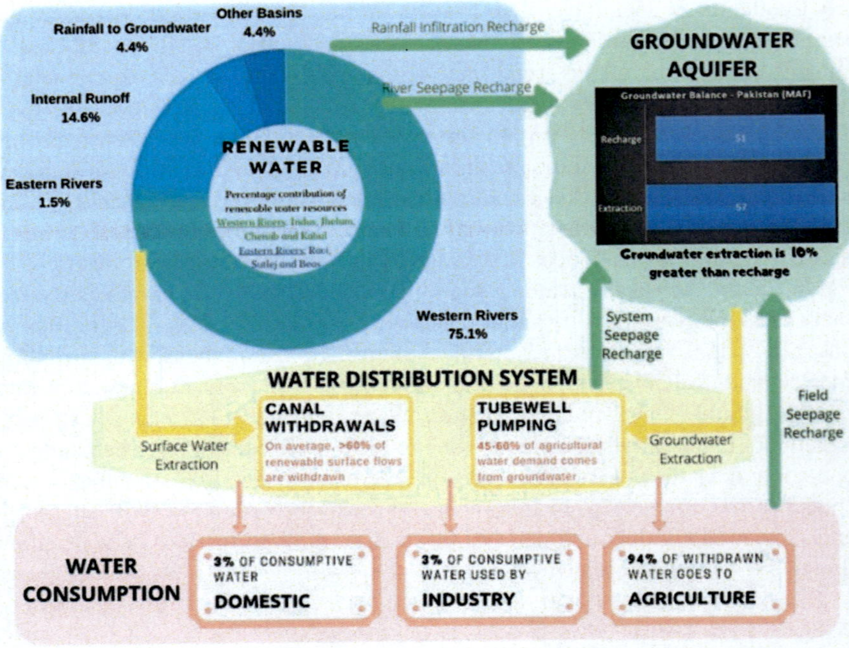

Fig. 4.2 Pakistan's water outlook: A broad perspective of water availability and consumptive use. The numbers provided in this figure are derived from Hassan (2016); Young et al. (2019) and authors' calculations based on simulations using the IBMR model (Yang et al. 2013)

decades; so much that annual groundwater extractions now are greater than the average annual recharge (see Fig. 4.2).

4.1.4 Beyond the Broad Perspective: Water Security at Local Scales

Section 1.3 (and Fig. 4.2) provides a cursory overview of water availability, access and use in Pakistan, and paints a generic picture of the socio-hydrology of the country. It alludes to the fact that groundwater is a key source of freshwater in the country that is being over-exploited. However, the national-scale discussion of Sect. 1.3 does not provide insights on water security at local scales within the country. For instance, no insights on urban water security can be drawn from Sect. 1.3 and Fig. 4.2. Local-scale water security insights require a local-scale analysis. Hence, in subsequent sections, we will discuss three local perspectives of water security in Pakistan, i.e.: (A) the South Punjab perspective; (B) the Karachi perspective; and (C) the Southern Sindh and the Indus Delta perspective. These perspectives have been selected to provide a varied discussion beyond the dominant bureau-centric discourse. We recognise that these perspectives alone are not representative of the entire country, and there are other equally, if not more, critical water security perspectives across Pakistan.

4.2 Perspective A: South Punjab

4.2.1 Context and Study Area

South Punjab is the epicentre of Pakistan's cotton and wheat production, with agriculture playing a central role in the socio-economics of the region. However, agricultural access to water is only possible via surface and groundwater irrigation. With an average annual rainfall of around 200 mm (Amin et al. 2018), South Punjab is predominantly an arid region. Hence, the economic, cultural and socio-hydrologic evolution of South Punjab is characterised by its inhabitants' deep relationship with rivers carrying waters from the upstream, canals diverting waters to arid (but rich) lands, and (more recently) electric tube-wells extracting groundwater to foster agro-economic growth.

The Indus Waters Treaty (IWT), a transboundary water-sharing agreement signed between India and Pakistan in 1960, brought a monumental shift in the socio-hydrology of South Punjab. According to IWT, consumptive rights for the Ravi and Sutlej Rivers (also called eastern rivers) were given to India, while consumptive rights for the Indus, Jhelum and Chenab Rivers were given to Pakistan (Qamar et al. 2019). The aftermath of IWT has led to a gradual recession of flows in Ravi and Sutlej (in Pakistan), so much so that, since 2000, inflows to Pakistan from these rivers has become negligible (excluding some monsoon months) (Young et al. 2019). The dramatic change in the flow regime of the Ravi and Sutlej has had a significant impact on socio-hydrology and water security of the eastern part of the South Punjab region (see the zone marked in red in Fig. 4.3) (Wescoat et al. 2018). Both rivers (in their natural regime) were primary sources of surface water access and groundwater recharge for South Punjab's dwellers and were also critical for the ecological health of the Indus Basin in South Punjab.

The South Punjab region is home to the largest proportion of the Indus Basin Irrigation System (IBIS) and encompasses approximately 30% of the irrigated area of IBIS. Development of such a massive irrigation system started causing major waterlogging issues within the South Punjab region in the mid to late twentieth century. Canal waters being diverted from the IBIS were inducing surplus recharge into the underlying unconfined aquifer. Government led tube-well installation campaigns were thus initiated in the 1960s to alleviate these waterlogging issues (Watto and Mugera 2016). A tube-well installation boom ensued (Wescoat et al. 2000) that coincided with the drying up of the Ravi and Sutlej and led to an extremely high dependence on groundwater for consumptive use in South Punjab. An in-depth understanding of this dependence and its implications on sustainable water use in the region is needed to better understand South Punjab's water security. We delve deeper into this understanding in subsequent subsections of Sect. 2.2.

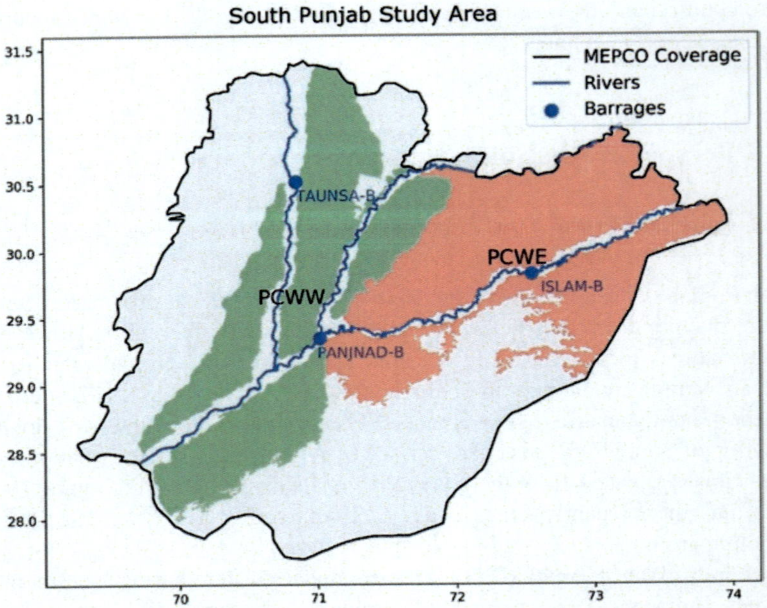

Fig. 4.3 Study area of South Punjab analysed in this study. This area is the 13 districts of Punjab that are serviced by MEPCO for electric supply

Figure 4.3 formally delineates the South Punjab region that is analysed in this study, in the context of understanding South Punjab's water security and related challenges. This region comprises of 13 districts: Multan, Muzaffargarh, Layyah, Dera Ghazi Khan, Rajanpur, Lodhran, Bahawalpur, Rahim Yar Khan, Khanewal, Sahiwal, Pakpattan, Vehari, and Bahawalnagar. As depicted in Fig. 4.3, this region is characterised as agriculture-centric (two agricultural zones are marked in red and green), with more than 50% land under cultivation and with direct access to canal waters.

4.2.2 *Agricultural Water Availability*

As mentioned earlier, agriculture is the primary water consuming sector in the South Punjab region (delineated in Fig. 4.3), where water is either accessed via the Indus Basin canal network, or from groundwater. Figure 4.4 provides an estimated overview of the temporal pattern of the water supply (with contribution from different supply sources) and demand balance of crops in the root zone. This estimate is derived from simulation results of the Indus Basin Model Revised (IBMR) (Yang et al. 2013, 2014).

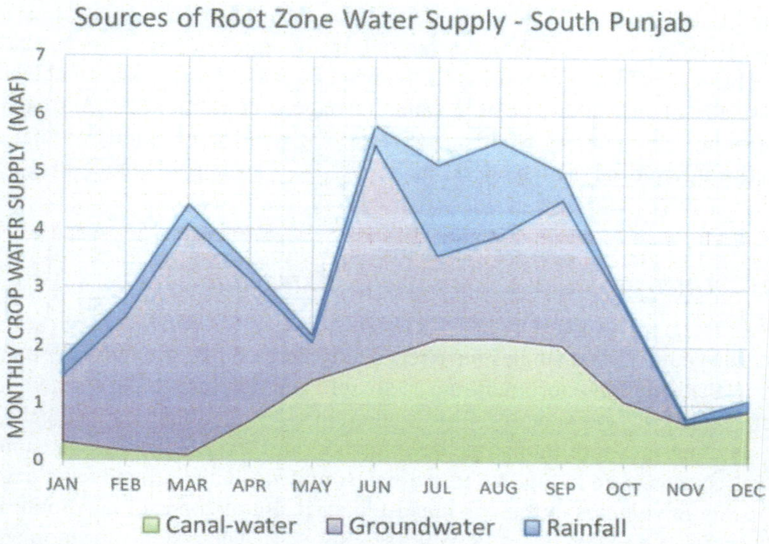

Fig. 4.4 Sources of crop water supply in South Punjab, estimated via simulation using the IBMR model. These estimates were made by simulating IBMR for the year 2017–18 (Apr–Mar)

IBMR is a large-scale river basin modelling and planning tool, designed to analyse the impact of structural and non-structural policy interventions within Pakistan's Indus Basin Irrigation System (IBIS). Developed by the Water and Power Development Authority (WAPDA) and the World Bank (Yu et al. 2013), IBMR is the de facto water resource planning tool used in Pakistan. The model is semi-distributed and divides the agricultural command area of IBIS into 12 spatial units, also called Agro-Climatic Zones (ACZs) (Yang et al. 2016; Yu et al. 2013). Two of these zones reside in the South Punjab study area delineated in Fig. 4.3 and are called Punjab Cotton Wheat East (PCWE) and Punjab Cotton Wheat West (PCWW), respectively.

A core feature of IBMR is its ability to simulate the supply of water to crops (i.e., the root zone) within an ACZ, from different supply sources (including direct precipitation/rainfall, groundwater, and canal-water irrigation. The model uses water balance equations to simulate supply to crops, under given agricultural production, and land-use constraints (e.g., minimum production of wheat in an ACZ is a model constraint).

Figure 4.4 provides a simulated estimation (via IBMR simulation) of the water supply sources of the PCWE and PCWW zones of the Indus Basin that are part of our South Punjab study area (see Fig. 4.3). The water supply estimations depicted in Fig. 4.4 are based on an IBMR model run with hydrologic and crop production data for the year 2017–18 (Apr–Mar) (MNFSR 2018).

It is evident from Fig. 4.4 that groundwater is the most critical water supply source for the agricultural economy and socio-economic well-being of South Punjab. It provides up to 90% of the crop water requirements during the peak winter growing season (Rabi)—a season that is also critical for the production of food crops, such as wheat. Even so, an in-depth understanding of the implications of such reliance on groundwater, and implementations of relevant policies about its sustainable use, are almost absent in South Punjab's water management framework. A fundamental question in this regard is whether the level of groundwater extraction that ensues from the agricultural demand depicted in Fig. 4.4 is sustainable in the long-run.

4.2.3 Unsustainable Groundwater Extraction

Numerous authors and studies have repeatedly iterated that groundwater extraction rates in Pakistan are exorbitant, unsustainable and a severe threat to the country's water security (Cheema et al. 2014; Laghari et al. 2012; Watto and Mugera 2016). The situation in South Punjab echoes this notion. Using IBMR, we expanded the analysis summarised in Fig. 4.4 to estimate/simulate annual groundwater extraction and recharge volumes in the two canal-fed agricultural zones of our South Punjab study region, i.e., PCWE and PCWW (see Fig. 4.4). The results are summarised in Fig. 4.5.

Figure 4.5 clearly shows that groundwater extraction in South Punjab is considerably greater than recharge, especially in the PCWE agricultural zone (marked in red in Fig. 4.3), where extraction is approximately 20–30% more than recharge. The huge gap between extraction and recharge in the PCWE zone is a direct consequence of unwarranted and extreme use of groundwater resources. The situation has been exacerbated by immense groundwater expansion in South Punjab, and in fact, in the entire country in the last few decades. This expansion continues to be

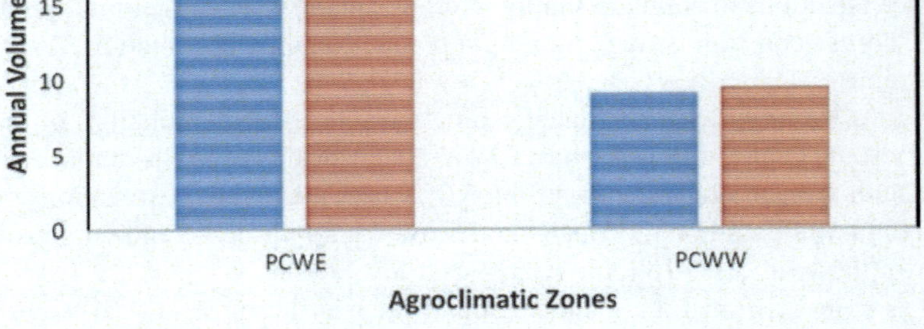

Fig. 4.5 Illustration of groundwater over-extraction in South Punjab (especially PCWE zone). Plots of estimated (via IBMR simulation) annual recharge and extraction in PCWE and PCWW zones of South Punjab

completely unregulated, with the number of installed private agricultural tube-wells increasing at a non-linear rate (MNFSR 2018; Qureshi et al. 2008; Siddiqi and Wescoat 2013; Watto and Mugera 2016).

4.2.4 Electric Tube-Wells and Policy Paradigm Shift

The lack of a comprehensive regulatory framework for ensuring sustainable extraction of groundwater is not the only threat to the water security of South Punjab. In fact, groundwater levels are threatened more by existing (and antiquated) policies implemented within the region that promote and foster over-extraction. For instance, electricity for agricultural use is heavily subsidised in South Punjab (Siddiqi and Wescoat 2013) by the Multan Electric Power Company (MEPCO), a distribution company that supplies electricity to the thirteen districts of South Punjab analysed in this study (see Fig. 4.3).

Subsidising electricity was a government strategy to maintain socio-economic and agro-economic growth of South Punjab, after shifts in water availability due to IWT and rapid population increase (Siddiqi and Wescoat 2013). Ready access to groundwater has since helped farmers in better managing water demands, given the uncertainties of surface water supplies (Qureshi et al. 2008). However, continued prevalence of electric subsidies has led to a rapid increase in cropping intensity and shifts in cropping choices, and is now a key cause of unsustainable and inequitable water extraction in the region. We will attempt to illustrate this by estimating the annual volume of groundwater extracted by electric tube-wells in South Punjab.

Annual electricity consumption data of South Punjab is used in this study to deduce an estimate of annual groundwater extracted by electric tube-wells in the region. The Appendix provides details of the method used in volume computation. MEPCO data of annual electricity consumption by electric tube-wells in South Punjab (2005–2017) is taken from the National Electric Power Regulatory Authority (NEPRA)'s state of industry reports (NEPRA 2018). A 50% pump efficiency and 15% transmission losses are assumed, as reported in International Resources Group (2015), and average depth to water table is assumed to be around 125 feet – a conservative estimate based on numerous reports and studies (Khan et al. 2016; Young et al. 2019) (see the Appendix for details regarding the method for extraction volume calculation).

Figure 4.6 plots annual groundwater extracted by electric tube-wells from years 2005–2017, as per estimates of this analysis. Trends show that groundwater extraction from electric tube-wells has significantly increased since 2005. Moreover, the extracted volume estimated in 2017 is considerably greater than the live storage capacity of the recently proposed Basha dam project (depicted by the red line in Fig. 4.6).

The magnitude of groundwater extraction from electric tube-wells in South Punjab (as estimated in Fig. 4.6) indicates that existing electricity subsidies for tube-well pumping in South Punjab should be urgently revisited. These subsidies promote unsustainable groundwater extraction and thus are a core source of water

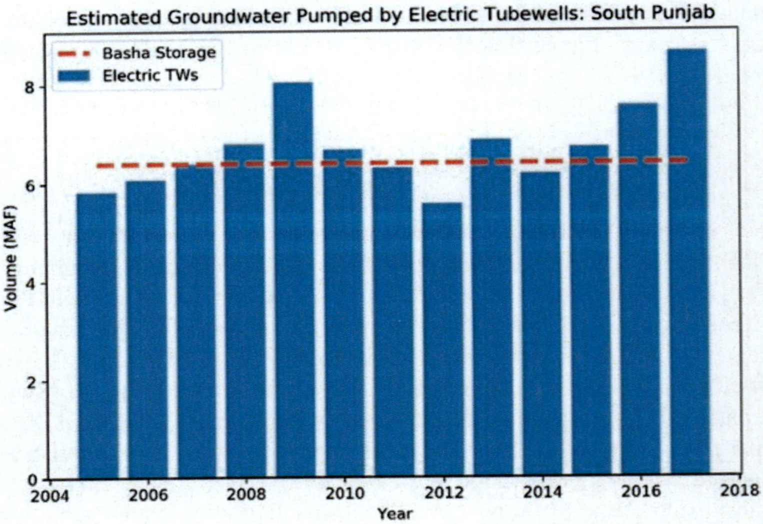

Fig. 4.6 Estimates of annual (2005–2017) volume in Million Acre Feet (MAF) of groundwater pumped by electric tube-wells in South Punjab

insecurity for the South Punjab region. Moreover, given that electric tube-wells are extremely costly to install, access to water resources from electric tube-wells is only limited to the economically affluent, large farmers.

According to Agricultural Statistics of Pakistan 2017 (MNFSR 2018), less than 15% of total installed tube-wells in Punjab are electric. If these numbers are assumed to be proportionate across sub-regions within Punjab, 15% of the total installed tube-wells in South Punjab that are electric accounted for approximately 25–30% of groundwater extraction in that region in 2017 (see Fig. 4.6). The number of installed electric tube-wells in the entire Punjab province was around 150,000 in 2017 (MNFSR 2018). Even if we assume that all electric tube-wells of Punjab (in 2017) were installed in South Punjab, the equity in water access within the region is extremely bleak, and to such an extent that less than 150,000 electric tube-well owners have unregulated access to groundwater and extracted more than 8 Million Acre Feet of water in 2017. This number (i.e., 8 MAF) is greater than the live storage volumes of Tarbela and Basha reservoirs, and accounts for approximately 15–20% of renewable freshwater available in South Punjab. The cumulative population of South Punjab was around 37 million in 2017, and less than 150,000 electric tube-well owners had access to and extracted approximately 15–20% of renewable freshwater in the region in 2017.

Our analysis of water security of South Punjab has three stark outcomes: (i) the agriculture-centric region is predominantly dependent on groundwater access for agro-economic progress and food security (see Fig. 4.4); (ii) groundwater extraction in South Punjab is unsustainably high and is greater than recharge (see Fig. 4.5); and (iii) only a fraction of the South Punjab population (less than 0.5%) has disproportionate access to freshwater resources (via electric tube-wells) enabled by agricultural subsidies that should be revisited urgently.

4.3 Perspective B: Karachi City

Pakistan is rapidly urbanising and by 2030, over half of its population will live in cities (United Nations 2019). This dramatic increase in urban population places an enormous stress on existing water infrastructure in cities. Nowhere is this effect more pronounced than in the coastal megacity of Karachi that serves as Pakistan's economic hub. Massive increase in Karachi's population coupled with deteriorating infrastructure has led to crippling water scarcity and water stress. Access to safe and reliable water is one of the major challenges that an urban dweller in Karachi faces on a daily basis.

4.3.1 Water Availability

Most of Karachi's water supply today is piped in from Keenjhar Lake, which has a catchment area of approximately 910 km^2, a usable storage capacity of 481 million cubic metres, and is fed by waters of the Indus River. Presently, a supply of approximately 500 million gallons per day (MGD) is pumped to Karachi from Keenjhar Lake through a variety of conduits. Transporting this vast volume of water requires a tremendous amount of energy and places a huge financial burden on Karachi's water utility, the Karachi Water and Sewerage Board (KWSB) (World Bank 2019). A second source of freshwater for Karachi, albeit much smaller relative to Keenjhar Lake, is the Hub Dam, which is completely reliant on precipitation to fill its reservoir. Given the significant spatial and temporal variability in precipitation, water supply from Hub Dam varies annually. Based on a yield analysis conducted in 1985, Hub Dam can supply approximately 75 MGD with a 95% reliability (JICA 2008).

Until recently, groundwater was another source of freshwater for Karachi with the wells of Dumlottee among the major water supply sources in the early to mid-twentieth century. The aquifer beneath the city is non-saline near sources of groundwater recharge but becomes saline deeper and further away from these recharge zones. With spiralling water demand in the 1990s and 2000s unmet by KWSB, many households installed on-premises tube-wells. However, due to the resultant over-utilisation of the aquifer, the water table has declined over the last several decades. Large scale pumping operations have created differential aquifer zones allowing saline and non-saline water to move laterally and vertically through the subsurface. Studies have shown that the aquifer located 1–5 km from the Arabian Sea is fully contaminated by saline water. Wastewater is also directly released at various points into surface and subsurface water, and rivers and underlying aquifers are getting recharged with contaminated water.

4.3.2 Water Demand

Much of the discourse regarding Karachi's water challenges focus on a perceived 'shortage' in supply compared to water demand, However, this perceived water demand in Karachi is not informed by credible empirical analysis and is based on unverified assumptions. Existing literature on Karachi's water system does not provide a sectoral breakdown of the city's water demand (domestic, commercial and industrial) nor does it provide insights into how this demand varies spatially (across various socio-economic and ethnic groups) and temporally (i.e., across seasons). Estimates for Karachi's water demand vary considerably from 1000 MGD – 1500 MGD. While not based on a thorough systematic study of water demand in the city, these estimates continue to be used today by water managers and planners alike for planning purposes.

4.3.3 Access

Access to domestic water in Karachi is highly unequal and is currently geared to provide water to the most affluent and politically connected consumers. In a majority of the neighbourhoods across the city, piped water supply is available for no more than a few hours every week, with no supply at all in many of the poorest neighbourhoods. Many factors contribute to this lack of access to piped water. Manipulation of water valves to direct water to higher-paying consumers (e.g., high-rise flats) has been recorded in several areas (Mustafa et al. 2017). In some cases, suction pumps are installed by households who can afford it. These powerful suction pumps divert water from the mains, which causes consumers' pipes further along in the system to dry up.

For the poorest households, often the only source of water is through deliveries from private water tankers, at a considerable expense for these citizens. Historically, many of these tankers obtain water from illegal fire hydrants, though recent efforts by the KWSB have led to a dismantling of several illegal hydrants and regularisation of private water vendors. Despite that, there exists great inequity with respect to the cost of water for domestic households. While political and socio-economic factors play a huge role in determining access to water in Karachi, so do geographic factors. The ad hoc manner in which the water infrastructure has been constructed to cope with the growth of the city has led to a haphazard network which leads to system inefficiency. In many cases, this is coupled with a lack of easily available documentation regarding the infrastructure which makes regular system maintenance and repair difficult.

A majority of water use in Karachi is for domestic purposes. With limited land availability and a lack of green spaces, landscaping water use is relatively low and is found generally in affluent neighbourhoods. In many cases, grey water is used for landscaping. Due to Karachi's position as a major trading port and Pakistan's economic hub, there is a significant industrial water demand. Over a third of water in Karachi is allocated for non-domestic use. Given the unreliable water supply, many of the industries have increased their use of treated wastewater and have shifted to water efficient processes.

4.3.4 Water Quality and Hazards

A large proportion of water in Karachi is untreated. Karachi currently has the infrastructure to treat only two-thirds of the piped water supply. Although two wastewater treatment facilities exist, they are non-operational and thus virtually all of Karachi's sewage is disposed untreated into the city's waterways and on the coast. Not only does this significantly degrade the ecosystems, but it also presents serious health hazards. Waterborne diseases in Karachi are estimated to lead to significant morbidity (e.g. diarrhoea), especially in the informal settlements which lack improved or even basic sanitation facilities (Young et al. 2019).

Unplanned growth of Karachi has led to riparian zones around regional rivers being greatly altered from their natural states, limiting the ability to mitigate high flows and limit flood effects. Along the banks of rivers in the city, densely populated slum settlements and industrial factories have encroached. In addition to these developments, there is significant sand and gravel excavation along the riverbeds, with rivers excavated up to 20 ft. deep. This alteration of the riverbeds allows stormwater to flow directly to the Arabian Sea without recharging the groundwater in the area, and increases the likelihood of major flooding events (Irfan et al. 2018).

4.4 Perspective C: Southern Sindh and the Indus Delta

Kotri Barrage, situated in Sindh, is the last major obstacle on the Indus River. The barrage diverts water to 1.3 million hectares of agricultural land in Southern Sindh (around 25% of the canal-fed agricultural area of Sindh) through four canal systems. The afore-mentioned agricultural land is the Southern Sindh study area of our analysis, is delineated in Fig. 4.7 below, and also includes the Indus Delta. Wheat and rice are the dominant crops planted in this area. Hence, it is also called the Sindh Rice Wheat South (SWRS) zone in this study (also called SWRS in Yang et al. 2013).

We have chosen SWRS as a local study in our analysis due to its unique socio-economic, ecological, and environmental importance within Pakistan and the Indus Basin. Since SWRS is the last agricultural zone of the Indus Basin it remains vulnerable to uncertain water availability, due to demand pressures in upstream agricultural areas. Moreover, SWRS also encapsulates the fifth-largest delta in the world, i.e., the richly biodiverse Indus Delta, home to valuable mangrove forests and a hub for fishing services (Renaud et al. 2013).

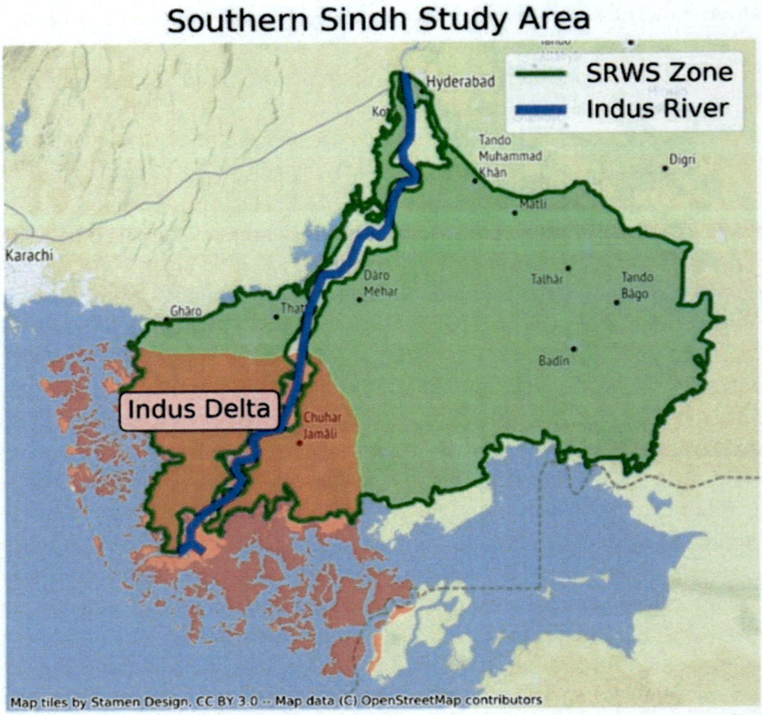

Fig. 4.7 The Southern Sindh area analysed in this study. This study area includes the south-most agricultural zone of the Indus Basin (marked in green) in Sindh, that includes the Indus Delta (delineated in red)

4.4.1 Uncertain Water Availability

Figure 4.8 provides an overview of the temporal (annual moving average) uncertainty in water withdrawals to SWRS. It can be noted that volumetric withdrawals deviate by around 20% from the average from 2001 to 2004. This variation is significantly higher than a 10% variation from the average observed at the upstream Sukkur barrage. Given that SWRS is the last agricultural zone within the Indus Basin, a higher uncertainty in water supply to SWRS is not surprising and may be attributed to uncertainties in demands and supplies at upstream diversion structures of the Indus Basin. However, this uncertainty may create a 'perceived' water security threat in Southern Sindh in future, given upstream inclinations towards development of new storage reservoirs and diversion canals (e.g., Basha reservoir and Jalalpur Canal).

A real water security challenge for SWRS is its lack of access to fresh groundwater. While farmers in most parts of Punjab and some parts of central and northern Sindh have access to fresh groundwater (Qureshi et al. 2008), this is not the case for farmers of southern Sindh and the SWRS zone. Consequently, uncertainties and

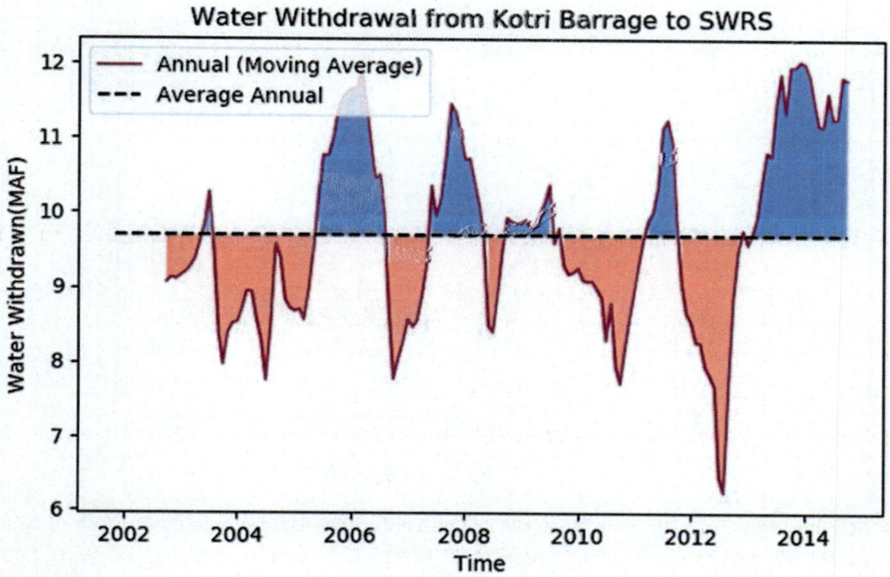

Fig. 4.8 12 month moving average of volume (MAF) of water withdrawn to SWRS zone from Kotri Barrage. (Source: NESPAK)

shortages in surface water withdrawals from Kotri (as depicted in Fig. 4.8) cannot be augmented via groundwater, and coupled with the arid nature of the region, pose severe challenges to the economic well-being of farmers in SWRS.

4.4.2 Water Security of the Indus Delta

Considered amongst the most vulnerable deltas of the world (Tessler et al. 2015), the Indus Delta comprises a significant portion of SRWS's land (see Fig. 4.7). Ranked sixth in the world on the list of deltas most vulnerable to seawater intrusion, the Indus Delta has suffered significant losses to its agricultural mudflats (Memon and Thapa 2011), biodiversity and mangrove forests (0.24 million hectares to 0.1 million hectares) in the past few decades (Renaud et al. 2013). These ecological losses are a consequence of narrow-sighted developments on and management of the Indus Basin Irrigation System that have led to significant flow reductions in Indus River downstream of Kotri and to the Arabian Sea, and have severely impacted the livelihoods of the delta dwellers (Memon and Thapa 2011; Renaud et al. 2013).

Figure 4.9 plots the flow duration curve (FDC) of Indus flows downstream of Kotri from daily data of years 2001 to 2014. An FDC is a plot of flow (y-axis) versus the probability of exceedance (%) of flow (computed from historical data). It is evident from Fig. 4.9 that, since 2001, the flow in Indus downstream of Kotri was

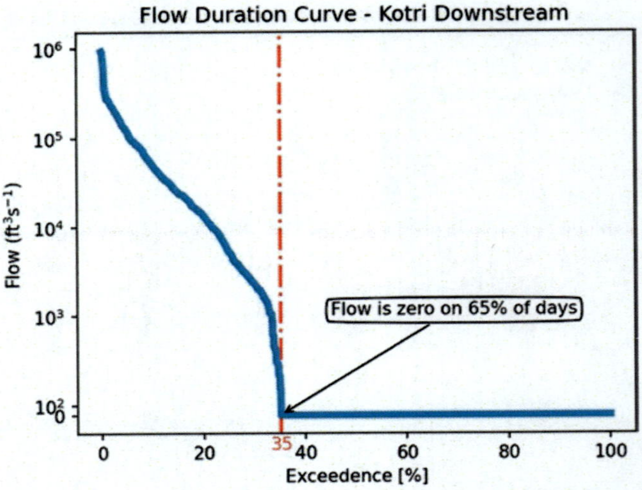

Fig. 4.9 Flow duration curve (FDC) of Indus flows to the Arabian Sea, downstream of Kotri Barrage for data from 2001 to 2014. (Data Source: NESPAK)

zero on around 65% of days in a year. This is a dramatic departure from the Indus River flow regime in prior decades. For instance, the average percentage of zero flow days downstream of Kotri from 1960–2000, was about 30%, whereas this percentage was 0% before 1960 (Memon and Thapa 2011).

The adverse impacts of the reduction in Indus flows downstream of Kotri on the population of the Indus Delta have also been immense. For instance, the dramatic reduction in downstream flows coupled with the uncertainty in water withdrawals to the canals of Kotri Barrage have diverted the occupation of paddy farmers and camel herders of the delta towards marine fishing. Paddy farming and camel herding was the primary occupation of more than 50% of deltaic households in 1990. This percentage reduced to less than 2% in 2011, when more than 85% of deltaic households shifted to marine fishing as a primary occupation (Memon and Thapa 2011). This monumental shift in the occupation of the Indus River's deltaic communities has significantly increased the stress on marine fishery resources.

The dramatically reduced river flows in the Indus Delta and consequent adverse impacts on society demand policy interventions that can minimise degradation of the Indus Delta and create pathways for delta restoration. One such policy measure that has been repeatedly ignored in policy circles is the allotment of minimum environmental flows downstream of Kotri. This need for allotment of minimum environmental flows is also documented in the 1991 Water Apportionment Accord (WAA) (Anwar and Bhatti 2018) when all provinces acknowledged that detailed studies are required to ascertain minimum environmental flows downstream of Kotri. One such official study was indeed conducted (14 years after signing of the WAA), where a minimum continuous flow of 5000 ft^3/s was proposed by a panel of experts (Gonzalez et al. 2005). However, and most unfortunately, it has been almost 30 years since the signing of the WAA, yet the intent to ascertain and implement minimum flows downstream of Kotri and the consequent intent to safeguard the water security of delta dwellers remain unfulfilled.

4.5 Concluding Thoughts

In order to develop future pathways for a water-secure Pakistan, we must understand the prevalent water security threats in the country, both at national and local scales. The analysis and discussion of this chapter initiated an assessment of the water security and related threats within three agent-based (i.e., centred on society and end-user) local perspectives of the country i.e., South Punjab, Karachi and Southern Sindh (including the Indus Delta). An overarching conclusion of our analysis is that more detailed data (especially urban data and regional/local data on water quality) should be acquired and shared publicly to better understand Pakistan's water security and related threats.

Within the context of South Punjab, our water security analysis depicts that exorbitant, unregulated, and unsustainable groundwater extraction is the most severe threat to South Punjab's water security. Annual groundwater extraction in South Punjab is at least 10% greater than annual recharge. The situation is worsened by existing electricity subsidies that promote high volumes of groundwater extraction via electric tube-wells. Moreover, less than 0.5% of South Punjab's population has access to these electric tube-wells that are consuming 15–20% of the region's annual renewable freshwater resource.

The analysis of Karachi's water security alludes to a severe lack of available information and data for a comprehensive assessment. However, it is evident that while Karachi's water demand is difficult to assess (accurately), access to water is highly unequal and tilted towards the affluent and politically influential consumers (a notion that is echoed in South Punjab as well).

The water security assessment of Southern Sindh indicates that strong policy interventions are needed (and are actually overdue) to ensure adequate environmental flows to the Indus Delta, to in-turn safeguard the water security of the dwellers of Southern Sindh. Such interventions were promised by state authorities in the past, but never implemented (Anwar and Bhatti 2018).

Aijaz and Akhter (2020) argue that the state-led evolution of water management in Pakistan has consistently viewed water as a 'quantifiable resource' only. With engineers and technocrats at the forefront of this evolution, large structural developments (like the Bhasha dam) continue to be the magic solutions for managing this 'quantifiable resource', while the plight of groundwater sustainability in South Punjab sees no policy intervention. Water security is thus poorly understood and only viewed through broad perspectives and simplistic correlations between water availability and population size. There is a dire need to move beyond such broad perspectives on water security that are dominated by a bureaucratic discourse, and disconnected from the water security challenges faced by civil society. Strong measures are imminently required to better understand the dynamics of water security threats in the country at fine scales, and to consequently develop and implement flexible, dynamic, politically inclusive, and sustainable policies and water management strategies.

Appendix

The following equation (Plappally and Lienhard 2012) is used to estimate annual groundwater extraction in South Punjab (with electricity supplied by MEPCO):

$$\text{Volume} = \frac{\text{Energy} * \text{eff} * (1 - \text{tloss})}{0.00273 * \text{depth} * 10^6}$$

In the above equation 'Volume' denotes the volume (m^3) of groundwater extracted by electric tube-wells, 'Energy' is the electricity (Kwh) consumed by electric tube-wells, 'eff' is an estimate of average efficiency of an electric tube-well pump, 'tloss' is average electricity transmission loss, and 'depth' is average depth to water table.

References

Aijaz A, Akhter M (2020) From building dams to fetching water: scales of politicization in the Indus Basin. Water 12(5):1351. https://doi.org/10.3390/w12051351

Akhbari M, Grigg NS (2013) A framework for an agent-based model to manage water resources conflicts. Water Resour Manag 27(11):4039–4052. https://doi.org/10.1007/s11269-013-0394-0

Amin A, Nasim W, Mubeen M et al (2018) Regional climate assessment of precipitation and temperature in Southern Punjab (Pakistan) using SimCLIM climate model for different temporal scales. Theor Appl Climatol 131(1–2):121–131. https://doi.org/10.1007/s00704-016-1960-1

Anwar AA, Bhatti MT (2018) Pakistan's water apportionment Accord of 1991: 25 years and beyond. J Water Resour Plan Manag 144(1):1–13. https://doi.org/10.1061/(ASCE)WR.1943-5452.0000831

Berglund EZ (2015) Using agent-based modeling for water resources planning and management. J Water Resour Plan Manag 141(11):1–17. https://doi.org/10.1061/(ASCE)WR.1943-5452.0000544

Cheema MJM, Immerzeel WW, Bastiaanssen WGM (2014) Spatial quantification of groundwater abstraction in the irrigated Indus Basin. Groundwater 52(1):25–36. https://doi.org/10.1111/gwat.12027

Cook C, Bakker K (2012) Water security: debating an emerging paradigm. Glob Environ Chang 22(1):94–102. https://doi.org/10.1016/j.gloenvcha.2011.10.011

Falkenmark M, Lundqvist J, Widstrand C (1989) Macro-scale water scarcity requires micro-scale approaches: aspects of vulnerability in semi-arid development. Nat Res Forum 13(4):258–267. https://doi.org/10.1111/j.1477-8947.1989.tb00348.x

FAO (2011) AQUASTAT - Country profile – Pakistan. http://www.fao.org/3/ca0403en/CA0403EN.pdf

Gonzalez FJ, Basson T, Schultz B (2005) Final report of international panel of experts for review of studies on water escapages below Kotri barrage. Federal Flood Commission Pakistan/Delft, The Netherlands

Hassan M (2016) Water security in Pakistan: issues and challenges. United Nations Development Programme Pakistan 3(4):1–34

International Resources Group (2015) Power Distribution Program 2010–2015. https://www.usaid.gov/documents/1865/power-distribution-program-2010-2015

Irfan M, Kazmi SJH, Arsalan MH (2018) Sustainable harnessing of the surface water resources for Karachi: a geographic review. Arab J Geosci 11(2):24. https://doi.org/10.1007/s12517-017-3365-6

JICA (2008) The study on water supply and sewerage system in Karachi in the Islamic Republic of Pakistan. Japan International Cooperation Agency (JICA)

Khan AD, Iqbal N, Ashraf M, Sheikh AA (2016) Groundwater investigations and mapping in the Upper Indus plain. PCRWR, Islamabad. http://www.pcrwr.gov.pk/publication.php?view_reports

Khan HF, Yang YCE, Xie H, Ringler C (2017) A coupled modeling framework for sustainable watershed management in transboundary river basins. Hydrol Earth Syst Sci 21(12):6275–6288. https://doi.org/10.5194/hess-21-6275-2017

Laghari AN, Vanham D, Rauch W (2012) The Indus Basin in the framework of current and future water resources management. Hydrol Earth Syst Sci 16(4):1063–1083. https://doi.org/10.5194/hess-16-1063-2012

Memon JA, Thapa GB (2011) The Indus irrigation system, natural resources, and community occupational quality in the delta region of Pakistan. Environ Manag 47(2):173–187. https://doi.org/10.1007/s00267-010-9569-0

Ministry of Water Resources (2018) Pakistan Water Charter. http://mowr.gov.pk/wp-content/uploads/2018/04/Pakistan-WATER-CHARTER.pdf

MNFSR (2018) Agricultural Statistics of Pakistan 2017–18. http://www.mnfsr.gov.pk/pubDetails.aspx

Mustafa D (2002) Theory versus practice: the bureaucratic ethos of water resource management and administration in Pakistan. Contemporary South Asia 11(1):39–56. https://doi.org/10.1080/0958493022000000035

Mustafa D, Gioli G, Karner M, Khan I (2017) Contested waters: subnational scale water conflict in Pakistan. United States Institute of Peace, Washington DC. www.usip.org

Nabi G, Ali M, Khan S, Kumar S (2019) The crisis of water shortage and pollution in Pakistan: risk to public health, biodiversity, and ecosystem. Environ Sci Pollut Res 26(11):10443–10445. https://doi.org/10.1007/s11356-019-04483-w

NEPRA (2018) State of Industry Reports. https://nepra.org.pk/publications/State of Industry Reports.php

Plappally AK, Lienhard VJH (2012) Energy requirements for water production, treatment, end use, reclamation, and disposal. Renew Sust Energ Rev 16(7):4818–4848. https://doi.org/10.1016/j.rser.2012.05.022

Qamar MU, Azmat M, Claps P (2019) Pitfalls in transboundary Indus Water Treaty: a perspective to prevent unattended threats to the global security. Npj Clean Water 2:22. https://doi.org/10.1038/s41545-019-0046-x

Qureshi AS, McCornick PG, Qadir M, Aslam Z (2008) Managing salinity and waterlogging in the Indus Basin of Pakistan. Agric Water Manag 95(1):1–10. https://doi.org/10.1016/j.agwat.2007.09.014

Renaud FG, Syvitski JPM, Sebesvari Z et al (2013) Tipping from the Holocene to the Anthropocene: how threatened are major world deltas? Curr Opin Environ Sustain 5(6):644–654. https://doi.org/10.1016/j.cosust.2013.11.007

Schewe J, Heinke J, Gerten D et al (2014) Multimodel assessment of water scarcity under climate change. Proc Natl Acad Sci U S A 111(9):3245–3250. https://doi.org/10.1073/pnas.1222460110

Siddiqi A, Wescoat JL (2013) Energy use in large-scale irrigated agriculture in the Punjab province of Pakistan. Water Int 38(5):571–586. https://doi.org/10.1080/02508060.2013.828671

Tessler ZD, Vörösmarty CJ, Grossberg M et al (2015) Profiling risk and sustainability in coastal deltas of the world. Science 349(6248):638–643. https://doi.org/10.1126/science.aab3574

UN WATER (2013) Water security and the global water agenda. UN University, Hamilton, Canada, UN Water Analytical Brief

United Nations (2019) World population prospects 2019. Department of Economic and Social Affairs, Population Division

van Steenbergen F, Kaisarani AB, Khan NU, Gohar MS (2015) A case of groundwater depletion in Balochistan, Pakistan: enter into the void. J Hydrol Reg Stud 4:36–47. https://doi.org/10.1016/j.ejrh.2014.11.003

Vörösmarty CJ, McIntyre PB, Gessner MO et al (2010) Global threats to human water security and river biodiversity. Nature 467(7315):555–561. https://doi.org/10.1038/nature09440

Watto MA, Mugera AW (2016) Groundwater depletion in the Indus Plains of Pakistan: imperatives, repercussions and management issues. Int J River Basin Manag 14(4):447–458. https://doi.org/10.1080/15715124.2016.1204154

Wescoat JL, Halvorson SJ, Mustafa D (2000) Water management in the Indus Basin of Pakistan: a half-century perspective. Int J Water Resour Dev 16(3):391–406. https://doi.org/10.1080/713672507

Wescoat JL, Siddiqi A, Muhammad A (2018) Socio-hydrology of channel flows in complex river basins: rivers, canals, and distributaries in Punjab, Pakistan. Water Resour Res 54(1):464–479. https://doi.org/10.1002/2017WR021486

World Bank (2019) Pakistan – Karachi Water and Sewerage Services Improvement Project (English). World Bank Group, Washington, DC. http://documents.worldbank.org/curated/en/695091561946580138/Pakistan-Karachi-Water-and-Sewerage-Services-Improvement-Project

Yang YCE, Cai X, Stipanović DM (2009) A decentralized optimization algorithm for multiagent system-based watershed management. Water Resour Res 45(8):1–18. https://doi.org/10.1029/2008WR007634

Yang YE, Brown CM, Yu WH, Savitsky A (2013) An introduction to the IBMR, a hydro-economic model for climate change impact assessment in Pakistan's Indus River Basin. Water Int 38(5):632–650. https://doi.org/10.1080/02508060.2013.830691

Yang YCE, Brown C, Yu W, Wescoat J, Ringler C (2014) Water governance and adaptation to climate change in the Indus River Basin. J Hydrol 519(PC):2527–2537. https://doi.org/10.1016/j.jhydrol.2014.08.055

Yang YCE, Ringler C, Brown C, Mondal AH (2016) Modeling the agricultural water–energy–food nexus in the Indus River Basin, Pakistan. J Water Resour Plan Manag 142(12):1–12. https://doi.org/10.1061/(ASCE)WR.1943-5452.0000710

Young W, Anwar A, Bhatti T et al (2019) Pakistan: getting more from water. World Bank, Washington, DC. https://openknowledge.worldbank.org/handle/10986/31160

Yu W, Yang YC, Savitsky A et al (2013) The Indus Basin of Pakistan: the impacts of climate risks on water and agriculture. The World Bank, Washington, DC. https://doi.org/10.1596/978-0-8213-9874-6

Zechman EM (2011) Agent-based modeling to simulate contamination events and evaluate threat management strategies in water distribution systems. Risk Anal 31(5):758–772. https://doi.org/10.1111/j.1539-6924.2010.01564.x

Chapter 5
Climate Change in the Mountains of Pakistan and its Water Availability Implications

Muhammad Naveed Anjum, Muhammad Jehanzeb Masud Cheema, Muhammad Zia ur Rahman Hashmi, Muhammad Azam, Arslan Afzal, and Muhammad Wajid Ijaz

Abstract Changes in precipitation and temperature directly affect the demand and supply of freshwater. This chapter presents and compares changes in these two climatic variables in the three major mountain ranges of Pakistan (Hindu-Kush, Karakoram, and Himalayan). The non-parametric Mann-Kendall, Sen's Slope Estimator, and Mann-Whitney U-test were applied to investigate changes in the precipitation and temperature time-series for the period from 1975 to 2014. The gauge weight method was used to estimate the areal average values of both climatic variables. The time-series datasets were classified into two equal periods to enable assessment of relative changes. Results indicated that the average annual temperature had an increasing trend in all mountainous ranges, with the highest warming trend in the Karakoram range (0.08 °C/decade), followed by Hindu-Kush (0.04 °C/decade) and the Himalayas (0.03 °C/decade). The results of seasonal trend analysis indicated the highest warming trend of the spring season in the Hindu-Kush range (0.33 °C/decade), followed by Karakoram (0.28 °C/decade) and the Himalayas (0.17 °C/decade). A cooling tendency of summer temperature was found in the Karakoram range (−0.10 °C/decade), followed by the Hindu-Kush range (−0.09 °C/decade), while a warming trend of summer temperature was found in the Himalayas. Overall, a decreasing trend of total precipitation was found in the Himalayas, with a significant decreasing trend of annual (38.90 mm/decade) and spring (14.88 mm/

M. N. Anjum (✉) · M. J. M. Cheema · M. Azam · A. Afzal
Faculty of Agricultural Engineering and Technology, Pir Mehr Ali Shah Arid Agriculture University, Rawalpindi, Pakistan
e-mail: naveedwre@uaar.edu.pk; mjm.cheema@uaar.edu.pk; muhammad.azam@uaar.edu.pk; arslanafzal@uaar.edu.pk

M. Z. u. R. Hashmi
Water Resources and Glaciology Section, Global Change Impact Studies Centre, Islamabad, Pakistan
e-mail: zia.hashmi@gcisc.org.pk

M. W. Ijaz
Environmental Protection Agency, Lahore, Pakistan

M. A. Watto et al. (eds.), *Water Resources of Pakistan*, World Water Resources 9,
https://doi.org/10.1007/978-3-030-65679-9_5

decade) precipitation. On the contrary, annual precipitation showed an increasing trend in the Hindu-Kush (13.88 mm/decade) and Karakoram (11.86 mm/decade). Overall, our investigation confirms that the climate of the region is already changing, with an increasing trend of temperature across all mountainous ranges and a decreasing tendency of precipitation in the Himalayas.

Keywords Climate change · HKH mountains · Northern Pakistan · Water resources · Mann-Kendall Test

5.1 Background

The high mountains of Hindu-Kush, Karakoram and the Himalayas (HKH) and the Tibetan Plateau, known as the third pole of our planet (Kehrwald et al. 2008; Smiraglia et al. 2007), supply water to all ten major rivers of Asia: the Indus, Ganges, Yangtze, Yellow, Tarim, Salween, Amudarya, Brahmaputra, Irrawaddy, and Mekong Rivers (Immerzeel et al. 2009; Rasul 2012). Hundreds of millions of people living in the catchments of these rivers and in the downstream areas rely on this water supply for drinking, domestic, industrial, irrigation, and hydropower production (Bocchiola et al. 2011; Bolch et al. 2012), and of all these river basins, the Indus Basin (Fig. 5.1) has been identified as the most depleted (Laghari et al. 2012). The economic sustainability of millions of people residing in the Indus Basin is predominantly dependent on agriculture, and is highly reliant on Indus waters for irrigated agriculture (FAO 2011). Almost all (95%) of the Indus River flow is being extracted by the Indus Basin Irrigation System (IBIS), one of the largest integrated irrigation systems of the world, to feed 145 million people of Pakistan (Laghari et al. 2012). The Indus River also services one third of Pakistan's hydropower demands (Pritchard 2017). Therefore, any future perturbations in the flow of the Indus River and its tributaries can lead to substantial impacts on the lives of people residing within the Indus Basin (Tahir et al. 2016).

It is well documented that the burgeoning agricultural and hydropower demands in Pakistan will put substantial pressure on available water resources in the headwaters of the Indus River (Bocchiola et al. 2011). Potential modifications in Indus River flows as a consequence of climate change will further exacerbate this critical situation. It is also reported that the concentration of absorbing atmospheric aerosols has increased in the HKH region (Bollasina et al. 2011), making freshwater resources in the cryospheric domain of the Indus Basin particularly vulnerable to climate change (Tahir et al. 2016).

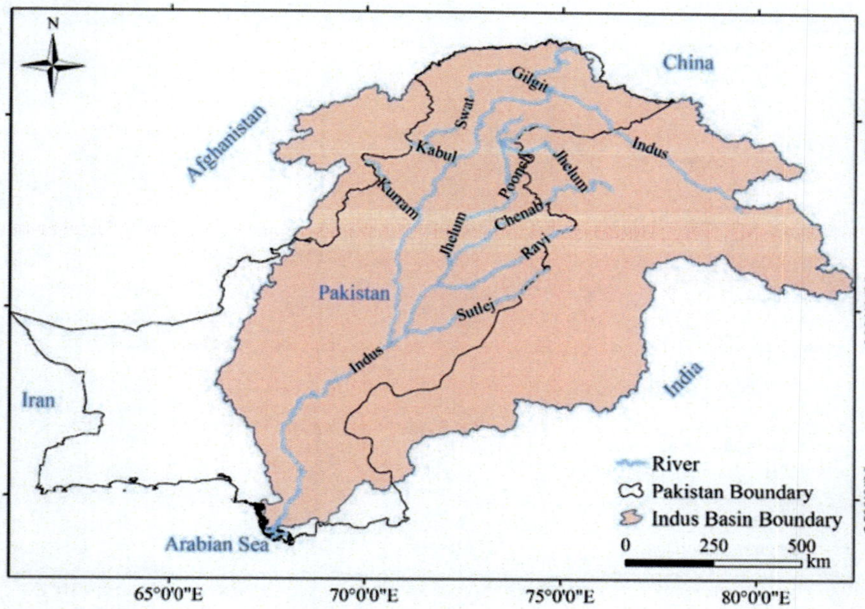

Fig. 5.1 Map of the Indus Basin and major tributaries

5.2 Climatology of Northern Pakistan

Overall, Pakistan has a diversified climate due to its areal extent, snow/glaciers covers, and heterogeneous topography. The Indian and Pacific Oceans, as well as the Eurasian continent, have a great influence on the climatic patterns of Pakistan. The region experiences the most powerful monsoon system of South Asia (Dahri et al. 2016). Climatologically, northern Pakistan is driven by the South Asian Atmospheric Circulation, which is associated with the extra-tropical cyclonic/anticyclonic circulation systems around low/high-pressure troughs in the winter season and monsoon evolutions in the summer seasons (Rajbhandari et al. 2014). Precipitation over the cryospheric region in northern Pakistan is thus largely influenced by these two principal weather systems of South Asia (Khan et al. 2015), and especially the heavy moisture-laden monsoon winds from the Arabian Sea, the Bay of Bengal, and the Indian Ocean. The summer monsoon circulation, also known as the Indian Summer Monsoon (ISM) circulation system is generated as a consequence of different heating between land and sea surfaces during the summer season (Ahmad et al. 2012; Palazzi et al. 2013). However, Western Disturbances (WD) advecting moisture from the Mediterranean and Caspian Sea as an extratropical frontal system cause precipitation over the cryospheric region of Pakistan in the winter and spring seasons (Dahri et al. 2016; Ministry of Water and Power 2010). Figure 5.2 shows routes of the westerlies and monsoon circulations as derived from a review of the literature (Ahmad et al. 2012; Dahri et al. 2016). Generally, the cryosphere in the Himalayan

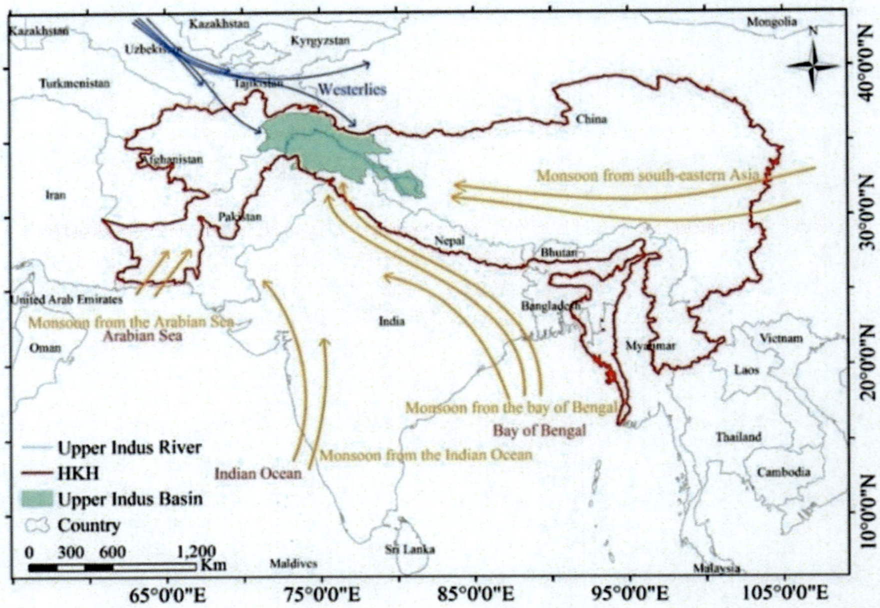

Fig. 5.2 Geographical stretches of the Upper Indus Basin (UIB) in the Hindukush-Karakoram-Himalayan mountains and routes of westerlies and monsoon circulations

range of Pakistan is mostly influenced by the monsoon system, the cryosphere in the Karakoram range is under the influence of both circulation systems (i.e., monsoon and westerlies), and the cryosphere in the Hindukush range is predominantly influenced by the westerlies circulation system.

5.3 Hydrology of Northern Pakistan

Understanding the hydrological processes in the headwaters of the Indus River is essential for investigating climate-induced impacts on river flow regimes. The hydrology of northern Pakistan is divided into three distinct regimes: precipitation (rainfall), snowmelt (nival), and glacier (glacial) melt regimes (Parry et al. 2017). The precipitation regime is mainly reliant on the variations in seasonality and intensity of the ISM and WD circulation systems. ISM is the main source of runoff in the catchments situated in the southern foothills of the Himalayas and downstream areas in the Indus plains (Archer et al. 2010). This regime is a major source of flooding in the downstream areas as it produces more intense runoff. The snowmelt dominated regime contributes about 35–40% of the total flow through the Indus Basin (Immerzeel et al. 2010; Savoskul and Smakhtin 2013). Runoff is fed by melting snow that fell during winter and spring seasons in the watersheds situated in the medium altitudes of the Himalayan and Hindukush mountains (Azmat et al. 2017; Tahir et al. 2011), with several researchers reporting a consistent negative temperature-runoff relationship (Archer et al. 2010; Immerzeel et al. 2009; Tahir et al. 2011). A slight increase in the seasonal temperature of this regime can lead to severe modifications in the runoff response of this regime (Archer et al. 2010; Tahir et al. 2016). The contribution of the glacial regime in the total water flow in the Indus Basin is estimated at about 25–35% (Immerzeel et al. 2010; Savoskul and Smakhtin 2013). The hydrological response of this regime is characterised by large variations because of its complex topographic and climatic conditions in the region (ICIMOD 2010). In this regime, studies (e.g. Archer et al. 2010) have indicated positive relationships between temperature and river flow (increase in temperatures leads to an increase in river flow) and negative correlations between precipitation and river flow in the summer season.

5.4 Cryospheric Profile of Pakistan

The northern highlands of Pakistan (NPK) host the country's greatest freshwater reserves in the form of perennial glaciers (covering an area of about 22,000 km^2). The coverage of these glaciers in the NPK was estimated using the Randolph Glacier Inventory (RGI) version 5. The spatial extents of glacier cover in Pakistan are presented in Fig. 5.3. Almost 28% of the total NPK area is covered by glaciers. Snow cover area in the NPK varies from more than 70% in winter to about 10% in summer seasons (Khan et al. 2015). The estimated stored water in the glacial reserves of the country is about 2738 km^3 (billion m^3) (Ashraf et al. 2014). Hence, meltwater of seasonal snow and glacier is the main contributor to the flow of the Indus River systems (the Indus River and its tributaries) (Ahmed et al. 2007; Tahir et al. 2016).

5.5 Climate Change in Pakistan

Availability of freshwater for agriculture, domestic and industrial uses in Pakistan is predominantly dependent on the flows of the Indus River and its tributaries. The river flows are composed of snow/glacier melt and precipitation runoff in the sub-basins of the Indus River system (Lutz et al. 2016). The snow and glaciers in the cryospheric domain of Pakistan serve as natural storage reservoirs that provide perennial supplies to the downstream areas of the country. According to Tahir et al. (2016), the contribution of snow or glacier meltwater in the total discharge of the watersheds situated in the Himalayan and Karakoram ranges are more than 60%. Pakistan is declared as a "water-stressed" country and its water resources are substantially susceptible to climate change (Hussain and Mumtaz 2014).

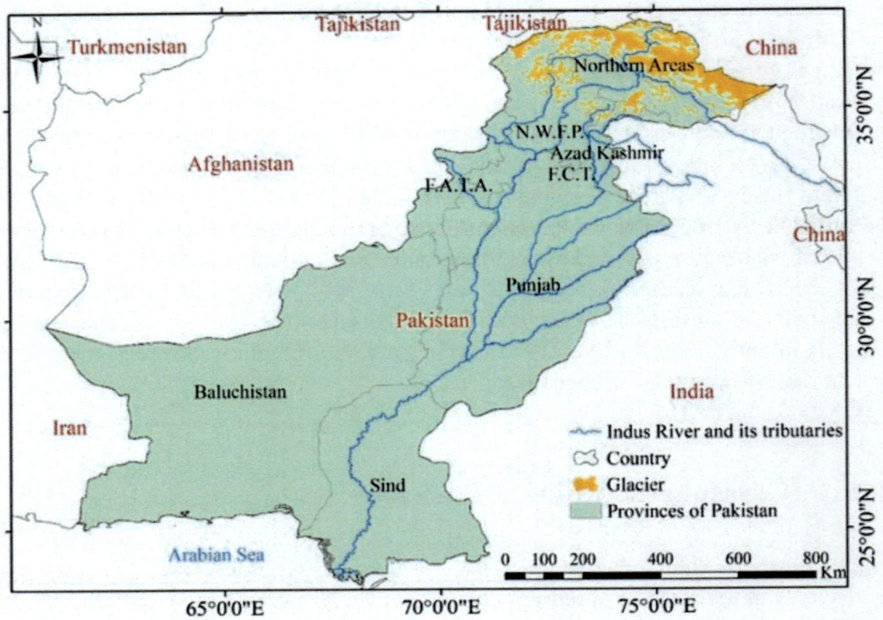

Fig. 5.3 Map of glacial cover extent in Pakistan

Almost 95% of canal irrigation water is supplied by the Indus River and its tributaries (Hussain and Mumtaz 2014). Snow and glacier meltwater are essential components of this supply, and thus critical to ensure the country's agricultural sustainability (Tahir et al. 2016). The headwaters of the Indus River system are inevitably connected with climate change, meaning there are serious implications for Pakistan's water resources from projected climate changes (Hussain and Mumtaz 2014). Climate change further exacerbates challenges for effective water resources management in the country due to its impacts causing haphazard snowmelt, change in glaciers mass balance, alterations in the westerlies and monsoon circulation systems, increase in extreme events (floods and droughts), seawater intrusion, and salinisation of coastal areas. These impacts also exacerbate water and food scarcity threats (Tahir et al. 2016). The scenario-based projections of several global climate models have revealed the possibility of more frequent extreme events (floods and droughts) and alterations in the spatial distribution of water resources of Pakistan (Garee et al. 2017).

Because Pakistan's economic sustainability is mainly dependent on the agricultural sector, these climate change-induced impacts and limited water resources are critical issues to understand and manage. The high-intensity precipitation events augmented with snow and glacier melt in the Swat, Jhelum, and Chenab tributaries of the Indus are the major sources of floods in the country's downstream areas (Ministry of Water and Power 2010). Sporadically, heavy moisture-laden monsoon winds adverting from the Arabian Sea, Indian Ocean and Bay of Bengal, and resultant low-pressure troughs, often lead to heavy downpours in the UIB. When these monsoon winds coincide with the westerlies over the Hindukush Range of UIB, it accentuates the destructive floods in the Swat basin's downstream areas (Ahmad et al. 2015). According to the Federal Flood Commission of Pakistan (Ministry of Water and Power 2010), that these extreme events occurred more or less simultaneously in the UIB is due to changes in the global climatic patterns. According to experts from the Intergovernmental Panel on Climate Change, World Meteorological Organisation, and World Climate Research Programme, extreme events are expected to become more frequent in northern Pakistan (Ahmad et al. 2015). The unprecedented floods of 2010 were also identified as being a result of climate change (Ministry of Water and Power 2010). Tahir et al. (2011) and Yaseen et al. (2015) studied the impacts of climate change on river flows of Hunza and Jhelum Rivers in Pakistan and found increased river flow in both watersheds. Both of these river watersheds are situated in the monsoon-dominated corridor of the Karakoram and Himalayas, respectively. Until now, implications of climate change for water resources availability in the westerlies dominated Hindukush range of Pakistan have not been comprehensively assessed (Kapnick et al. 2014; Tahir et al. 2016).

5.5.1 Changes in Precipitation and Temperature across the HKH Mountains of Pakistan

Available literature indicates that most studies of climate change in Pakistan have considered the entire Upper Indus Basin (UIB) in general, even though the three mountainous ranges (Hindu-Kush, Karakoram, and the Himalayas) in the UIB are experiencing different climatic systems. In this study, therefore, changes in the two main climatic variables (precipitation and temperature) across the three ranges of the HKH mountains were analysed separately. Figure 5.4 shows the three mountain range areas and the locations of installed weather stations. To examine and compare historical changes in precipitation and temperature, the daily observations of weather stations installed in the three respective mountain areas were collected from the Pakistan Meteorological Department (PMD) and Water and Power Development Authority (WAPDA) for the 40-year period from 1975 to 2014 – except that temperature data of most weather stations in the Karakoram range were only available from 1995 to 2014, meaning that analysis of temperature variation in that range was done for that 20-year period only. Relative changes in the climatic variables were assessed with reference to the baseline period. For this purpose, the time-series data of precipitation and temperature were divided into two time periods, i.e., 1975–1994 and 1995–2014. Monthly, seasonal, and annual values of precipitation and temperature were derived from their daily time series data. Areal average values of climatic variables were obtained using gauge weights in the respective mountain ranges. Historical changes in the climatic variables were also assessed for the four seasons

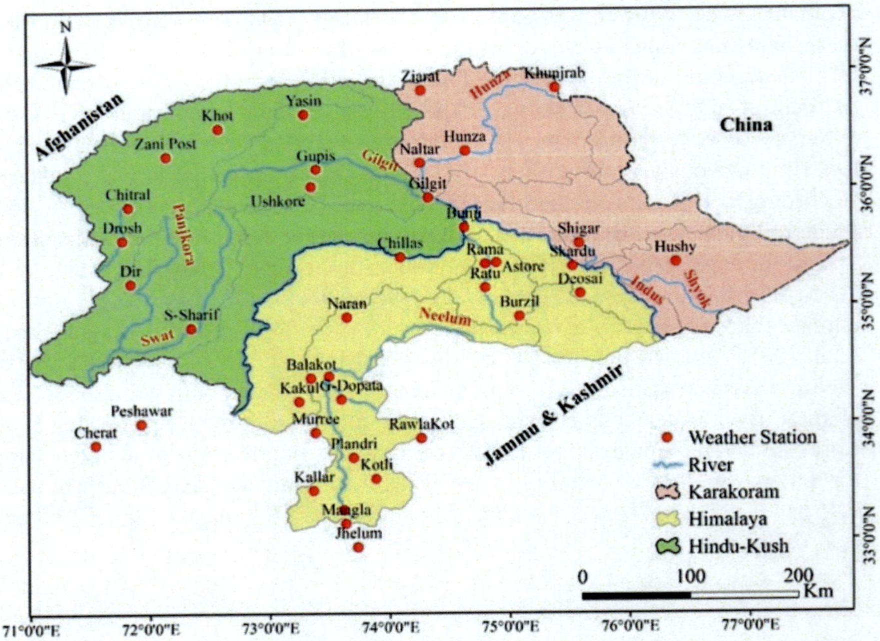

Fig. 5.4 Spatial extents of Hindu-Kush, Karakoram, and Himalayan mountains and locations of installed weather stations in northern Pakistan

of winter (December to February), spring (March to April), summer (May to September), and autumn (October to November).

The non-parametric Mann-Whitney U-test was used to examine the significance of relative changes (at 95% confidence level) in precipitation and temperature data in the second period (1995–2014) compared with the first period (1975–1994). Direction and magnitude of the trends in times-series of areal average precipitation and temperature data were analysed using non-parametric Mann-Kendall and Sen's Slope estimator. The Mann-Kendall (MK) test is approved by the World Meteorological Organisation for analysing trends in climatic and hydrological variables.

Table 5.1 shows the comparisons of the relative changes in the precipitation and temperature in the HKH mountains of Pakistan. It is clear from the results that the average annual and seasonal precipitation in the Hindu-Kush range slightly increased in the second period (1995–2014) as compared with the reference period (1975–1994), except for the spring season in which precipitation slightly decreased (−6.2%). The respective values of increase in average annual, winter, summer, and autumn precipitation were 1.6%, 4.5%, 6.1%, and 2.6%. However, according to the Mann-Whitney U-test, the change in precipitation in the Hindu-Kush range was not significant (at 5% level). In the case of Karakoram range, the average autumn precipitation decreased by 60.7% in the second period as compared with the first

Table 5.1 Comparison of relative changes (%) in areal average annual and seasonal precipitation and temperature in the HKH mountains of Pakistan

		Precipitation					Temperature				
No.	Range	Annual	Winter	Spring	Summer	Autumn	Annual	Winter	Spring	Summer	Autumn
1	Hindu-Kush	1.6	4.5	−6.2	6.1	2.6	1.5	7.6	**6.4**	−0.3	1.4
2	Karakoram	13.9	25.6	16.4	24.3	−60.7	0.5	2.1	6.2	−0.3	−0.3
3	Himalaya	**−10.8**	−6.6	**−19.1**	−7.8	−22.4	1.2	0.6	4.4	0.2	2.2

Bold values indicate significant changes at 95% confidence level based on the Mann-Whitney U-Test period, while average areal precipitation increased by 13.9%, 25.6%, 16.4%, and 24.3% for the annual, winter, spring, and summer timescales respectively. The overall areal average precipitation in the Himalayas decreased, with decreases of 10.8%, 6.6%, 19.1%, 7.8%, and 22.4% in the second period for annual, winter, spring, summer, and autumn timescales, respectively. Results of the Mann-Whitney test showed that the decrease in the annual and spring precipitation in the second period was significant (at 95% level) with reference to the first.

Analysis of the areal average time series data of annual and seasonal temperatures showed that all of the mountainous ranges experienced warmer temperatures in the second period as compared with the first. In the case of the Hindu-Kush range, the summer season was cooler in the second period as compared with the first, whereas all other seasonal and annual temperatures were higher, including a significant (6.4%) increase in the areal average spring temperature. In the case of the Karakoram range, the summer and autumn average temperatures decreased slightly (0.3% in both seasons), while annual and all other seasonal temperatures increased in the second period. The Himalayas also experienced higher temperatures for annual and seasonal timescales in the second period as compared with the first. However, the cooling or warming of the Karakoram and Himalayan ranges in the second period were not significant at the 95% confidence level.

Long-term changes in time series of areal average precipitation for annual and seasonal time scales in the HKH mountains over 1975–2014 are presented in Fig. 5.5. Results of the MK test indicated an overall decreasing tendency of precipitation in the Himalayas. It was found that the average annual and spring precipitation decreased significantly (38.90 mm/decade and 14.88 mm/decade, respectively) in this range, with both decreasing trends found to be significant at the 95% confidence level. A slight and statistically non-significant increase (2.44 mm/decade) in winter precipitation was found in the Himalayas. An overall increasing tendency of precipitation was found in the Hindu-Kush and Karakoram ranges. The increase in average annual precipitation was higher in the Hindu-Kush range compared with the Karakoram range, being 13.77 and 11.86 mm/decade respectively. The distribution of MK test (Z) and Sen's Slope Estimator (Q) values for annual and seasonal timescales are provided in Table 5.2.

Figure 5.6 presents the trends in average annual and seasonal temperatures in the Hindu-Kush and Himalayan mountains in 1975–2014, while trends in temperature in the Karakoram range are based on the period of 1995–2014. It was found that the average temperature of the summer season in the Hindu-Kush range decreased, although the trend was not significant at 95% confidence level. The rate of decrease in summer temperature was –0.09 °C/decade. On the other hand, an increasing tendency in annual and all other seasonal temperatures was detected across the Himalayas. The increasing trend in spring temperature of 0.33 °C/decade was statistically significant, and is consistent with the 0.34 °C/decade warming rate identified in northwestern China (Li et al. 2013). In the case of the Karakoram range, the average winter and summer temperatures showed decreasing tendency, with decreasing rates of minus 0.10 °C/decade in both ranges, whereas average annual, spring, and autumn temperatures showed increasing tendencies. None of these

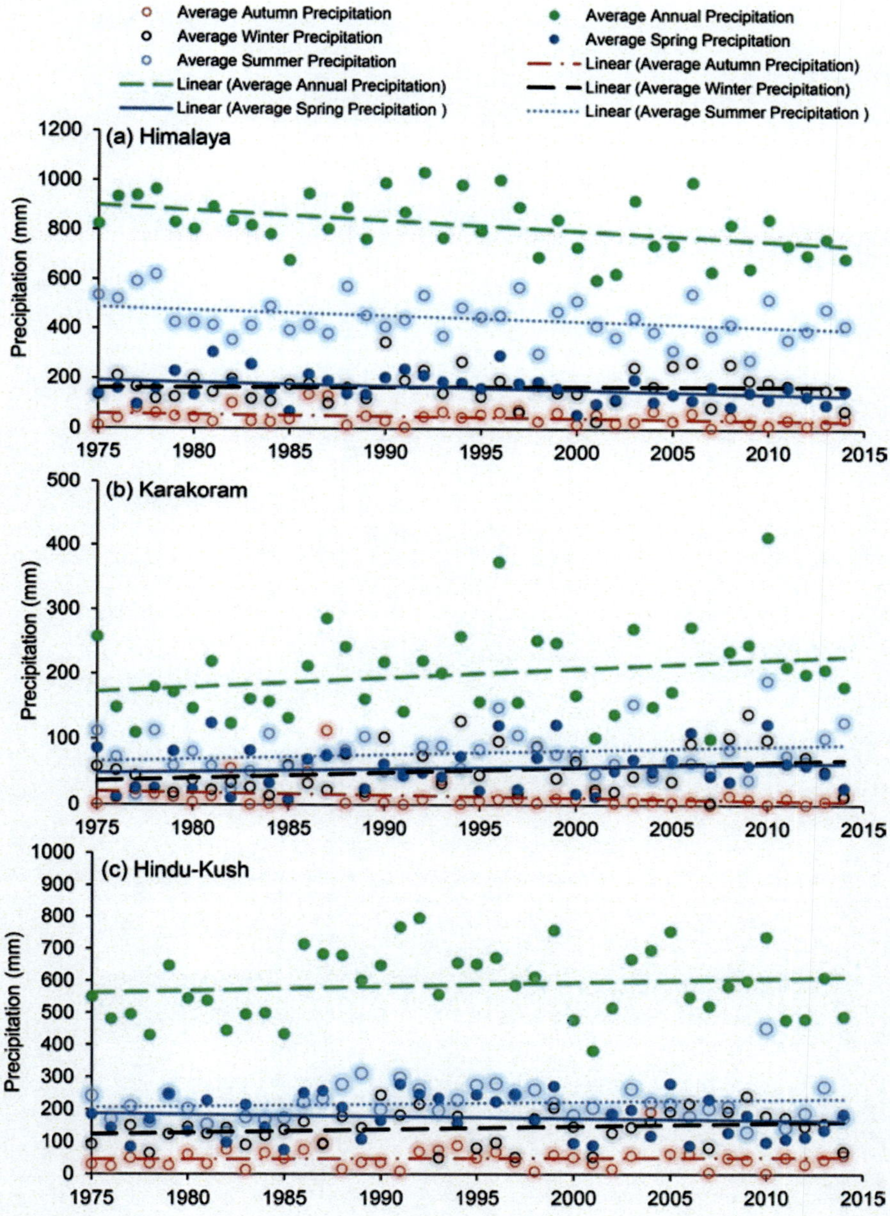

Fig. 5.5 Historical trends of precipitation in the Himalayan, Karakoram, and Hindu-Kush mountains for annual and seasonal timescales

Table 5.2 Distribution of Z and Q values on annual and seasonal scales in the Hindu-Kush, Karakoram, and Himalaya mountains

		Precipitation					Temperature				
Range		Annual	Winter	Spring	Summer	Autumn	Annual	Winter	Spring	Summer	Autumn
Hindu-Kush	Z	0.83	1.76	−0.57	0.34	0.03	0.48	1.21	**2.09**	−0.87	0.05
	Q	13.77	11.25	−5.94	2.83	0.14	0.04	0.15	0.33	−0.09	0.01
Karakoram	Z	1.34	1.34	0.48	0.71	−0.87	0.62	−0.16	1.78	−0.42	0.03
	Q	11.86	7.08	2.22	2.99	−0.81	0.08	−0.10	0.28	−0.10	0.02
Himalaya	Z	**−2.44**	0.15	**−1.97**	−1.74	−1.32	0.45	0.48	1.13	0.13	0.94
	Q	−38.90	2.44	−14.88	−19.74	−4.80	0.03	0.06	0.17	0.02	0.10

Bold values indicate a significant trend at 95% significance level

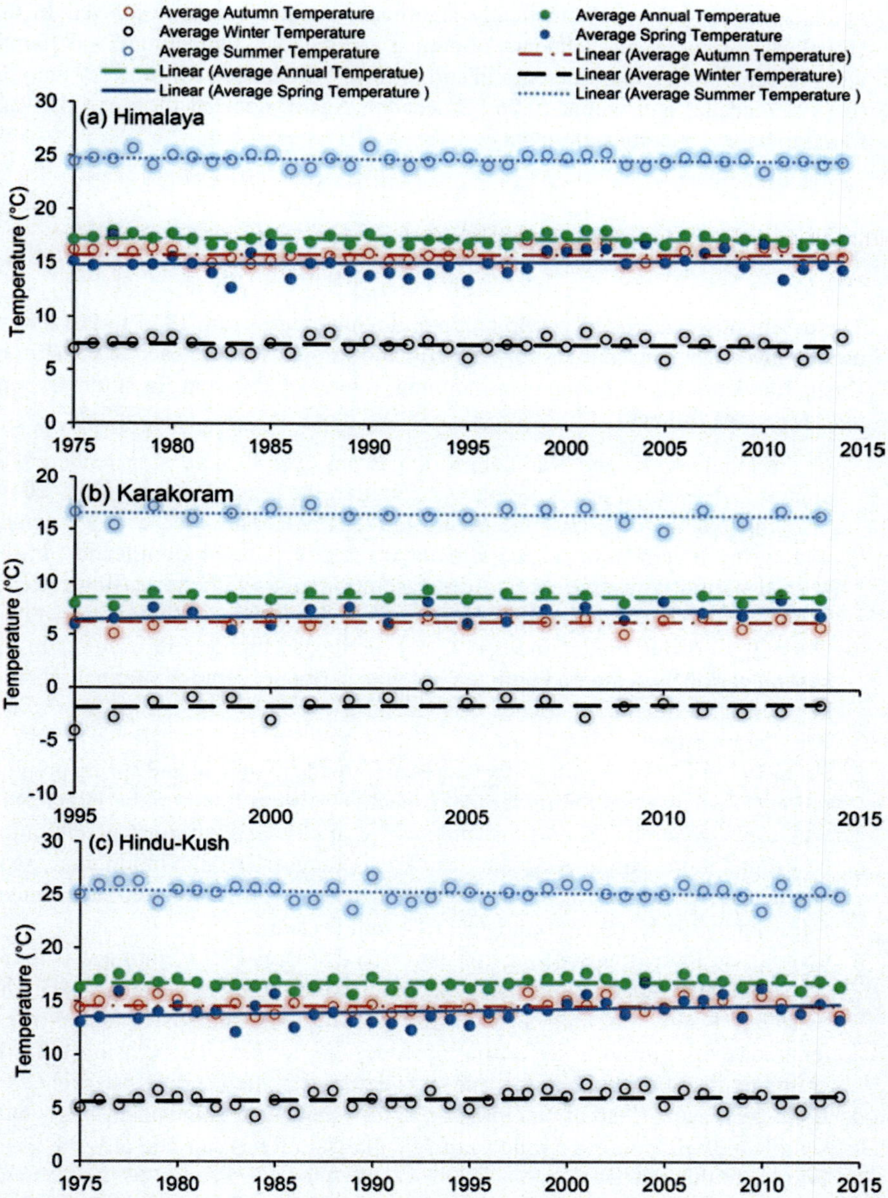

Fig. 5.6 Temperature trends in the Himalayan, Karakoram, and Hindu-Kush mountains for annual and seasonal timescales

trends were found to be statistically significant at 95% confidence level. In the Himalayas, an increasing tendency of annual and seasonal temperature was found, with the maximum rate of increase in spring (0.17 °C/decade), followed by autumn (0.10 °C/decade) and winter (0.06 °C/decade). Again, none of these trends were found to be statistically significant.

5.6 Concluding Remarks

The in situ observations of precipitation and temperature from 1975 to 2014 were used to investigate the changes for both climatic variables across each of the Hindu-Kush, Karakoram, and Himalayan mountain ranges of Pakistan. In summary, our findings are as follows:

1. In the Hindu-Kush and Karakoram ranges, an overall increasing tendency of annual and seasonal precipitation was found in the second period (1995–2014) as compared with the baseline period (1975–1994). However, average annual and spring precipitation in the Himalaya range decreased significantly in the second period with reference to the baseline period, with decreasing rates of annual and spring precipitation being minus 10.80% and 19.10% respectively.
2. Overall, all of the mountainous ranges experienced higher temperatures in the second period as compared with the baseline period, except for summer where a decreasing temperature trend was identified across the Hindu-Kush and Karakoram ranges.
3. From 1975 to 2015, significantly decreasing trends of annual and spring precipitation (38.90 and 14.88 mm/decade respectively) were found in the Himalayas, whereas a generally increasing tendency of annual and seasonal precipitation was found in the Hindu-Kush and Karakoram ranges. The maximum increase in annual precipitation (13.77 mm/decade) was found in the Hindu-Kush range, followed by the Karakoram range (11.86 mm/decade).
4. Overall, an increasing tendency of annual and seasonal temperature was found across all three mountainous ranges, except for the summer season in the Hindu-Kush and Karakoram ranges and winter season in the Karakoram range, where temperatures showed a decreasing tendency. A significant increasing trend in spring temperature (0.33 °C/decade) was detected in the Hindu-Kush range.
5. If the decreasing trend of precipitation in the Himalayan range persists, it would significantly influence the inflows of Mangla Dam Reservoir and also the water balance of this range.
6. If the decreasing trend of summer temperature and warming trend of other seasons persists, it will significantly affect the timing of the volume of water flows, which will have significant implications for downstream areas.

Acknowledgments The authors would like to thank the Pakistan Meteorological Department (PMD) and Water and Power Development Authority (WAPDA) Pakistan for providing climatic data.

References

Ahmad S, Nishii K, Tamura T et al (2012) Characteristics of climatological tropospheric conditions during pre-monsoon and matured phases of Pakistan summer monsoon. J Jpn Soc Civil Eng 56:1–6. https://doi.org/10.2208/jscejhe.68.I_157

Ahmad I, Tang D, Wang T et al (2015) Precipitation trends over time using Mann-Kendall and Spearman's rho tests in Swat River basin, Pakistan. Adv Meteorol 431860:2015. https://doi.org/10.1155/2015/431860

Ahmed A, Iftikhar H, Chaudhry GM (2007) Water resources and conservation strategy of Pakistan. Pak Dev Rev 46(4):997–1009

Archer DR, Forsythe N, Fowler HJ, Shah SM (2010) Sustainability of water resources management in the Indus Basin under changing climatic and socio economic conditions. Hydrol Earth Syst Sci 14(8):1669–1680. https://doi.org/10.5194/hess-14-1669-2010

Ashraf A, Roohi R, Naz R, Mustafa N (2014) Monitoring cryosphere and associated flood hazards in high mountain ranges of Pakistan using remote sensing technique. Nat Hazards 73(2):933–949. https://doi.org/10.1007/s11069-014-1126-3

Azmat M, Liaqat UW, Qamar MU, Awan UK (2017) Impacts of changing climate and snow cover on the flow regime of Jhelum River, Western Himalayas. Reg Environ Chang 17(3):813–825. https://doi.org/10.1007/s10113-016-1072-6

Bocchiola D, Diolaiuti G, Soncini A et al (2011) Prediction of future hydrological regimes in poorly gauged high altitude basins: the case study of the upper Indus, Pakistan. Hydrol Earth Syst Sci 15(7):2059–2075. https://doi.org/10.5194/hess-15-2059-2011

Bolch T, Kulkarni A, Kääb A et al (2012) The state and fate of Himalayan glaciers. Science 336(6079):310–314. https://doi.org/10.1126/science.1215828

Bollasina MA, Ming Y, Ramaswamy V (2011) Anthropogenic aerosols and the weakening of the South Asian summer monsoon. Science 334(6055):502–505. https://doi.org/10.1126/science.1204994

Dahri ZH, Ludwig F, Moors E et al (2016) An appraisal of precipitation distribution in the high-altitude catchments of the Indus basin. Sci Total Environ 548–549:289–306. https://doi.org/10.1016/j.scitotenv.2016.01.001

FAO (2011) AQUASTAT – country profile – Pakistan. http://www.fao.org/3/ca0403en/CA0403EN.pdf

Garee K, Chen X, Bao A et al (2017) Hydrological modeling of the Upper Indus Basin: a case study from a high-altitude glacierized catchment Hunza. Water 9(1):1–20. https://doi.org/10.3390/w9010017

Hussain M, Mumtaz S (2014) Climate change and managing water crisis: Pakistan's perspective. Rev Environ Health 29(1–2):71–77. https://doi.org/10.1515/reveh-2014-0020

ICIMOD (2010) Islamic Republic of Pakistan: glacial melt and downstream impacts on Indus dependent water resources and energy. Final report prepared by International Centre for Integrated Mountain Development (ICIMOD) for Asian Development Bank Project RETA 6420-PAK. ICIMOD, Kathmandu, Nepal

Immerzeel WW, Droogers P, de Jong SM, Bierkens MFP (2009) Large-scale monitoring of snow cover and runoff simulation in Himalayan river basins using remote sensing. Remote Sens Environ 113(1):40–49. https://doi.org/10.1016/j.rse.2008.08.010

Immerzeel WW, Beek LPH, Van Bierkens MFP (2010) Climate change will affect the Asian water towers. Science 328(5984):1382–1385. https://doi.org/10.1126/science.1183188

Kapnick SB, Delworth TL, Ashfaq M et al (2014) Snowfall less sensitive to warming in Karakoram than in Himalayas due to a unique seasonal cycle. Nat Geosci 7(11):834–840. https://doi.org/10.1038/ngeo2269

Kehrwald NM, Thompson LG, Tandong Y et al (2008) Mass loss on Himalayan glacier endangers water resources. Geophys Res Lett 35(22):2–7. https://doi.org/10.1029/2008GL035556

Khan A, Naz BS, Bowling LC (2015) Separating snow, clean and debris covered ice in the Upper Indus Basin, Hindukush-Karakoram-Himalayas, using Landsat images between 1998 and 2002. J Hydrol 521:46–64. https://doi.org/10.1016/j.jhydrol.2014.11.048

Laghari AN, Vanham D, Rauch W (2012) The Indus basin in the framework of current and future water resources management. Hydrol Earth Syst Sci 16(4):1063–1083. https://doi.org/10.5194/hess-16-1063-2012

Li BF, Chen YN, Li WH et al (2013) Spatial and temporal variations of temperature and precipitation in the arid region of Northwest China from 1960–2010. Fresenius Environ Bull 22(2):362–371

Lutz AF, Immerzeel WW, Kraaijenbrink PDA et al (2016) Climate change impacts on the upper Indus hydrology: sources, shifts and extremes. PLoS One 11(11):1–33. https://doi.org/10.1371/journal.pone.0165630

Ministry of Water and Power (2010) Annual flood report. Office of the Chief Engineering Adviser, Federal Flood Commission, Ministry of Water and Power, Government of Pakistan, Islamabad, Pakistan

Palazzi E, Von Hardenberg J, Provenzale A (2013) Precipitation in the Hindu-Kush Karakoram Himalaya: observations and future scenarios. J Geophys Res Atmos 118(1):85–100. https://doi.org/10.1029/2012JD018697

Parry J-E, Osman H, Terton A et al (2017) The vulnerability of Pakistan's water sector to the impacts of climate change: identification of gaps and recommendations for action. International Institute for Sustainable Development, Canada and COMSATS Institute of Information Technology, Islamabad, Pakistan. https://www.iisd.org/publications/vulnerability-pakistans-water-sector-impacts-climate-change-identification-gaps-and

Pritchard HD (2017) Asia's glaciers are a regionally important buffer against drought. Nature 545(7653):169–174. https://doi.org/10.1038/nature22062

Rajbhandari R, Shrestha AB, Kulkarni A et al (2014) Projected changes in climate over the Indus River Basin using a high resolution regional climate model (PRECIS). Clim Dyn 44(1–2):339–357. https://doi.org/10.1007/s00382-014-2183-8

Rasul G (2012) Climate data modelling and analysis of the Indus ecoregion. World Wide Fund for Nature – Pakistan, Karachi

Savoskul O, Smakhtin V (2013) Glacier systems and seasonal snow cover in six major Asian river basins: hydrological role under changing climate. IWMI Report No. 150. International Water Management Institute (IWMI), Colombo, Sri Lanka. https://www.iwmi.cgiar.org/publications/iwmi-research-reports/iwmi-research-report-150/

Smiraglia C, Mayer C, Mihalcea C et al (2007) Ongoing variations of Himalayan and Karakoram glaciers as witnesses of global changes: recent studies on selected glaciers. In: Baudo R, Tartari G, Vuillermoz E (eds) Mountains witnesses of global changes research in the Himalaya and Karakoram: Share-Asia Project. Developments in Earth Surface Processes, 10(6): 235–247. https://doi.org/10.1016/S0928-2025(06)10026-7

Tahir AA, Chevallier P, Arnaud Y, Ahmad B (2011) Snow cover dynamics and hydrological regime of the Hunza River basin, Karakoram Range, Northern Pakistan. Hydrol Earth Syst Sci 15(7):2275–2290. https://doi.org/10.5194/hess-15-2275-2011

Tahir AA, Adamowski JF, Chevallier P et al (2016) Comparative assessment of spatiotemporal snow cover changes and hydrological behavior of the Gilgit, Astore and Hunza River basins (Hindukush–Karakoram–Himalaya region, Pakistan). Meteorog Atmos Phys 128(6):793–811. https://doi.org/10.1007/s00703-016-0440-6

Yaseen M, Khan K, Nabi G et al (2015) Hydrological trends and variability in the Mangla. Sci Int (Lahore) 27(2):1327–1335

Chapter 6
Pakistan's Water Resources in the Era of Climate Change

Ghulam Hussain Dars, Bakhshal Khan Lashari, Mehran Sattar Soomro, Courtenay Strong, and Kamran Ansari

Abstract Pakistan, the sixth most populous country, has been one of the most vulnerable countries to the harmful impacts of climate change. The climate risk index (CRI) includes Pakistan in the top ten nations based on exposure and vulnerability to extreme events. Specific challenges amplifying the adverse effects of climate change include rapid population increase, increasing temperatures, more frequent and intense extreme events, glacier melting, and sea-level rise. Pakistan's water resources are largely dependent on one river, the Indus. The Indus River, being dependent on glacial and snowmelt and precipitation, is highly sensitive to climate change. Global and regional climate models are highly certain about the increase in temperature but uncertain about future precipitation patterns in most of the country, which could severely affect the water resources in the future. An improved understanding of the adverse impacts of climate change on water resources is therefore needed. Adaptation measures need to be identified and actively pursued to ensure limited freshwater resources are used and managed sustainably.

Keywords Indus River · Pakistan · Water resources · Climate change impacts · Glaciers

6.1 Geographical Location

Pakistan extends from 24°N to 37°N in latitude and 61°E to 76°E in longitude (Fig. 6.1). Pakistan has a geographic area of 79.61 million hectares (Mha), and only 21.50 Mha (27% of the total area) is under cultivation (Qureshi 2005). The

G. H. Dars (✉) · B. K. Lashari · M. S. Soomro · K. Ansari
US Pakistan Center for Advanced Studies in Water, Mehran University of Engineering and Technology, Jamshoro, Sindh, Pakistan
e-mail: ghdars.uspcasw@faculty.muet.edu.pk; kansari.uspcasw@faculty.muet.edu.pk

C. Strong
Department of Atmospheric Sciences, University of Utah, Salt Lake City, UT, USA
e-mail: court.strong@utah.edu

M. A. Watto et al. (eds.), *Water Resources of Pakistan*, World Water Resources 9,
https://doi.org/10.1007/978-3-030-65679-9_6

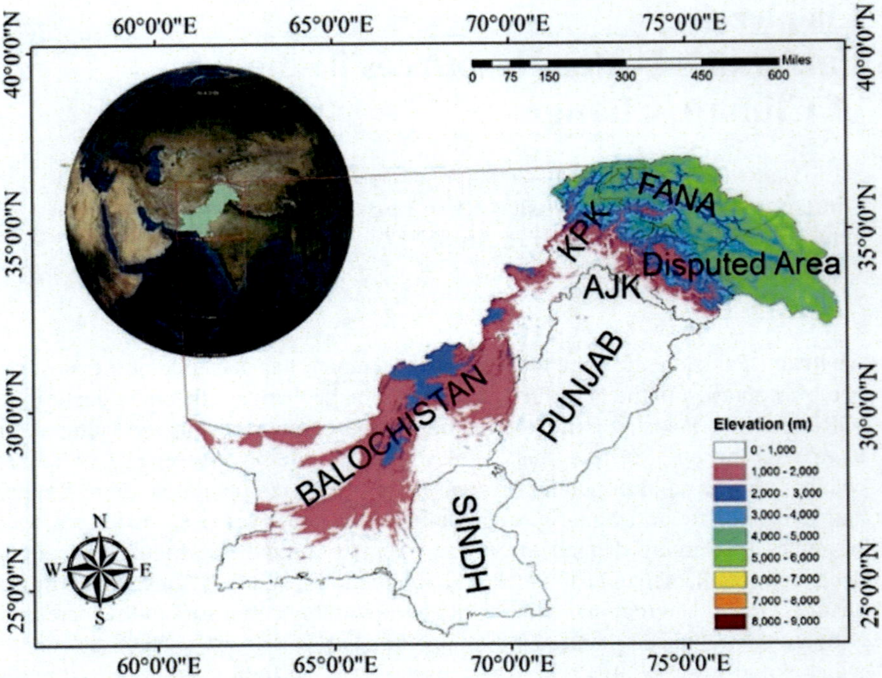

Fig. 6.1 Study area with elevation

geography of Pakistan is divided into three major areas: the Northern Highlands, Indus River Plain, and Balochistan Plateau.

The Northern Highlands of Pakistan consist of Hindukush, Himalayan, and Karakoram (HKHK) ranges, including the famous peaks K2 and Nanga Parbat. The flat Indus plain is made up of alluvium, deposited mainly by the Indus River and its tributaries. The Balochistan Plateau is an arid, mountainous, and desert area, located in the southwest of the country (Ahmed et al. 2015).

6.2 Situation Analysis

Pakistan relies heavily on the Indus Basin Irrigation System (IBIS), one of the world's most extensive. Surface water in the Indus Basin is generated by flows through the Indus and its tributaries – Jhelum, Chenab and Indus Rivers on the west, and Sutlej, Ravi, and Beas Rivers on the east. The total length of the Indus River is 2900 km, with a drainage area of 966,000 km^2 (Qureshi 2011). The Indus Basin (IB) is arid to semi-arid, with surface water flows fed by glacial and snowmelt and monsoon rainfall. The Indus Basin contains two major reservoirs, Mangla and Terbela, 12 inter-linked canals, 16 barrages, 45 canal commands, with a combined length of

59,000 km, and more than 120,000 watercourses with a combined length of 107,000 km (Qureshi 2005). The Indus Basin in Pakistan provides around 180 billion cubic metres (BCM) with 165 BCM from western rivers, i.e., Indus, Chenab, and Ravi, and remaining from eastern rivers, i.e., Beas, Sutlej, and Ravi. Out of 180 BCM, 128 BCM is withdrawn as surface water, while 52 BCM as groundwater withdrawal (Qureshi 2011). Another report states that more than 90% is diverted to canals for irrigated agriculture, and the remaining water is provided to the domestic and industrial sectors (WWF-Pakistan 2007).

Pakistan's per capita water storage capacity is only 150 m^3, which is lower than 220 m^3 in India, 460 m^3 in Morocco, 6000 m^3 in the US, and 2200 m^3 in China (Qureshi 2011). In addition, Pakistan can only store water for 30 days in the IB, which is far less than Colorado and Orange River Basins, having storage capacities of 900 days and 500 days, respectively (Briscoe et al. 2005). Pakistan's population is estimated to increase to 250 million by the year 2025 and the urban population is expected to increase to 52% by 2025 (Qureshi 2011). These increases will place immense pressure on the management of already limited water resources. Water demand for agriculture, domestic, and industrial sectors will rise by 10% of the total available water resources, and per-capita water availability will fall below 800 m^3 by the year 2025 (WWF-Pakistan 2007).

Despite having low storage capacity, Pakistan could not build more reservoirs due to a lack of financial resources, mistrust among provinces, and other governance issues. Besides, the storage capacity of existing major reservoirs is diminishing due to silting, i.e., approximately 5 million acre-feet (MAF) out of 18.5 MAF or 27% of the total live storage up to the year 2018 (WAPDA 2018). The rapid population growth and less water storage capacity would not only increase the water demand by all sectors but could also trigger inter-provincial water disputes. Moreover, poor efficiency of the irrigation system, water theft, and lack of operation and maintenance of old water infrastructure is further aggravating water management problems.

Historically, the growth rate of tube-wells in the Punjab province is higher than in other provinces, i.e., around 100,000 in 1970 and 800,000 in 2010 (Watto and Mugera 2016). Currently, the number of tube-wells has gone above one million, and most of them are used for irrigation (Yu et al. 2013). The development of tube-wells inclined until 2005 in Pakistan overall, but has slightly declined after 2005. Similarly, the groundwater share for irrigation has increased from 10 BCM to 60 BCM from 1965 to 2005, but has gradually decreased since 2005 (Watto and Mugera 2016). In Pakistan, the sustainability of major crops, including cotton, wheat, rice, and sugarcane, depends upon groundwater. Groundwater fulfils more than 50% of the total irrigation requirements of Pakistan, making the agricultural sector the largest consumer of groundwater (Qureshi et al. 2008). However, in the most populous province, i.e., Punjab, 90% of inhabitants depend on groundwater resulting in over-exploitation of the resource. The main sources of groundwater recharge of the IB are the precipitation and seepage from unlined canals. Approximately 45% of replenishable groundwater is recharged from canal seepage, 26% from return flows, 21% from rainfall, and just 6% from the river (van Steenbergen and Gohar 2005).

The water table of fresh groundwater in most areas of the aquifer is falling due to over-extraction.

Irrigation application efficiency of the current but outdated irrigation system is only 40% (Qureshi 2011). Despite low irrigation efficiencies, most farmers grow water-intensive crops, which has affected crop productivity severely. The per hectare production of wheat in Pakistan is 17% lower than India and 65% lower than China. The average production of rice in Pakistan is 2415 kg/ha compared with 2490 kg/ha in India (Watto and Mugera 2016). The average crop yield also varies in Pakistan; for instance, farmers get an average wheat production of 2267 kg/ha to 3847 kg/ha, and rice production from 1756 kg/ha to 3545 kg/ha (Qureshi et al. 2004). The factors that contribute towards the low crop productivity include salinisation, outdated irrigation practices, and above all, water shortage. Therefore, Pakistan needs to improve its crop-water productivity by adopting new methods and techniques of on-farm irrigation and water management.

6.3 Major Climatic Threats

Global warming due to climate change will alter the water cycle functioning in Pakistan and increase the frequency and intensity of extreme events (Dars et al. 2017). The global climate risk index (GCRI) 2019 has placed Pakistan in the top ten countries most affected by climate change (Eckstein et al. 2017). The GCRI 2019 has analysed various regions and countries which were affected by extreme events such as floods, droughts, and heatwaves for the last 20 years (1998 to 2017). The Indus Basin (IB) is one of the world's most sensitive regions to the impacts of climate change being highly dependent on glacial and snowmelt water supplies coming from the Upper Indus Basin (Immerzeel et al. 2010). The water resources in the IB are also expected to be affected by climate change through warming temperatures and increasingly erratic precipitation patterns, as described in detail below. The lower reaches of IB are most vulnerable to floods in the monsoon season causing significant casualties and economic losses. The following are the major water and climate-related issues in Pakistan.

6.3.1 Temperature Rise

Several studies have been conducted to analyse the temperature changes over this region. Sheikh et al. (2009) investigated the temperature trend for Pakistan using the Climate Research Unit (CRU) gridded data and showed a warming of 0.6 °C over the past century. Chaudhary et al. (2009) analysed temperature trends in Pakistan and found that annual maximum temperature had risen by 0.87 °C and minimum temperature by 0.48 °C between the years 1960 to 2007. Similarly, Farooqi et al. (2005) analysed Pakistan Meteorological Department (PMD) weather station data for the time frame of 1951 to 2001 and has projected temperature rise ranging from 0.6 °C to 1 °C. In another study, Iqbal and Zahid (2014) projected mean temperature in summer in the South Asia region by using 24 Global Climate Models (GCMs) under two Representative Concentration Pathways (RCPs), i.e., RCP4.5 and RCP8.5. The study projected that the mean temperature in Pakistan could increase to 4.38 °C under high emission scenarios from 2011 to 2100. Su et al. (2016) found a consistently increasing trend in mean annual temperature over the entire Indus Basin. Under RCPs 2.6, 4.5 and 8.5, they projected that the annual average temperature might rise, respectively, by 1.21 °C, 1.93 °C and 2.71 °C during the mid-century (2046–2065) while it could increase by 1.1 °C, 2.49 °C, and 5.19 °C in the late century (2081–2100).

We also downscaled ten global climate models (GCMs) for mean temperature over the entire Indus Basin and Pakistan using quantile mapping and delta downscaling method following Mosier et al. (2018). The ten GCMs used were CCSM4, CESM1-BGC, MIROC5, ACCESS1-3, INM-CM4, CSIRO-Mk3-6-0, MPI-ESM-LR, GFDL-CM3, BCC-CSM1-1, and CanESM2. The Mean Monthly Error (MME) averaging technique (i.e., simple arithmetic mean) was used in this study. These changes were estimated for the historical reference period (1960–1990). The changes in spatial patterns of seasonal temperatures for the future period (2040–2070) relative to the baseline period (1960–1990) are given in Fig. 6.2. It can be observed that the annual mean temperature has a consistently increasing trend throughout the basin. In the IB, the annual mean temperature is projected to increase by 2 °C under RCP4.5 and 2.66 °C under the RCP8.5 scenario. These projections also show a consistent increase in the temperature in the IB in the winter season by 2.1 °C under RCP4.5 and 2.8 °C under the RCP8.5 scenario. Similarly, the summer temperature is also projected to increase by 1.93 °C under RCP4.5 and 2.53 °C under RCP8.5 scenarios. However, in the summer season, the Upper Indus Basin (UIB) and western parts of the IB are projected to experience greater increases in temperature under both scenarios.

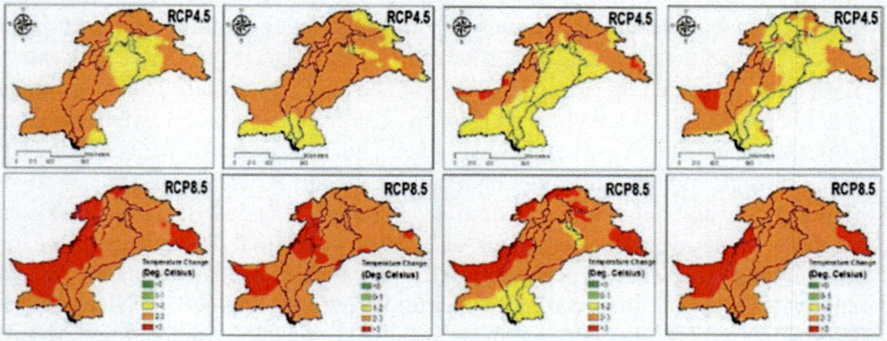

Fig. 6.2 Projected changes (°C) in temperature for 2040–2070 relative to the historic period (1960–1990). From left to right the figures represent winter, spring, summer, and autumn seasons

The changes in spatial patterns of mean annual, winter, and summer temperatures across Pakistan follow similar increasing trends of the IB. However, the western parts of Balochistan province are projected to experience larger increases in temperature under the RCP4.5 scenario. It can be seen that the annual mean temperature in Pakistan is projected to increase by 2.14 °C under RCP4.5 and 2.8 °C under the RCP8.5 scenario. The winter temperature in Pakistan may increase by 2.3 °C under RCP4.5 and 2.9 °C under the RCP8.5 scenario. Similarly, the summer temperature is also projected to be rising by 2.06 °C under RCP4.5 and 2.76 °C under RCP8.5 scenarios. However, a lower increase in the summer temperature is projected in Sindh and Punjab provinces under the RCP4.5 scenario.

6.3.2 *Erratic Precipitation Patterns*

Precipitation is an essential parameter in evaluating climate water balance. Pakistan receives its monsoon precipitation from the Bay of Bengal and winter precipitation through westerly winds in the winter season. Most of the southern part of the country lies under arid to semi-arid climate conditions, with annual rainfall varying from 200 mm in southern areas to 1600–2000 mm in northern areas (Farooqi et al. 2005). However, the precipitation patterns are observed to be changing with time. For instance, Chaudhary et al. (2009) have presented long-term precipitation time series, i.e., from 1901 to 2007, showing that from 1900 to 1940, the precipitation had gradually decreased from 600 mm to 400 mm. However, after 1940, an increase of 133 mm in precipitation was observed. This study showed an increase of 61 mm, 22 mm, and 21 mm in the annual, monsoon, and winter precipitation, respectively. They also analysed projected future precipitation trends over all of Pakistan up to the middle of the twenty-first century. They showed mixed trends of precipitation over various regions of Pakistan under A2 and A1B scenarios. Rajbhandari et al. (2015) also projected climate changes over the Indus Basin using the PRECIS Regional Climate Model (RCM). They projected an increasing precipitation trend in UIB while a decreasing trend in the Lower Indus Basin (LIB). Su et al. (2016) projected annual precipitation trends over Indus Basin under three RCPs, i.e., RCP 2.6, 4.5 and 8.5 for the mid (2046–2065) and late (2081–2100) twenty-first century. Their study projected precipitation patterns to be highly uncertain. The mean annual precipitation in the mid-century will rise by 3.2%, 0.1%, and 6.2%, whereas, in the late-century, it would increase by 5.6%, 4.0%, and 7.8% under RCP 2.6, 4.5 and 8.5.

We downscaled the same ten global climate models (GCMs) listed previously to model future precipitation projections over the entire Indus Basin and Pakistan, using the same averaging technique and reference period. The results show that precipitation in the IB will be highly uncertain in the future. The changes in spatial patterns of seasonal precipitation for the future period (2040–2070) relative to the baseline period (1960–1990) are given in Fig. 6.3. It can be observed that average annual precipitation in the Indus Basin is projected to slightly increase under both scenarios. Under the RCP4.5 and RCP8.5 scenario, the mean annual precipitation is

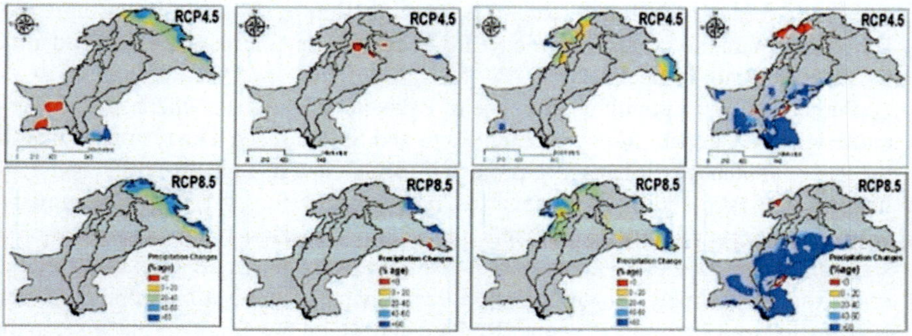

Fig. 6.3 Projected changes (in percentages) in precipitation for 2040–2070 relative to the historic period (1960–1990). From left to right the figures represent winter, spring, summer, and autumn seasons. Grey grid cells indicate locations where the absolute uncertainty between ensemble projections is greater than the projected changes

projected to increase by 7% and 12%, respectively in the IB. The projections do not show any increase in the winter (December–January–February) precipitation under both scenarios in the IB. The spring (March–April–May) precipitation in the IB is projected to decrease by 4% under the RCP4.5 scenario with no change under the RCP8.5 scenario. However, the summer (June–July–August–September) precipitation is projected to increase by 10% and 15% under RCP4.5 and RCP8.5 scenarios, respectively. Similarly, a significant increase of 29% and 43% in the autumn (October–November) precipitation is projected in the IB.

Most of Pakistan follows the same projections as the IB. In Pakistan, the annual precipitation is projected to increase by 7% and 10% under the RCP4.5 and RCP8.5 scenarios, respectively. The projections do not show any increase in the winter (DJF) precipitation under both scenarios in Pakistan. The spring (MAM) precipitation in Pakistan is projected to decrease by 4% under the RCP4.5 scenario with no change under the RCP8.5 scenario. However, the summer (JJAS) precipitation is projected to increase by 11% and 17% under RCP4.5 and RCP8.5 scenarios, respectively. Similarly, a significant increase of 18% and 36% in the autumn (ON) precipitation is projected in Pakistan. It seems that the precipitation patterns in the IB and Pakistan are projected to be shifting more from winter to autumn season.

The Northern Areas of Pakistan also come under the federal jurisdiction of Pakistan. This region has great significance regarding hydrological and environmental aspects, and this region consists of the world's largest glaciers and several highest peaks. More importantly, the entry point of the Indus River in Pakistan lies in this region. Therefore, precipitation changes in the future will significantly affect the flow of the Indus River. The results show that the mean annual precipitation is projected to increase in the northern region of Pakistan by 14% and 19% under the RCP4.5 and RCP8.5 scenarios, respectively. The winter precipitation is projected to increase by 17% under the RCP4.5 and by 22% under the RCP8.5 scenario. In the spring season, a slight increase of 3% is projected under both scenarios. The precipitation in the summer season is likely to increase by 14% and 18% under RCP4.5

and RCP8.5 scenarios, respectively. The precipitation in the autumn season is projected to remain the same under the RCP4.5 scenario and increase by 17% under the RCP8.5 scenario, respectively.

Punjab, the most populous province of Pakistan, is the place where major rivers, namely Beas, Chenab, Indus, Jhelum, Ravi, and Sutlej, meet. The results show that the mean annual precipitation is projected to increase in Punjab by 6% and 13% under the RCP4.5 and RCP8.5 scenarios, respectively. However, the winter precipitation is projected to remain the same under both scenarios. In the spring season, a significant decrease of 13% and 9% is likely to occur under RCP4.5 and RCP8.5 scenarios, respectively. However, a significant increase in the summer precipitation is projected by 12% and 53% under RCP4.5 and RCP8.5 scenarios. Likewise, in the autumn season, the precipitation is projected to increase by 25% and 75% under RCP4.5 and RCP8.5, respectively.

Sindh, being the lower riparian province, is highly vulnerable to the severe impacts of climate change. It faces floods in case of excess rainfall and droughts in case of low rainfall. The results show that in Sindh, mean annual precipitation is projected to increase by 22% and 27% under RCP4.5 and RCP8.5 scenario, respectively. The winter precipitation is likely to increase by 20% under the RCP4.5 scenario and remain the same under RCP8.5. In the spring season, the precipitation is not likely to change under RCP4.5 and increase by 25% under the RCP8.5 scenario. However, a significant increase in precipitation is projected by 27% and 24% under RCP4.5 and RCP8.5 scenarios, respectively, in the summer season. In the autumn season, a huge increase in precipitation is projected by the models, up to 67% under RCP8.5.

Balochistan is the largest province of Pakistan by land size, having an area of 347,000 km^2. Its climate is diverse, ranging from extremely hot summers on the plains in the east to freezing winters in the northern mountainous areas. The results show that the mean annual precipitation is projected to increase in Balochistan province by 8% and 17% under RCP4.5 and RCP8.5 scenarios, respectively. However, the winter precipitation is projected to decrease by 5% under RCP4.5 and 11% under RCP8.5. In the spring season, the precipitation is not projected to change under either scenario. However, the summer precipitation is likely to increase by 15% and 23% under the RCP4.5 and RCP8.5 scenarios, respectively. In the autumn season, precipitation is projected to increase by 50% and 75% under RCP4.5 and RCP8.5, respectively.

The Khyber Pakhtunkhwa (KPK) is one of the four provinces of Pakistan and is located in its northwest. It hosts several catchments contributing to the Indus River. The results show that the mean annual precipitation is projected to increase slightly in KPK province by 4% and 10% under RCP4.5 and RCP8.5 scenarios, respectively. The winter precipitation is likely to increase by 5% under the RCP8.5 scenario with no change under RCP4.5. However, in the spring season, precipitation is likely to decrease by 3% under both scenarios. The summer precipitation is projected to increase by 15% and 23% under RCP4.5 and RCP8.5 scenarios, respectively. In the autumn season, a huge precipitation increase is projected, amounting to 50% and 75% under RCP4.5 and RCP8.5, respectively.

Azad Jammu and Kashmir (AJK) is the second-largest contributor of irrigation water supply and hydropower generation in the country, also known as a green energy base for Pakistan. The results show that in the AJK, the mean annual precipitation is projected to increase slightly by 3% and 7% under the RCP4.5 and RCP8.5 scenarios, respectively. However, no change in winter precipitation is projected under RCP8.5 with a slight decrease of 4% under the RCP4.5 scenario. In spring, the precipitation is projected to decrease by 9% and 6% under RCP4.5 and RCP8.5, respectively. The summer precipitation is projected to significantly increase by 51% and 59% under RCP4.5 and RCP8.5 scenarios, respectively. The precipitation in the autumn season shows mixed patterns. The precipitation is projected to decrease by 6% under RCP4.5 but increase by 16% under RCP8.5.

The Islamabad Capital Territory (ICT) is the capital of Pakistan, having a subtropical climate. The results show that the mean annual precipitation is projected to increase slightly in the ICT region by 4% and 8% under RCP4.5 and RCP8.5 scenarios, respectively. However, the winter precipitation is projected to decrease slightly by 4% under RCP4.5 but increase by 2% under the RCP8.5 scenario. In the spring season, a significant decrease of 12% and 8% will occur under RCP4.5 and RCP8.5 scenarios. The summer precipitation is projected to increase by 8% and 12% under RCP4.5 and RCP8.5 scenarios, respectively. In the autumn season, the precipitation is projected to not change under RCP4.5 but increase by 17% under the RCP8.5 scenario.

In the Federally Administered Tribal Areas (FATA) the mean annual precipitation is projected to increase slightly by 6% and 8% under RCP4.5 and RCP8.5 scenarios, respectively. However, the winter precipitation is projected to decrease by 3% under the RCP4.5 scenario with no change under RCP8.5. In the spring season, a slight decrease of 8% and 6% will occur under RCP4.5 and RCP8.5 scenarios, respectively. The summer precipitation is projected to significantly increase by 18% and 26% under RCP4.5 and RCP8.5 scenarios, respectively. In the autumn season, the precipitation is projected not to change under RCP4.5 but increase by 7% under the RCP8.5 scenario.

6.3.3 *Extreme Events (Floods and Droughts)*

Extreme events (floods and droughts) are caused either by intense and huge rainfall or deficit of rainfall. Floods in Pakistan usually occur due to heavy rainfalls during the monsoon, which is sometimes augmented by snowmelt flows in the rivers. According to the Federal Flood Commission Report (2010), from 1950 to 2009, Pakistan has faced a total financial loss of US$20 billion with an affected area of 407,132 km^2. The worst floods of 2010 alone destroyed an area of 160,000 km^2 with a financial loss of US$10 billion. Heavy monsoon rainfall in Sindh, Balochistan, KP, and the lower Punjab made the rivers overflow, causing severe destruction in areas from Gilgit-Baltistan to Kotri (the last structure on the Indus River) and downstream from there.

Drought is one of the most frequently occurring extreme events in Pakistan. It occurs in about four out of ten years (Jamro et al. 2019). Sindh, Punjab, and Balochistan have experienced droughts for the past few decades due to low rainfall. Jamro et al. (2019) analysed the droughts in Pakistan using the standardised precipitation evapotranspiration index (SPEI). They found a statistically significant trend in the increase of droughts in Sindh and Balochistan; and found no statistically significant trend of droughts for the northern areas of Pakistan.

6.3.4 Glacier Melting

The Hindukush-Himalayan region, also known as the Asian Water Tower (Immerzeel et al. 2010), alone covers an area of 4 million km^2, and spanning over eight countries. The HKH region plays an essential role in the climate system of the region, and it is a sensitive indicator of global climate change. The Indus Basin consists of 18,495 glaciers covering an area of 21,192 km^2 with ice-reserves of 2696 km^3 (Bajracharya and Shrestha 2011). The glacial and snowmelt in the UIB contribute more than 80% of the flow to the Indus River (Hewitt et al. 1989), with the Karakoram range alone contributing more than 50% (Immerzeel et al. 2010). Global warming has caused glaciers all over the world to recede, including in the Himalaya and Hindukush mountain ranges.

One-sixth of the world population depends on water from snow and glacier-melt (Barnett et al. 2005). Climate change has threatened the freshwater reserves and has increased the risk of hazards in the mountainous region and downstream (Bajracharya and Shrestha 2011). The water supply provided by the UIB is critical to the country's economy. Pakistan is largely dependent on the Indus River for its agriculture, energy production, and domestic water supply. Rising temperatures and changes in precipitation patterns will alter the seasonal snow cover, glacier volume, and runoff characteristics affecting the water availability both upstream and downstream of the Indus Basin (Lutz et al. 2014).

Despite these significant risks faced by Pakistan and its people, Pakistan's mountainous region is one of the most data-scarce. An inadequate number of hydrometeorological measuring gauges are installed, and the generated data are insufficient to monitor glacier dynamics and thus devise policies.

6.3.5 Sea Level Rise and Seawater Intrusion in the Indus Delta

Global warming will cause melting ice sheets and glaciers and thermal expansion of the oceans, ultimately leading to rises in sea level all around the globe (IPCC 2013). Due to rising sea levels, coastal areas will increasingly experience adverse impacts such as coastal flooding and erosion, loss of wetlands and estuaries, saltwater intrusion, rising water tables, infrastructure damage, loss of agricultural productivity, and socio-economic impacts.

The Indus Delta is located in the southeast of Pakistan and covers an area of about 6000 km^2 (Kalhoro et al. 2016). It comprises 17 major creeks, numerous minor creeks, and mudflats, with an estuary covering an area of 36 km^2. The fully active part of the delta has reduced in size by 10% (ADB 2005). Earlier, the Indus River provided 40 million tons of silt per year to the delta, and it was one of the largest sediment-providing rivers of the world (Nasir and Akbar 2012). But unfortunately, today, the lower reaches of the river are highly affected by reduced and highly intermittent flow of water downstream of Kotri. The reduction of water and sediment loads has resulted in land erosion and seawater intrusion (Kalhoro et al. 2016).

In the dry season, the seawater has the potential to intrude up to 84 km upstream of the delta and the average land erosion rate is 0.179 to 0.0135 km/year (Kalhoro et al. 2016). Most of the creeks of the delta are filled with saline water from the sea and the subsurface land of the delta is highly saline affected due to seawater intrusion. The salinity in creeks is higher than the salinity of seawater itself (Anwar et al. 2014). The salinity of water in the coastal areas off Karachi has reached 37,000 ppm, and has increased to 42,000 ppm in the tidal creeks (Chandio and Anwar 2009). Various studies have reported impacts from seawater intrusion differently. According to one report (Ministry of Finance 2011), seawater intrusion has degraded 12% of the total cultivatable land of Sindh province. With 0.486 Mha of Indus Delta degraded, around 0.25 million people have had to migrate. A study by Chandio et al. (2011) reported that seawater intrusion is deteriorating 80 acres of land per day. The sub-surface intrusion of seawater has degraded 1.15 Mha or 88% of the entire delta, and only 0.15 Mha or 11% is unaffected (Siyal 2018). Thus, it is obvious that seawater intrusion has deteriorated the ecosystem, groundwater resources, and agricultural productivity in the deltaic region.

The Indus Delta is likely to be affected by anthropogenic interventions upstream, including large scale hydraulic engineering, withdrawal of water for hydropower and irrigation, and climate change. Therefore, it is essential to better understand the processes and mechanisms of seawater intrusion by conducting robust studies with better accuracy and integrity of the environment, agriculture, and socio-economic conditions of the region.

6.4 Conclusions and the Way Forward

Freshwater is a finite resource, and climate change is posing a potential threat to the water resources of the country, including surface and groundwater. Due to a higher rate of population growth expected for Pakistan in the future, 40% more food would be required by 2025, while per capita water availability is projected to decline to 600 m^3 by the year 2025 (Qureshi 2011). The water management problems and challenges differ from place to place and scale to scale. Therefore, an integrated approach needs to be adopted to resolve these issues. Pakistan can tackle water management and climate change challenges by following these measures and strategies:

- Raise awareness about climate change/water management.
- Give high priority to climate change adaptation through, for example, water-saving technologies, recycling and reuse of water, increases in water storage capacity, better regulation of groundwater extractions, and adequate drainage provisions.
- Strengthen early warning systems.
- Provide incentives to progressive farmers.
- Reinforce knowledge/data sharing.
- Enhance governance through, for example, water pricing, capacity building, institutional building, and improving coordination among provinces/organisations.
- Improve water quality.
- Promote renewable energy.
- Grow heat resistant crops.

References

ADB (2005) Sindh coastal and inland community development project, interim report

Ahmed K, Shahid S, bin Harun S et al (2015) Assessment of ground-water potential zones in an arid region based on catastrophe theory. Earth Sci Inf 8:539–549. https://doi.org/10.1007/s12145-014-0173-3

Anwar MM, Chandio NH, Bhalli MN (2014) Economic deprivation of Indus river delta, Sindh, Pakistan: causes and suggestions. Sci Int 26(2):885–890

Bajracharya SR, Shrestha BR (2011) The status of glaciers in the Hindu Kush-Himalayan region. International Centre for Integrated Mountain Development (ICIMOD), Kathmandu, Nepal

Barnett TP, Adam JC, Lettenmaier DP (2005) Potential impacts of a warming climate on water availability in snow-dominated regions. Nature 438:303–309. https://doi.org/10.1038/nature04141

Briscoe J, Qamar U, Contijoch M et al (2005) Pakistan's water economy: running dry. World Bank, Washington, DC

Chandio NH, Anwar M (2009) Impacts of climate on agriculture and its causes: a case study of Taluka Kamber, Sindh, Pakistan. Sindh Univ Res J 41(2):59–64

Chandio NH, Anwar MM, Chandio AA (2011) Degradation of Indus delta, removal mangroves forestland its causes: a case study of Indus river delta. Sindh Univ Res J 43(1):67–72

Chaudhary QZ, Mahmood A, Rasul G, Afzaal M (2009) Climate change indicators of Pakistan. Pakistan Meteorological Department Technical Report No. PMD-22/2009, Islamabad, Pakistan

Dars GH, Najafi MR, Qureshi AL (2017) Assessing the impacts of climate change on future precipitation trends based on downscaled CMIP5 simulations data. Mehran Univ Res J Eng Technol 36(2):385–394

Eckstein D, Künzel V, Schäfer L (2017) Global climate risk index 2018: who suffers most from extreme weather events? Weather-related loss events in 2016 and 1997 to 2016. Germanwatch e.V. https://germanwatch.org/en/download/20432.pdf

Farooqi AB, Khan AH, Mir H (2005) Climate change perspective in Pakistan. Pak J Meteorol 2(3):11–21

Hewitt K, Wake CP, Young GJ, David C (1989) Hydrological investigations at Biafo Glacier, Karakoram Range, Himalaya; an important source of water for the Indus River. Ann Glaciol 13:103–108. https://doi.org/10.3189/S0260305500007710

Immerzeel WW, Beek LPH, Van Bierkens MFP (2010) Climate change will affect the Asian water towers. Science 328(5984):1382–1385. https://doi.org/10.1126/science.1183188

IPCC (2013) Climate change 2013: the physical science basis. Working Group I contribution to the Fifth Assessment Report of the Intergovernmental Panel on Climate Change. IPCC, Geneva

Iqbal W, Zahid M (2014) Historical and future trends of summer mean air temperature over South Asia. Pak J Meteorol 10(20):67–74

Jamro S, Dars GH, Ansari K, Krakauer NY (2019) Spatio-temporal variability of drought in Pakistan using Standardized Precipitation Evapotranspiration Index. Appl Sci 9(20):4588. https://doi.org/10.3390/app9214588

Kalhoro NA, He Z, Xu D et al (2016) Vulnerability of the Indus River delta of the North Arabian Sea, Pakistan. Glob NEST J 18(3):599–610

Lutz AF, Immerzeel WW, Shrestha AB, Bierkens MFP (2014) Consistent increase in High Asia's runoff due to increasing glacier melt and precipitation. Nat Clim Chang 4:587–592. https://doi.org/10.1038/NCLIMATE2237

Ministry of Finance (2011) Economic survey of Pakistan 2010–11. Finance Division, Ministry of Finance, Revenue and Economic Affairs, Government of Pakistan. http://www.finance.gov.pk/survey_1011.html

Mosier TM, Hill DF, Sharp KV (2018) Update to the Global Climate Data package: analysis of empirical bias correction methods in the context of producing very high resolution climate projections. Int J Climatol 38:825–840. https://doi.org/10.1002/joc.5213

Nasir SM, Akbar G (2012) Effect of River Indus flow on low riparian ecosystems of Sindh: a review paper. Rec Zool Surv Pak 21:86–89

Qureshi AS (2005) Climate change and water resources management in Pakistan. In: Mirza MMQ, Ahmad QK (eds) Climate change and water resources in South Asia. Taylor & Francis, London, pp 197–230

Qureshi AS (2011) Water management in the Indus Basin in Pakistan: challenges and opportunities. Mt Res Dev 31(3):252–260. https://doi.org/10.1659/mrd-journal-d-11-00019.1

Qureshi AS, Akhtar M, Sarwar A (2004) Role of changing energy pricing policies on groundwater development in Pakistan. J Appl Irrig Sci (Zeitschrift für Bewässerungswirtschaft) 39:329–342

Qureshi AS, Mccornick PG, Qadir M, Aslam Z (2008) Managing salinity and waterlogging in the Indus Basin of Pakistan. Agric Water Manag 95(1):1–10. https://doi.org/10.1016/j.agwat.2007.09.014

Rajbhandari R, Shrestha AB, Kulkarni A et al (2015) Projected changes in climate over the Indus River basin using a high-resolution regional climate model (PRECIS). Clim Dyn 44:339–357. https://doi.org/10.1007/s00382-014-2183-8

Sheikh MM, Manzoor N, Adnan M et al (2009) Climate profile and past climate changes in Pakistan. Global Change Impact Studies Centre Research Report No. 1, Islamabad, Pakistan

Siyal AA (2018) Climate change: assessing impact of seawater intrusion on soil, water and environment on Indus delta using GIS and remote sensing tools. US-Pakistan Center for Advanced Studies in Water (USPCAS-W), MUET, Jamshoro, Pakistan

Su B, Huang J, Gemmer M et al (2016) Statistical downscaling of CMIP5 multi-model ensemble for projected changes of climate in the Indus River Basin. Atmos Res 178–179:138–149. https://doi.org/10.1016/j.atmosres.2016.03.023

van Steenbergen F, Gohar S (2005) Groundwater development and management. In: Briscoe J, Qamar U (eds) Pakistan water economy running dry - background papers. Oxford University Press, Oxford

WAPDA (2018) WAPDA annual report 2017–18. Pakistan Water and Power Development Authority, Lahore, Pakistan

Watto MA, Mugera AW (2016) Groundwater depletion in the Indus Plains of Pakistan: imperatives, repercussions and management issues. Int J River Basin Manag 14(4):447–458. https://doi.org/10.1080/15715124.2016.1204154

WWF-Pakistan (2007) Pakistan's waters at risk: water and health related issues in Pakistan and key recommendations. WWF-Pakistan, Lahore, Pakistan

Yu W, Yang YC, Savitsky A et al (2013) The Indus Basin of Pakistan: the impacts of climate risks on water and agriculture. World Bank Publications, Washington, DC

Chapter 7
Unlocking Economic Growth Under a Changing Climate: Agricultural Water Reforms in Pakistan

Stephen Davies and William Young

Abstract Irrigation dominates water use in Pakistan, but from an economic perspective, irrigation water use efficiency and productivity remain very low. Conventional agronomic practices and irrigation water application methods, deteriorating irrigation infrastructure, and agricultural subsidies all act to repress a desire by farmers to improve their irrigation water use efficiency. We explore the performance of irrigation in the national economy out to 2055 using a computable general equilibrium model for Pakistan. We adopt a moderate population growth scenario and assume plausible rates of productivity growth by sector, and then explore how removal of agricultural subsidies and changing diets may affect irrigated crop choices and economic productivity under different rates of climate warming. We find that without critical reforms or efficiency improvements, water demand could easily exceed supply by 2055. However, with appropriate investments and reforms, water scarcity need not prevent Pakistan from ensuring food security and from reaching upper-middle income status by 2050. Critical to this transformation is ensuring that a fraction of the water saved through irrigation efficiency improvements is available to support faster growth in the industrial and service sectors, and a fraction is protected for environmental purposes.

Keywords Economic modelling · Water resources policy · Water productivity · Irrigation efficiency · Climate change

S. Davies (✉)
International Food Policy Research Institute, Islamabad, Pakistan
e-mail: S.Davies@cgiar.org

W. Young
Global Water Practice, World Bank Group, Washington, DC, USA
e-mail: W.Young@worldbank.org

M. A. Watto et al. (eds.), *Water Resources of Pakistan*, World Water Resources 9,
https://doi.org/10.1007/978-3-030-65679-9_7

7.1 Introduction

Hydrologically and agriculturally, Pakistan is dominated by the Indus Basin. The Upper Indus has its headwaters in China and then flows northwest through Jammu and Kashmir before turning sharply to exit the mountains through Khyber Pakhtunkhwa. To the east, the Jhelum, Chenab, Ravi, and Sutlej (and its tributary the Beas) flow from India into the Pakistan province of Punjab. The extensive Indus floodplain is closely connected to alluvial aquifers extending across 16 million hectares (Mha), of which 6 Mha are fresh (mostly in Punjab) and remainder saline (mostly in Sindh). The Indus exits through an extensive delta system to the Arabian Sea.

Pakistan has experienced two decades of steady, although not dramatic, economic growth to reach US$1,500 per capita and has significantly increased food supply, more than keeping pace with the population that has reached 208 million (United Nations 2017). Although food supply has increased, food security remains a serious challenge, including challenges in food distribution and variety. While food production currently exceeds demand, deficiencies in food diversity, procurement, storage, and distribution undermine food security (Kirby et al. 2017; Hussain and Routray 2012). Food access is thus uneven, and malnutrition is high among certain groups, as 47% are food insecure through undernourishment and micronutrient deficiencies (Davies et al. 2018; WFP 2018). Thus, agriculture in Pakistan must meet both increased grain demand from population and income growth and diversify production (Hayat et al. 2016). At the core of these challenges is the need to obtain better performance from Pakistan's water resources.

To become truly food and water secure, Pakistan must substantially raise per capita income and concurrently ensure that sufficient water is recovered from irrigation improvements to support faster industrial and service sector growth. Additionally, some water needs to be reallocated to meet environmental needs. These reallocations must occur amid continued population growth and climate change, both of which increase pressure to retain water in agriculture.

The average annual resource (surface inflows and groundwater recharge) of the Indus Basin in Pakistan is estimated to be around 218 billion cubic metres (BCM), with withdrawals of 187 BCM (Young et al. 2019). However, this withdrawal total includes a significant double counting because much of the groundwater withdrawal is first withdrawn as surface water (diversions into irrigation canals) and then leaked to groundwater from canals and fields. Around 70% of groundwater withdrawals are supported by canal leakage and irrigation drainage, the remainder being rainfall and river recharge (Laghari et al. 2012). Adjusting for this double counting suggests a net annual withdrawal of around 136 BCM (Young et al. 2019).

Water use is heavily dominated by irrigation, which accounts for over 95% of withdrawals. An approximate water balance for the Indus (Fig. 7.1) highlights the high level of water use, the large water losses – including natural losses, the recycling of water between surface and groundwater – and the very low level of residual basin outflows. However, Pakistan's total water resource is somewhat uncertain, as

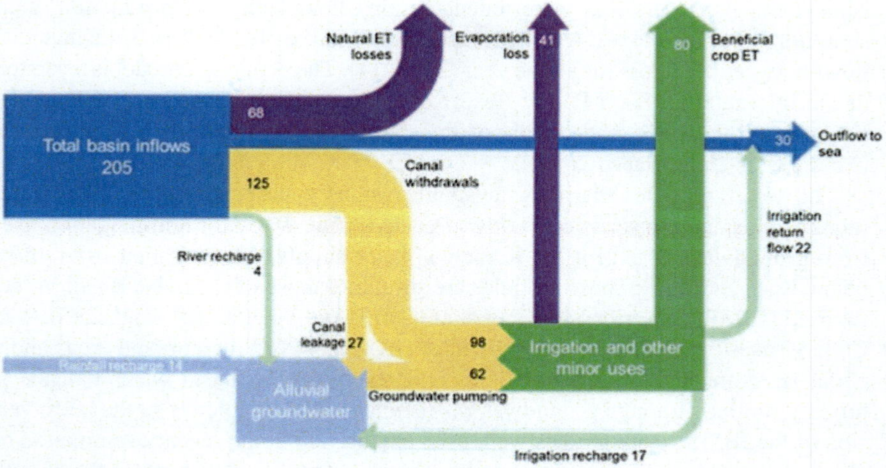

Fig. 7.1 Indus Basin Water Balance from Young et al. (2019)

monitoring is incomplete, surface-groundwater exchanges are poorly quantified, and robust water accounting has not been established.

While a diverse mix of crops is grown in Pakistan, around three-quarters of the area and two-thirds of the value comes from wheat, rice, sugarcane, and cotton. These four crops account for around 80% of agricultural water consumption but currently contribute less than 5% to total GDP, a share that is in decline. Thus, while agriculture employs 43% of the labour force, and while irrigation heavily dominates water use, the relatively small contribution of irrigation to GDP suggests that, contrary to common perception, Pakistan should not be considered an "irrigation economy".

In this chapter, we summarise results from a computable general equilibrium (CGE) model that explores whether Pakistan can reach upper-middle-income status by 2050 and specifically whether water scarcity represents a serious impediment to growth (CGE-W). We consider the water required to ensure food security, as well as the water required to improve the environmental health of the river and delta. On the demand side, we consider population growth and urbanisation, climate change and changing diets. We conclude with examples of how the analytical results can guide key policy reforms and investments for a water-secure Pakistan.

7.2 Modelling Framework

The CGE-W model consists of a series of modules, including an annual economy-wide component, and modules computing water demand, water basin management, and water allocation. CGE-W starts with the whole-economy model, with consumers, producers and government entities linked through production, consumption, trade, and taxes. A Social Accounting Matrix (SAM) defines the financial flows between all actors for a base year (2013–14). The economic model is a version of the International Food Policy Research Institute (IFPRI) standard CGE model (Löfgren et al. 2002), and details of the model can be found in related articles (Young et al. 2019; Saeed 2017).

The water modules distribute available water to crops based on crop water requirements and the stress on yields, after deducting water for non-irrigation uses from groundwater – except for Karachi's water supply, which is met from "flow below Kotri Barrage". These modules are similar to those of the Indus Basin Model Revised (IBMR), described by Yang et al. (2013) and Yu et al. (2013). Like IBMR, CGE-W uses twelve agroecological zones (ACZs), 44 canal command areas and a nodal structure including major dams and reservoirs. Its basin water balance is broadly similar, although groundwater is handled more completely in the latest version of the IBMR. Both models were used extensively in the recent examination of climate change by Yu et al. (2013), which highlighted the relative strengths of both approaches. CGE-W does not consider the water resources of the Makran Coast or the Kharan Desert hydrological units in Balochistan, outside of the Indus Basin.

The economic model is initially run for a given year, after updating exogenous trends like productivity, international prices, and tax rates, to obtain outputs by sector and allocation of land to various crops. Next, the water demand module, described in more detail below, calculates water demand for crops, industry, households, and livestock, with the latter three being determined before water is made available for agricultural uses. Available supply is then allocated across canals, regions, and crops to minimise water stress, subject to flow routing equations allocating water through the Pakistan Indus canal system. These are described completely in Young et al. (2019), and Davies et al. (2016a, b).

Subsequently, the water stress module evaluates a measure of yield stress for every crop, given changes in temperature and rainfall. New yields are then calculated and applied to the CGE model, which is solved again for a final equilibrium, but with the allocation of land to crops fixed, since farmers cannot change their cropping decisions after planting. Groundwater pumping is allowed only in non-saline areas, and non-irrigation demands are met solely from groundwater and assumed to be fully consumptive. In reality, a large fraction of non-irrigation water is flowing back to groundwater or surface water, albeit in a more polluted state. These steps are explained in more detail below.

7.2.1 *The Water Demand Module*

Four components are determined in the water demand module, and together provide the total demand for water. Industrial water demand, identified by agroecological zone (z) and month (m), is proportional to the square root of industrial GDP (see Eq. 7.1 below). Similarly, livestock and household demands are proportional to the square roots of livestock GDP and aggregate household expenditures respectively (7.2), (7.3). These demands, of course, are growing more slowly than the related economic activities, and are met from groundwater before making groundwater available for irrigation.

$$\text{Industrial and Services Demand}: wci_{z,m} = qwi_{z,m} * \left(\sum_i qai_i\right)^{0.5} \tag{7.1}$$

$$\text{Domestic}\left(\text{household}\right)\text{Demand}: wcd_{z,m} = qwd_{z,m} * \left(\sum_h eh_h\right)^{0.5} \tag{7.2}$$

$$\text{Livestock Demand}: wcl_{z,m} = qwl_{z,m} * \left(\sum_l qal_l\right)^{0.5} \tag{7.3}$$

Irrigation water demand is a function of areal irrigation demand ($iwd_{z,c,m}$) less soil moisture for the month ($subirr_{z,c,m}$), multiplied by the number of acres of each commodity within a zone (7.4).

$$\text{Irrigation Demand}: wir_{z,m} = \sum_c \left(\max\left(\left(iwd_{z,c,m} - subirr_{z,c,m}\right), 0\right) * Acres_{z,c}\right) \tag{7.4}$$

Areal irrigation demand is based on known crop water requirements less effective rainfall, while soil moisture is based on water table depth (exogenous and fixed). Crop areas are given by zone but weighted by canal commands within a zone, which makes surface water demand go to a given canal command within the zone.

7.3 Scenarios and Key Assumptions

The baseline scenario ("Business as Usual" – BAU) simulates average per capita income growth near the current rate to reach US$3,030 by 2055. A Reaching Upper Middle Income (RUMI) scenario is defined and modelled to explore the plausibility, especially from a water management perspective, of Pakistan attaining a per capita income of US$6,000 by 2055. This would require an annual GDP per capita growth rate of 4.0%, or 0.5% above the best recent performance by a group of comparator countries reviewed in Young et al. (2019). Thus, this is a "stretch goal" for Pakistan that illustrates the importance of water security to economic growth and can inform macro decisions on investments and reforms in the water sector.

Variants of both BAU and RUMI are explored (Table 7.1). The BAU and RUMI base cases include moderate climate change: a 1 °C rise in mean annual temperature by 2055, similar to recent rates of warming across Pakistan. Faster warming variants explore a 3 °C increase by 2055. For the faster warming RUMI variant, the impacts of different consumer preferences are explored. As incomes and education improve, dietary preferences typically shift from cereals, fats, and sugar, to include more protein, fruit, and vegetables, affecting agricultural water use. Two other RUMI-Hi

Table 7.1 Summary of scenarios modelled

Scenario	Description	Annual GDP growth	Productivity growth	Consumer preferences	Climate change
BAU-Lo	Business as usual; the current rate of climate warming.	2.26% to reach US$3,036 per capita by 2055.	Agriculture 1.0%; Livestock 1.0%; Industry 0.6%; Services 0.8%	No change	The baseline rate of climate warming: 1 °C increase in mean annual temperature by 2047.
BAU-Hi	Business as usual; the faster rate of climate warming.	2.26% to reach US$3,036 per capita by 2055.	Agriculture 1.0%; Livestock 1.0%; Industry 0.6%; Services 0.8%	No change	The faster rate of climate warming: 3 °C increase in mean annual temperature by 2047.
RUMI-Lo	Accelerated economic growth; the current rate of climate warming.	3.99% to reach US$6,028 per. capita by 2055	Agriculture 2.1%; Livestock 1.9%; Industry 1.2%; Services 2.0%	No change	The baseline rate of climate warming: 1 °C increase in mean annual temperature by 2047.
RUMI-Hi	Accelerated economic growth; the faster rate of climate warming.	3.98% to reach US$6,013 per. capita by 2055	Agriculture 2.1%; Livestock 1.9%; Industry 1.2%; Services 2.0%	No change	The faster rate of climate warming: 3 °C increase in mean annual temperature by 2047.
RUMI-Hi-Diet	Accelerated economic growth; the faster rate of climate warming; diet shifts.	3.82% to reach US$5,854 per capita by 2055	Agriculture 2.1%; Livestock 1.9%; Industry 1.2%; Services 2.0%	More meat, dairy, and fruit.	The faster rate of climate warming: 3 °C increase in mean annual temperature by 2047.
RUMI-Hi-Reform	Accelerated economic growth; the faster rate of climate warming; agricultural policy reforms and trade shifts.	3.93% to reach US$5,819 per capita by 2055.	Agriculture 2.1%; Livestock 1.9%; Industry 1.2%; Services 2.0%	No change	The faster rate of climate warming: 3 °C increase in mean annual temperature by 2047.

(continued)

Table 7.1 (continued)

Scenario	Description	Annual GDP growth	Productivity growth	Consumer preferences	Climate change
RUMI-Hi-Eflows	Accelerated economic growth; the faster rate of climate warming; flows to the delta.	3.98% to reach US$6,013 per. capita by 2055	Agriculture 2.1%; Livestock 1.9%; Industry 1.2%; Services 2.0%	No change	The faster rate of climate warming: 3 °C increase in mean annual temperature by 2047.

variants explore (i) agricultural price policy reform and changes in international trade, and (ii) additional environmental flows to the delta.

The objective of the modelling is to examine how drivers of change (population growth, changing consumer preferences, climate change) affect water-dependent economic outcomes. The main assumption is that on the demand side, preferences are set outside the model, as elasticities vary by household type but not across years. Actual household expenditures are determined by the prices of goods and household income, determined within the model. On the supply side, production is affected by exogenously set productivity growth rates for each commodity and key inputs (e.g., land and labour). The population is not modelled directly but is incorporated as labour force growth, and its effects emerge as expanded household expenditures. The nature of the water balance assumed in the CGE-W and its climate change assumptions are presented below.

After these parameters are set, the model solves for commodity supply, household incomes, business profits, and domestic prices. International export/import prices are fixed, except for the simulation of altered textile and rice prices. When products have large trade positions, reactions to external price changes can be large. In the base year, textiles and rice have such large trade positions.

7.4 Indus Basin Water Balances in the CGE-W

Accounting for water inflows, withdrawals, and outflows (surface and groundwater) across the Indus Basin of Pakistan underpins the modelling. Figure 7.1, adapted from Young et al. (2019), shows the average basin water balance as compiled from multiple data sources. This is similar to the water balance adopted in the CGE-W, but with differences because of data sources, averaging periods and assumptions. The largest differences are on the supply-side: Young et al. (2019) include estimated runoff within Pakistan to get its resource total (in addition to inflows from upstream countries), which adds to total availability, but this is offset using the most recent inflows from the eastern tributaries, where flows have declined as a result of development upstream in India. Also, compared to values in Fig. 7.1, the CGE-W is calibrated to a specific year rather than to an average value.

The model does not maintain a groundwater balance and so does not explicitly account for the flows to and from surface water shown in Fig. 7.1; it thus has a different treatment of losses. Figure 7.1 combines canal leakage and all other losses (watercourse and field-level) into a single term, while CGE-W combines watercourse losses with canal losses but treats field-level groundwater losses explicitly. As it ignores the contribution of these "losses" to groundwater supply, CGE-W does not need to include the "natural ET" losses in the basin to ensure closure of the water balance, and outflows are often significantly higher than in reality.

7.4.1 CGE-W Climate Change Assumptions

To model climate change effects, we used results from General Circulation Models (GCMs) in Yu et al. (2013) for precipitation, temperature, and meltwater inflows. These results indicate that by the 2050s, temperatures could average about 3 °C higher across Pakistan. The range of estimates and variances for precipitation was very high, so we confined the climate simulations to temperature increases. Yu et al. (2013) concluded that multiyear trends in meltwater inflows were unlikely to systematically increase or decrease in the coming decades. Climate change is expected to shift the seasonal timing of meltwater, which was not simulated here but could be assessed in future analyses.

The historical inflows used to create a synthetic future series for 2014 to 2047 were taken from Robinson and Gueneau (2014) from 1975 to 2008, and from Davies (2016b) for 2009 to 2014. The historical sequence was spliced to create the future series, such that the final ten years, when droughts were experienced, were switched with the middle ten years to explore drought-recovery mid-scenario and ensure end-of-scenario conditions were more representative of the longer-term trend.

7.5 Future Water Demand and Use

This section examines CGE-W results for agricultural and non-agricultural water uses. Young et al. (2019) show that new industrial and domestic demands will be an increasing challenge for water resource management, as, by the time Pakistan reaches its 100th anniversary, domestic demand could grow by 3.8 BCM while industrial demand including livestock could increase by 5.7 BCM. Agricultural water demand also increases, both to meet growing food demand and because of climate warming. Under BAU, growth in irrigation water consumption is not affected by the degree of climate warming and is essentially unconstrained by water availability (Fig. 7.2). Under RUMI-Lo, irrigation water consumption is lower than BAU because of higher economic growth, which relies on the reallocation of water

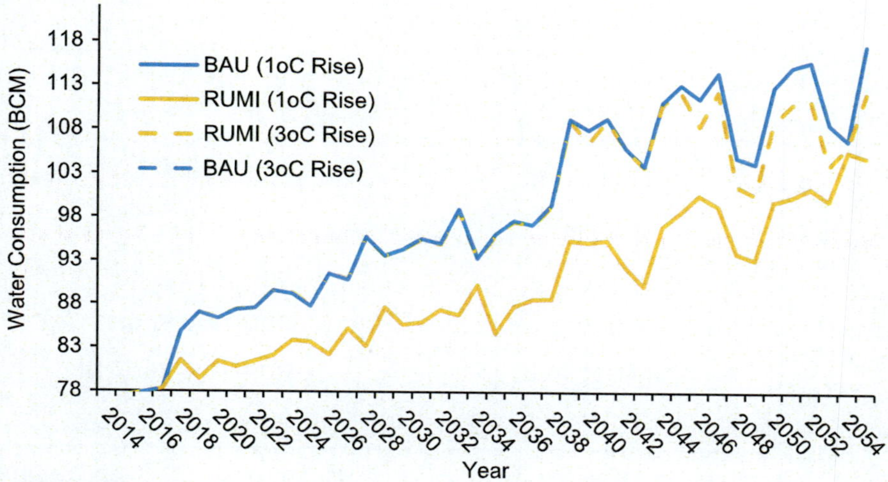

Fig. 7.2 Modelled irrigation water use (BCM) under BAU and RUMI, for both slow and faster climate warming

away from agriculture, and less warming reduces demand within the sector. Irrigation water consumption is similar under BAU and RUMI-Hi until 2037 when BAU irrigation water consumption outstrips RUMI-Hi because non-agricultural demands from faster economic growth limit irrigation water use.

As with surface water, groundwater is allocated away from agriculture more quickly under RUMI scenarios (Fig. 7.3). BAU and RUMI-Hi exhibit similar groundwater use in the first half of the simulations, after which the fraction used in irrigation declines consistently and faster under RUMI. Groundwater helps to balance water shortages in drier years until around 2030 when greater industrial and household water demands reduce the role of groundwater as an irrigation drought-buffer. Under lower climate warming, this buffering role continues longer under RUMI. Industrial and domestic use is not fully consumptive, so there will be opportunities for wastewater reuse in agriculture. Untreated wastewater is, however, often too polluted for safe use, so careful, detailed economic and technical analyses of wastewater treatment and reuse options are required.

7.5.1 *Crop-Wise Water Use*

Around four-fifths of irrigation water is used for wheat, rice, sugarcane, and cotton. Under BAU-Lo the water used by three of these crops increases (Fig. 7.4) – with expected year-to-year variations reflecting water availability – while rice irrigation use slowly declines. Competition for irrigation water varies by season. Rice and cotton are *Kharif* crops, while wheat and sugarcane are either fully or partially *Rabi*

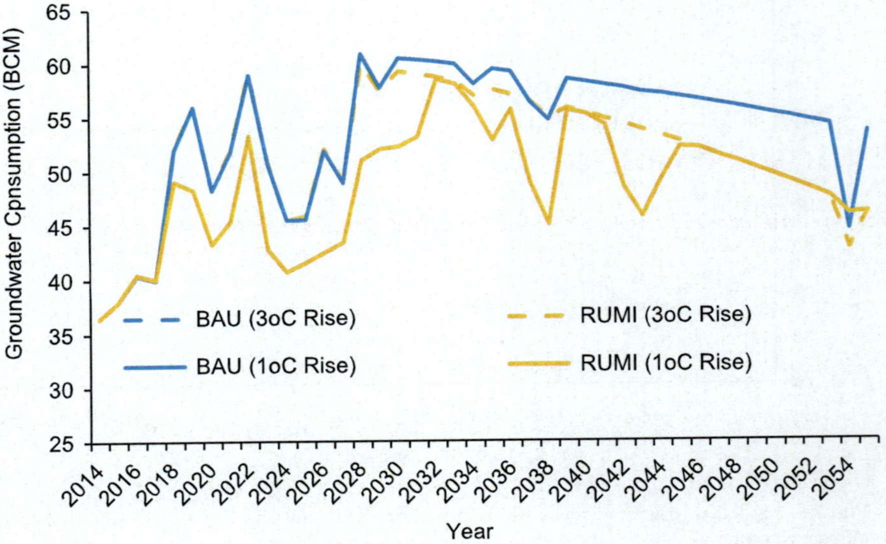

Fig. 7.3 Groundwater use for irrigation (BCM) under BAU and RUMI, for both slow and faster climate warming

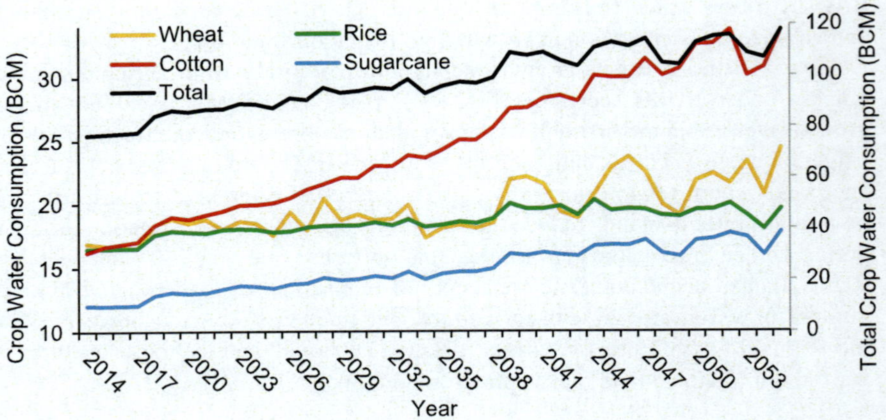

Fig. 7.4 Modelled water use (BCM) for major crops (left-hand axis) and total water use (right-hand axis) under BAU-Lo

crops. Nearly half of current rice production is exported, so with steady international prices, water for cotton during *Kharif* becomes an increasingly better option than rice. Cotton production supports exports of yarn, cloth, and garments, whose higher value leads to a transfer of water from rice under BAU. Wheat is grown in *Rabi* when water shortages are greater, and so wheat irrigation use is more sensitive

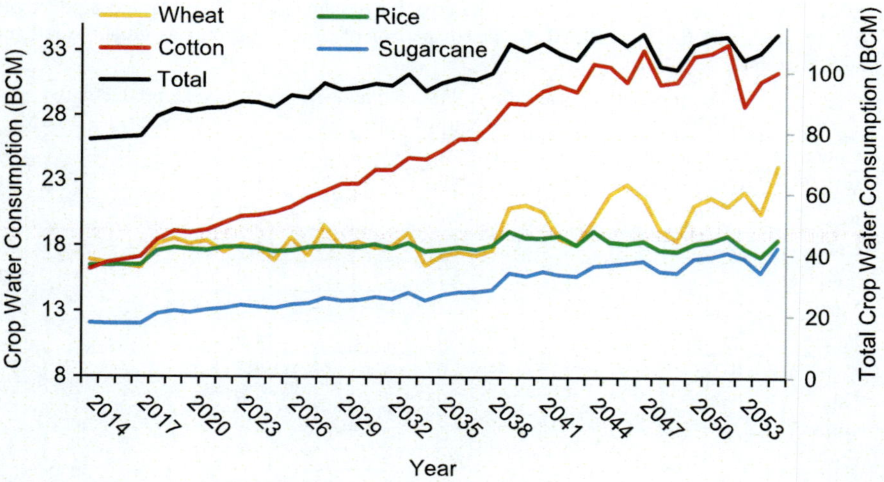

Fig. 7.5 Modelled water use (BCM) for major crops (left-hand axis) and total water use (right-hand axis) under RUMI-Hi

to precipitation and temperature than for the *Kharif* crops and is thus more variable across years.

Under RUMI-Hi (see Fig. 7.5), total irrigation water use by 2055 is more than 5 BCM lower than under BAU because greater economic activity reallocates water out of agriculture, and faster warming forces more production tradeoffs. Total water use rises steadily with expanding production until about 2038, when further expansion becomes supply-limited. With an approximate supply constraint of 109 BCM, water use for cotton stabilises and water use for rice declines to meet rising domestic food demand through increased wheat production.

In reality, the irrigation supply constraint is likely to be reached sooner than modelled because of existing inefficiencies in field-level water management. The growing demand for industrial and household water, and the impacts of warming on demand in all sectors, means that around 10% of irrigation withdrawals (approximately 12 BCM) must be reallocated to other sectors in the next few decades. This is a significant challenge given limited good quality groundwater and increasing crop water demands. A mix of policy reforms, improved management, and infrastructure and technology investments will be required to facilitate this shift.

The economic productivity of water varies by crop (Table 7.2). While the irrigation literature often cites values for water in irrigation, they are often based on gross production values, ignoring the contribution of other inputs, as pointed out by Scheierling (2014) and European Union (2016). Davies et al. (2016a) used regression analysis, which controlled for other inputs, to show the additional output gained for each increment of water input for *Rabi* and *Kharif* crops in Pakistan. Combining estimates from this prior work with crop water consumption from CGE-W permits

Table 7.2 Economic value of water by the crop for baseline year (2013/14) and productivity growth rates under BAU and RUMI, with a comparison of climate warming assumptions (Hi and Lo Scenarios)

	Water productivity Baseline (US$/m³)	Water productivity growth rate (%), BAU		Water productivity growth rate (%), RUMI	
Climate warming		Lo	Hi	Lo	Hi
Wheat	0.012	2.0	1.6	3.4	3.0
Rice	0.009	2.1	1.7	3.4	3.1
Cotton	0.013	2.2	1.8	3.7	3.2
Sugarcane	0.013	2.3	1.9	3.4	3.0
Maize	0.032	2.1	1.6	3.8	3.3
Potato	0.006	2.4	1.9	2.9	2.5
Vegetables	0.013	1.6	1.1	3.3	2.9
Other crops	0.081	2.0	1.6	3.3	3.4
Fruit	0.174	2.0	1.6	3.7	3.2
Average	0.034	2.2	1.8	3.5	3.1

the value of water in crop production to be estimated separately from the contributions of fertiliser, labour, and land (Table 7.2, first column[1])

Water productivity gains to 2055 are 60–70% higher under RUMI than under BAU, highlighting the relationship between water productivity and income, and accordingly pointing out that reaching upper-middle-income requires much faster increases in water productivity. The forecasted BAU values are about 10% lower than estimates from 1980 to 2013 in Young et al. (2019), which were 2.1% per year in Punjab and 3.4% in Sindh (2.4% overall). Table 7.2 shows that water productivity drops under faster warming, as the same output requires more water. Higher temperatures reduced water productivity by about 0.4% per year in both scenarios.

7.6 Connecting Economic Productivity and GDP Growth with Water Productivity

Underpinning the results in Table 7.2 is an assumption that water resources are sufficient to accommodate the higher growth under RUMI. To test this assumption, the net surface water flows generated in the CGE-W (with a trend) were first regressed against calculated productivity growth, or total factor productivity (TFP) values from the World Bank (Ghosh and Kraay 2008), which showed that TFP increased

[1] For comparison, we used the reported costs for 6 hours of tube-well use (Rs. 2000) in 2015 (Punjab Agricultural Marketing Information Service). Assuming a 15 hp pump delivers 1000 gallons per hour, the cost per cubic metre was Rs. 1.46, which is similar to the value found for major crops in the CGE-W.

by 0.2% for each additional 1% added to net surface water flows.[2] To assess the economic impact, we increased TFP by 0.6% under BAU-Lo, assuming 6.9 BCM could be used to more effectively support TFP growth.[3] This delivered a US$3.9 billion gain in GDP by 2055, or $0.57/m^3, confirming that the expansion of water resources historically had a significant positive effect on economic growth. As this outcome includes agricultural benefits, shifting water to non-agricultural uses would probably yield even greater returns. Of course, limited growth in water resources would slow GDP growth.

Next, the broader link between economic growth and required levels of water productivity is examined via a comparison of the sources of productivity growth in comparator countries. CGE-W's growth drivers are based on macroeconomic categories, including expansions of capital and labour (with human capital included), as well as TFP (Atiyas and Bakis 2013; Van Der Eng 2009; Bosworth et al. 2007). In Table 7.3, TFP growth for Pakistan and four comparator countries is examined over the last three decades, as are the contributions to that growth from the drivers noted above.[4] Pakistan's growth was the lowest of the five countries from 1980 to 2010, while India's was about 50% above Egypt (the second-best performer) and was generally more than double the other countries. At India's historical growth rate, productivity doubles in 36 years, while at Pakistan's historical growth rate, this doubling requires 87 years. The consistently good performance across decades in India is one of the main reasons for the divergence in economic outcomes.

For Pakistan under BAU, average TFP across all sectors was 0.82, just about the cumulative historical value for Pakistan (Table 7.3). To reach upper-middle-income

Table 7.3 GDP per capita and average annual growth rate, annual Total Factor Productivity growth rate and average contributions to growth in Pakistan and comparator countries

	GDP		Annual Total Factor Productivity growth (%)				Average contributions to economic growth (%)		
	Per capita (2010 US$)	Annual growth (%)	The 1980s	The 1990s	The 2000s	1980-2010 mean	Capital	Labour	TFP
Pakistan	1200	1.9	2.3	0.1	0.3	0.83	33	56	11
India	1900	3.5	1.4	1.7	3.3	2.01	23	51	26
Egypt	2700	2.6	2.9	1.3	0.0	1.40	43	26	31
Indonesia	4000	3.5	0.6	0.0	2.4	0.89	70	34	−4
Turkey	11,100	2.6	0.8	−0.2	3.2	1.08	41	40	19

[2] The data on surface water flows used the inflows at the rim stations less the outflows below Kotri Barrage. The surface water withdrawals as calculated in the CGE-W included significant double counting, and so were not the best series.

[3] TFP and water are linked in CGE-W indirectly, as agricultural water affects agricultural yields, which then have effects on crops and, through multipliers, on the rest of the economy.

[4] Decadal TFP values from Atiyas and Bakis (2013); factor contributions over various years from Atiyas and Bakis (2013) for Turkey; Van der Eng (2009) for Indonesia; Bosworth et al. (2007) for India; and The Conference Board (2014) for Egypt.

($6,000 per person) by 2055, average growth must be more than doubled, and be higher than any comparator country except India. Additionally, all countries had decades of poor performance, with Pakistan having negligible productivity growth during two of the three periods. Thus, policies emphasising consistent performance are important, so raising Pakistan's rate after 1990 to 1.0% would make their growth rates equivalent to that of Egypt.

In addition to more than doubling overall TFP, reaching upper-middle-income will require an increase of about 50–60% in water productivity. Thus, the likelihood of reaching the RUMI income level may be more constrained by general TFP growth than by water management improvement. But gains may be harder in the water sector, given political economy issues. Nonetheless, awareness and political interest are continually rising, so there is a basis for optimism.

7.7 Dietary Change

Faster income growth under RUMI allows increased consumption, and will likely also drive changes in consumer preferences. Young et al. (2019) showed how absolute expenditures on food evolve as Pakistani households get richer, with wheat declining in absolute household expenditure with income, while expenditures on fruit, meat and milk rise significantly. Here, we compare RUMI-Hi with a RUMI-Hi-Diet scenario, implemented using altered demand elasticities in the model's linear expenditure system (Nganou 2005). The latter has minimal growth in wheat consumption and increased expenditure on vegetables, livestock, and milk.

The modelling suggests that these changes may have a large impact on irrigation water use, with some unexpected outcomes. Figure 7.6 compares water use in RUMI-Hi (implied by the zero line) versus RUMI-Hi-Diet, which sees total irrigation water drop by up to 2 BCM in 2054. However, sugarcane and cotton water use

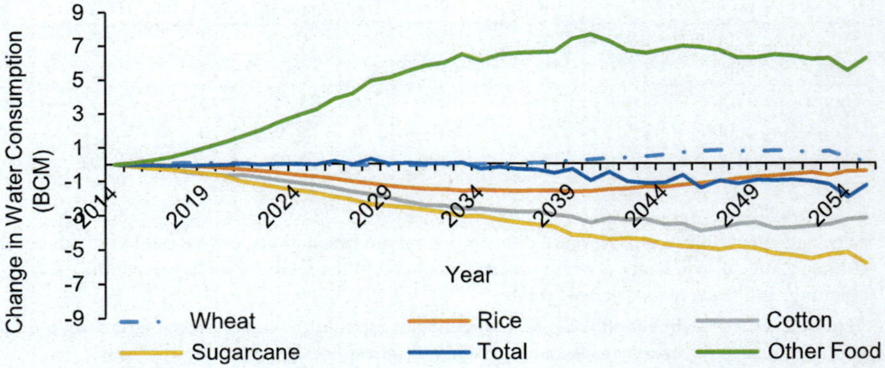

Fig. 7.6 Modelled difference in irrigation water use (BCM) for major crops and other food for RUMI-Hi-Diet versus RUMI-Hi (zero line)

Table 7.4 Modelled annual growth rate (%) in commodity consumption by scenario to 2055

	BAU-Hi	RUMI-Hi	RUMI-Hi-Diet
Wheat	2.3	3.6	3.4
Rice	2.7	4.0	3.9
Fruit and vegetables	2.9	4.3	4.2
Meat	3.5	5.2	5.2
Dairy	3.9	5.7	5.1
Sugar	2.8	4.2	3.1

decline much faster, by 5.8 and 3.2 BCM respectively. This released water goes mostly to "other food" crops (fruits and vegetables, and oilseeds), which adds over 7.0 BCM in 2040, and wheat, which after 2038 consumes more water. This later upward trend is also found in rice and cotton, so water use in "other food" crops declines. Cotton water consumption levels off after a fairly steep decline at about 3 BCM lower than the initial level, while rice water consumption declines more than one MAF but then grows like wheat. Therefore, the model simulations suggest that full shifts out of cereals, even with better diets, should not be assumed. These outcomes are affected by the varying price responses and trade patterns that arise from altered diets.

Table 7.4 summarises the dietary changes in the model's results. Population growth is set at 1.3% for all scenarios, so per capita growth occurs because all reported growth rates exceed this rate. Given base elasticities and growth in household expenditures in BAU, only meat and dairy consumption increase by more than 3.0% annually. Under RUMI-Hi, consumption increases for all commodities are higher. The changed elasticities under RUMI-Hi-Diet have large dietary consequences. Sugar consumption growth slows by over 1.0% annually, and both rice and wheat slow somewhat, although perhaps less than expected; per capita expenditure on other more nutritious commodities remains steady. The per capita availability of wheat, for example, grows much faster to 2055 than in the forecast of Kirby et al. (2017), demonstrating the large effects of yield and water productivity in the RUMI simulations. Thus, with faster economic growth and changing preferences, consumption is expected to shift towards a more nutritious and diversified diet, while also improving water security if appropriate agricultural reforms and irrigation investments are made. These effects are discussed below.

7.8 Critical Agricultural Reforms

With nearly 80% of irrigation withdrawals used for just four crops, policies and outlooks for these crops can affect total irrigation water use and the economy. Two oft-discussed mechanisms to rationalise water use in Pakistan are reforms of the wheat procurement program and indicative sugarcane prices, which, until recently, have kept their domestic prices well above international prices. Cotton (with

Table 7.5 2055 Water use by major crop under RUMI-Hi variants and percent change under scenario variants relative to the base case

	RUMI-Hi (Annual BCM in 2055)	No W/S support + LEP (BCM change)	No W/S support + LEP/Diet (BCM change)	No W/S support + LEP/ Diet/E-flows (BCM change)
Wheat (W)	24.2	−0.9	0.1	−0.2
Rice (R)	18.5	−0.1	−0.4	−3.1
Cotton (C)	31.5	−2.7	−3.2	−2.4
Sugarcane (S)	17.9	−1.9	−5.8	−5.8
Other Crops	20.2	4.8	8.1	6.4
Total/Net BCM	112.3	−1.0	−1.2	−5.1

textiles) and rice are key exports for Pakistan, but this dominance depends on international prices and is affected by domestic productivity growth and demand.

Simulations by Young et al. (2019) showed that reforming price policies for wheat and sugarcane (noted as W/S support in Table 7.5) improves water security, with only minor impacts on economic growth. They also simulated falling export prices for rice and cotton (LEP in Table 7.5), which reduced GDP because of fewer exports and lower international prices. The combined effects are shown here for a longer period, which leads to a reduction in irrigation water consumption of 1.0 BCM per year (Table 7.5). This impact is spread across all four crops but is mostly from cotton and sugarcane, with water consumption for these two crops declining by 4.6 BCM. However, the production of high-value crops such as fruits and vegetables expands, so the net reduction in agricultural water consumption is small. If diets change as well, significantly less water is consumed by the four main crops, especially sugarcane (with a 5.8 BCM reduction), but that is offset by increased production of other crops (including fruit and vegetables), where consumption is 8.1 BCM more than under RUMI-Hi. These changes deliver multiple benefits, as nutrition improves, and water is available for higher-valued uses. If environmental flows are added, irrigation water consumption declines by a further 3.9 BCM to support those uses.

Table 7.6 shows the impacts of RUMI-Hi scenarios on economic growth. The economic losses from reforms and lower international commodity prices are spread throughout the economy except in agriculture, likely due to rising prices of commodities. Agricultural GDP rises overall because of increased production from other crops, livestock and lower than expected declines in wheat. Thus, overall GDP drops by just 0.4% relative to the RUMI-Hi base case. With changes in diet added, demand shifts away from staple agricultural commodities and agricultural GDP declines by 7.2%, given the importance of staples in production and processing,

Table 7.6 2055 GDP by major crop under RUMI-Hi and percent change under scenario variants relative to the base case

	RUMI-Hi (US$ in 2055)	No W/S support + LEP (% change)	No W/S support + LEP/Diet (% change)	No W/S support + LEP/ Diet/E-flows (% change)
Ag sector	411	2.6	−7.2	−7.4
Ag processing	264	−3.6	−4.2	−4.0
Ag exports	170	−29.4	−32.9	−32.1
Ag imports	101	−10.9	−22.8	−22.8
Industry sector	221	−0.8	1.3	1.1
Service sector	1076	−0.7	1.7	1.7
Total GDP	1972	−0.4	−1.0	−1.1

while the industrial and services sectors expand moderately.[5] Adding environmental flows only causes GDP to decline by an additional 0.1%, as negative impacts in agriculture and industry are largely offset by lower trade deficits and growth in services.

It is estimated that about 12 BCM will need to be reallocated from agriculture by 2055 to meet growing non-agricultural demands, which, with policy reform, diet and trade changes, can be met. Moreover, under slower climate warming, it would be easier to shift the required water as irrigation demand would be lower. The model retains water in agriculture mostly because non-agricultural requirements are met first from groundwater and no value is placed on flows below Kotri Barrage. However, the model puts water first into high-value agriculture, which, if correct, makes reallocation of water out of agriculture complex. In reality, these transitions could be achieved by pricing irrigation water, supported by other regulatory measures and investments.

CGE-W does not consider social outcomes but can indicate economic costs of insufficient water for industry and services. Not all services and industries are heavily water-dependent but establishing new industries in Pakistan is slow and costly, so anything that constrains growth can impose large penalties. A hypothetical 5% reduction in industry and service productivity growth (which just decreases average TFP growth from 0.8 to 0.77 in services for BAU Lo) was modelled to assess the consequences of water-constrained growth. A GDP loss of US$84 billion by 2055 was observed, equivalent to 90% of the GDP contributions of Pakistan's major crops. Thus, losses associated with inadequate water services to the industrial and service sectors can justify major investments in urban water supply and reallocation of water away from irrigation. Policies and investments that hold water in

[5] While not attempted here, with higher elasticities for industrial goods and services, the released resources from agriculture might lead to gains in GDP.

agriculture will ultimately impose large economic costs on Pakistan, in addition to creating greater social costs of these policies.

7.9 Improving Environmental Management

The CGE-W allows limited exploration of improved environmental management. We explore an increased flow below Kotri Barrage, which is critical to sustaining the ecosystems of the Indus Delta, as well as to contribute to Karachi's water supply. Although the 1991 interprovincial Water Allocation Accord recognises the importance of environmental flows, it does not specify either agreed-upon environmental objectives or environmental flow allocations. Although not scientifically robust, widely accepted or implemented, prior work has suggested a constant environmental flow below Kotri of 5000 m^3/s (4.44 BCM annually and 0.37 BCM per month). This flow is rarely achieved, especially at a monthly level, as other uses are given priority (Amir and Habib 2015). During the 2000–2001 drought, flows below Karachi were insufficient to even meet Karachi's demands, let alone provide an environmental flow for the Delta.

Karachi's current bulk volumetric supply is sufficient for the existing demand, but service provision is very poor because of leakage, theft and other mismanagement. Karachi demand will, however, grow significantly, creating imperatives for increased flow below Kotri Barrage and further reducing flows to the Indus Delta, especially in dry years. Regardless of the actual requirements, Amir and Habib (2015) argue that environmental water demands will increase considerably because of climate warming and the need to counter increased seawater intrusion.

To explore the implications of increased consumptive and environmental water needs in the lower basin, RUMI variants with different demands below Kotri Barrage were modelled. While water use in Karachi and for environmental purposes probably yield far higher values than water used for irrigation, this cannot be assessed directly using CGE-W. CGE-W can, however, evaluate the foregone value of agricultural production from shifts in water allocation (Table 7.7).

Table 7.7 Annual value of lost agricultural production as a result of increasing flow below Kotri Barrage under RUMI-Hi

Level of demand	Source of demand		Agricultural production loss (net present value US$ billion)	Agricultural production loss (2053–55 average, US$ billion)
	Environmental	Other uses[a]		
Current	4.4	5.9	0.54	0.85
Moderate increase	8.9	8.1	1.09	1.29
Major increase	11.8	10.4	2.55	2.38

[a]Domestic and industrial uses

If the current demand below Kotri was fully met, the annual lost production is US$0.54 billion (net present value [NPV]) and an average of US$0.85 billion over 2053–55. A moderate increase in the flow below Kotri (combining the recommendations of the Karachi Water and Sewerage Board and an environmental flow slightly above the Amir and Habib [2015] estimate) would double the NPV loss, or increase the annual loss in 2053–55 by almost one and half times; a major flow increase generates even greater agricultural losses.

The higher values of water for Karachi and the environment would more than compensate for these losses, so mechanisms to achieve greater end-of-system flows should be explored and implemented. The loss estimates are probably conservative because CGE-W uses water for irrigation efficiently, and so the modelled flows below Kotri are higher than in reality, leading to a lower opportunity cost. The current low field-level irrigation efficiency ties up water, so greater withdrawal reductions would in actuality be required to meet below-Kotri requirements, with greater agricultural impacts.

7.10 CGE-W Support to Reforms for a Water Secure Pakistan

In the final chapter of Young et al. (2019), the main institutional, legal, policy, and infrastructure dimensions needed for a water-secure Pakistan are outlined, noting that under BAU water quality and security would decline, and environmental degradation and groundwater depletion would increase. However, per capita income could reach the upper-middle level, even given projected water scarcity and climate change, given appropriate reforms and investments. Here, we illustrate uses of the CGE-W results to support three areas with large effects: (1) long-term strategic basin planning; (2) irrigation and drainage services; and (3) inter-sectoral water allocation.

7.10.1 Strategic Planning

Young et al. (2019) identified basin-scale planning as key to guide long-term sustainable economic development and create a water-secure Pakistan. Such planning would consider environmental flows, inter-provincial water sharing, and inter-sectoral developments. The recently passed National Water Policy anticipates a National Water Council could do such planning, along with strengthening the technical capacity and legislative foundation of the Indus River System Authority to transform it into a basin management organisation.

The CGE-W provides a broad framework for many strategic planning issues in the water sector. The simulations with temperature and groundwater variation identified water balance consequences and effects on water productivity. This chapter illustrated how agricultural water requirements vary with climate change, and how non-agricultural water demand grows with higher income. Additionally, the economic and water use effects from the choice of crops, the outlook in international markets and selected policies were examined, as was the agricultural output foregone if put into environmental uses.

CGE-W can provide ongoing assessments of factors that limit economic growth. The chapter showed that a doubling of overall TFP is needed to reach RUMI per capita income, but only 50–60% growth in water productivity would be needed over 30 years. These insights, especially when provided in an ongoing and systematic way, can help guide strategic planning: they can sensitise programs to set TFP targets in the water sector; similar simulations can track effects of dietary change and environmental uses over time to help set evidence-based goals; and simulations can point to the possible consequences of limited data on groundwater, non-beneficial uses or benefits of environmental flows, among others.

7.10.2 *Improvements in Irrigation and Drainage Services*

Improving irrigation and drainage services is one of the most impactful reform areas proposed in Young et al. (2019) and is likely to be one of the most complex. Moreover, to improve farm income and food security, and to allow reallocation of water to other sectors, this is an area for urgent reform. While many of the recommendations of Young et al. (2019) are micro-oriented, the results here provide important contexts for those efforts including: (i) identifying limits for water use in agriculture and their determinants (faster economic growth and climate change); (ii) monitoring the tendency for water to remain in agriculture even when wheat and sugarcane policies alter incentives; and (iii) raising awareness of the water use impacts of dietary change and textiles and rice export changes.

7.10.3 *Inter-sectoral Water Allocations*

Young et al. (2019) suggest that while addressing inter-sectoral water allocations is a complex but perhaps less urgent activity, it is one that has among the greatest likely impacts. Policies that define allocation options must be promulgated, and the institutional context needs to become broader, so that irrigation departments become water resource planning and management institutions that consider inter-sectoral allocation and use of water, including for the environment. Of course, part of making inter-sectoral transfers successful is to have better water productivity both within agriculture and outside that sector. The CGE-W addressed many of the related issues to give quantitative perspectives. The topics such as the amount of water that needs to be moved out of agriculture in the face of climate change, income growth and perhaps continued high grain demand were explored. Additionally, CGE-W simulations showed the intertwined effects of agricultural policy reform, dietary change and environmental flows on water use and economic growth.

7.11 Conclusions

We explored the economic performance of Pakistan's water resources in the face of continued population and economic growth, with effects of temperature increases reflected. The issues are examined using CGE-W simulations considering two rates of income growth, high and low levels of climate warming, changes in dietary preferences, and a range of agricultural and water policy options. We assess the plausibility of Pakistan reaching upper-middle-income status by 2055, which would require average economic growth rates to be more than double, and water productivity to increase by more than 50%, making this a significant stretch goal but one that has been matched by other similar countries.

At present, Pakistan already has, by international levels, very large water withdrawals from the Indus River, and significant imbalances in use, as four major crops using 80% of irrigation withdrawals just contribute less than 5% to GDP. This small contribution of irrigation to GDP means Pakistan should not be considered an "irrigation economy", and other uses will have a much higher value.

Domestic water demand is expected to grow by 3.8 BCM by 2055, and industrial demand by 5.7 BCM. Under faster climate warming, surface irrigation water consumption is not affected by the rate of economic growth until around 2037, after which irrigation water consumption is constrained because of demands in other sectors. Groundwater irrigation consumption is not affected by the rate of economic growth until around 2030, and thus, until then, groundwater remains an important buffer in water-scarce years. After 2030, groundwater irrigation consumption declines because of increased non-agricultural demand, and this buffering capacity is lost.

Reforming price policies for wheat and sugarcane improves water security, with only minor impacts on economic growth. Irrigation water use is reduced, primarily for sugarcane and cotton, but that reduction is partly offset by an increase in water consumption by high-valued agriculture. Dietary changes could have significant impacts on water consumption, reallocating water away from staple crops, causing a decline in overall agricultural GDP which, however, is mostly offset by production increases in the industrial and services sectors. The potential costs of not reallocating water away from irrigation are large, where a 5% reduction in industry and service productivity growth – a hypothetical consequence of water-constraints – generates a loss equivalent to 90% of the GDP contribution of Pakistan's four major crops.

Allocating environmental water to maintain the Lower Indus and its delta is becoming increasingly urgent, and additionally, water demand for Karachi is expected to grow considerably. While increasing flows below Kotri Barrage reduce the value of agricultural production by an estimated US$1–2 billion (NPV), the much higher value of water for Karachi and the environment would more than compensate for these agricultural production losses. More research is required to define realistic targets for flows below Kotri Barrage and guide a measured approach to manage these flows with a program of agricultural policy reforms and irrigation modernisation, which would help minimise agricultural production losses.

Improving water security in Pakistan is likely to require establishing long-term strategic basin planning, improving irrigation and drainage services, and making significant inter-sectoral water allocations. CGE-W analyses suggest that about 12 BCM needs to be relocated away from irrigation, but that challenges exist because grain demand may well remain higher than expected. Agricultural policy reform allows water to move to other uses, but without explicit institutional and policy reforms and targeted infrastructure development, water will remain in agriculture, albeit possibly supporting more high-value uses than currently. Reallocating water out of agriculture prematurely could however push up prices for staple products and create food security challenges. This could be tracked with regular CGE analyses to inform a program of strategic policy reforms.

References

Amir P, Habib Z (2015) Estimating the impacts of climate change on sectoral water demand in Pakistan. Action on Climate Today. https://cyphynets.lums.edu.pk/images/Readings_concluding.pdf

Atiyas I, Bakis O (2013) Aggregate and sectoral TFP growth in Turkey: a growth accounting exercise. Working Paper 2013-1The Competitiveness Forum, Sabanci University, Istanbul Turkey

Bosworth B, Collins S, Virmani A (2007) Sources of Growth in the Indian Economy. NBER Working Paper 12901. National Bureau of Economic Research, Cambridge MA. February 2007. Accessed at https://www.nber.org/system/files/working_papers/w12901/w12901.pdf

Davies S, Saeed W, Majeed O et al (2016a) Economic analysis of DBDP using an integrated CGE-W model. In: Hartel P, Trouille B, Bello J, Chinbat D (eds) Report 13: draft economic analysis for the environmental and social impact assessment for Diamer Basha Dam Project. MwH Americas, Chicago

Davies S, Gueneau A, Mekonnen DK et al (2016b) Irrigation and water management in the Indus Basin: infrastructure and management strategies to improve agricultural productivity. In: Spielman DJ, Malik SJ, Dorosh PA, Ahmad N (eds) Agriculture and the rural economy in Pakistan: issues, outlooks, and policy priorities. University of Pennsylvania Press, Philadelphia, pp 117–170

Davies S, Soofi S, Turab A et al (2018) A strategic review of food security and nutrition in Pakistan. World Food Programme, Islamabad

European Union (2016) Assessing the environmental and economic value of water: review of existing approaches and applications to the Armenian context. Final report for the EU Water Initiative, Brussels. https://www.oecd.org/env/outreach/AM%20Water%20Value.pdf

Ghosh SR, Kraay A (2008) Measuring growth in total factor productivity. PREM Notes No 42. World Bank, Washington, DC. https://openknowledge.worldbank.org/handle/10986/11423

Hayat N, Hussain A, Yousaf H (2016) Food demand in Pakistan: analysis and projections. South Asia Econ J 17(1):94–113

Hussain A, Routray J (2012) Status and factors of food security in Pakistan. Int J Dev Issue 11(2):164–185
Kirby M, Ahmad MD, Mainuddin M et al (2017) Agricultural production, water use and food availability in Pakistan: historical trends, and projections to 2050. Agric Water Manag 179:34–46
Laghari AN, Vanham D, Rauch W (2012) The Indus basin in the framework of current and future water resources management. Hydrol Earth Syst Sci 16:1063–1083. https://doi.org/10.5194/hess-16-1063-2012
Löfgren H, Harris RL, Robinson S (2002) A standard Computable General Equilibrium (CGE) model in GAMS. Microcomputers in Policy Research 5. IFPRI, Washington, DC
Nganou JP (2005) Estimation of the parameters of a Linear Expenditure System (LES) demand function for a small African economy. RePEC Archive, Munich, Germany. https://mpra.ub.uni-muenchen.de/id/eprint/31450
Robinson S, Gueneau A (2014) Economic evaluation of the Diamer-Basha dam: analysis with an integrated economic/water simulation model of Pakistan. Pakistan Strategy Support Program Working Paper 14. IFPRI, Washington, DC. http://ebrary.ifpri.org/cdm/ref/collection/p15738coll2/id/128002
Saeed WS (2017) A 2010-11 social accounting matrix for Pakistan. International Food Policy Research Institute, Islamabad, Pakistan (Unpublished manuscript)
Scheierling SM, Treguer DO, Booker JF, Decker E (2014) How to assess water productivity? Looking for water in the agricultural productivity and efficiency literature. Policy Research Working Paper 6982. World Bank Global Practice Group, Washington, DC
The Conference Board (2014) Growth accounting and Total Factor Productivity country details, 1990–2013. The Conference Board Total Economy Data Base. http://www.conference-board.org/data/economydatabase/
United Nations (2017) World population prospects: the 2017 revision. Department of Economic and Social Affairs, Population Division, New York
Van Der Eng P (2009) Total Factor Productivity and economic growth in Indonesia. Working Paper No. 2009/01. Arndt Corden Division of Economics, Australian National University College of Asia and the Pacific, Canberra
WFP (2018) Pakistan: food security bulletins, January 2018. World Food Programme (WFP), Rome, Italy. https://www.wfp.org/publications/pakistan-food-security-bulletins-2018
Yang YE, Brown CM, Yu WH, Savitsky A (2013) An introduction to the IBMR, a hydro-economic model for climate change impact assessment in Pakistan's Indus River Basin. Water Int 38(5):632–650. https://doi.org/10.1080/02508060.2013.830691
Young W, Anwar A, Bhatti T et al (2019) Pakistan: getting more from water. World Bank, Washington, DC. https://openknowledge.worldbank.org/handle/10986/31160
Yu W, Yang YC, Savitsky A et al (2013) The Indus Basin of Pakistan: the impacts of climate risks on water and agriculture. The World Bank, Washington, DC. https://doi.org/10.1596/978-0-8213-9874-6

Chapter 8
Sustainability Analysis of Irrigation Water Management in Punjab, Pakistan

Muhammad Abid, Mohsin Hafeez, and Muhammad Arif Watto

Abstract Water management in the irrigation-dominated Indus Basin of Pakistan is under pressure to ensure equitable, long-term, stable and flexible water supplies for meeting crop water demands, growing non-agricultural water demands (domestic and industrial supplies), and minimising adverse environmental impacts of one of the largest irrigation systems in the world. In this chapter, we focus on the irrigation system in Punjab by carrying out a sustainability analysis of its current irrigation water application methods. Cai et al.'s (Sustainability analysis for irrigation water management: concepts, methodology, and application to the Aral Sea region. Environment and production technology division, discussion paper no. 86, International Food Policy Research Institute, Washington, DC, 2001) analytical framework is used, which comprises indicators of risk and vulnerability, environmental system integrity, and economic acceptability and equity. The analysis suggests that irrigation water management in Punjab is currently unsustainable due to declining surface water supplies and excessive pressure on groundwater to support intensive agriculture and increasing demand from non-agricultural uses. Furthermore, climate change projections suggest reduced overall water availability leading to reduced crop productivity. Groundwater exploitation, unsustainable irrigation and agricultural practices, and industrial effluents are affecting water quality and worsening the overall health of the Indus Basin and its ecosystem. The cost of irrigation water management is economically not viable due to the high level of subsidies for technological interventions at the farm level and minimal water charges. The gap between collected water charges and overall operation and maintenance costs has reached USD 76 million. Water productivity in the Punjab is one

M. Abid (✉)
Centre for Climate Research and Development, COMSATS University Islamabad, Islamabad, Pakistan
e-mail: muhammad.abid@comsats.edu.pk

M. Hafeez
International Water Management Institute (IWMI), Lahore, Pakistan
e-mail: m.hafeez@cgiar.org

M. A. Watto
University of Agriculture Faisalabad, Sub-campus Depalpur, Okara, Pakistan

M. A. Watto et al. (eds.), *Water Resources of Pakistan*, World Water Resources 9,
https://doi.org/10.1007/978-3-030-65679-9_8

of the lowest in the South Asia region due to use of traditional irrigation practices with low irrigation and application efficiency. Equitable distribution of water in the province has become a big challenge for water managers given increasing water allocation conflicts, especially between upstream and downstream water users. We thus suggest adopting an approach that is more inclusive of all major stakeholder interests keeping in view the competing inter-sectoral water demands in future and the ongoing challenges of climate change, urbanisation and economic growth. Such efforts are required to improve water use efficiency as well as equity in the distribution of water among users.

Keywords Irrigation water management · Water use efficiency · Indus Basin · Sustainability analysis · Punjab · Pakistan

8.1 Introduction

Pakistan is predominantly an arid country that relies heavily on irrigation to sustain its agricultural sector. Pakistan withdraws about 184 km^3 surface water and groundwater, or 78% of the total annual available water resources for irrigation uses (Young et al. 2019). They predict that water demand, without effective management and reforms, could increase by nearly 60% by 2047 with the largest increase being the irrigation sector. Pakistan's cropped area increased from 11.6 to 22.68 million hectares (Mha) between 1947 and 2018 due to farm mechanisation and improvements in irrigation water availability (Ministry of Finance 2019). However, irrigation water application and management practices are criticised for being inadequate and inefficient especially given the impacts of the changing climate. Extreme climatic variability (drought/floods) highlights the importance of managing water resources for greater sustainability in Pakistan than ever before. Therefore, significant improvements in water management is warranted, particularly in the agriculture sector (SBP 2017). Similarly, recent statistics show that the productivity of agriculture sector needs to be improved. Pakistan's irrigated agriculture contributes around US$22 billion to the national GDP with less than 5% contribution to GDP from the four major crops (wheat, rice, sugarcane, and cotton), which account for 80% of the total water use in irrigation (Young et al. 2019). Many studies report that the water productivity of all these crops remain significantly lower than the global averages which is mainly due to poor irrigation water use and management practices at farm level (Kahlown and Kemper 2007; Watto and Mugera 2016).

Punjab province has the highest irrigated area and is the largest producer of major crops, making it the largest user of freshwater resources among Pakistan's other provinces (Abid et al. 2015). Irrigation in Punjab is sourced from surface water and groundwater. Surface water is available through the Water Apportionment Accord of 1991 and is managed by the Punjab Irrigation Department (PID). Groundwater, however, remains unregulated and has become a significant source of irrigation for the agriculture sector given increased water demands from increasing cropping intensity and inadequate and variable access to surface water supplies (Watto and Mugera 2015).

As a precursor to the implementation of the National Water Policy (NWP) (Ministry of Water Resources 2018), the Government of Punjab developed and approved the Punjab Water Policy (PID 2018) on December 29, 2018. The policy constituted a Punjab Water Council under the chairmanship of the Chief Minister of Punjab, as well as a Punjab Water Policy Implementation Committee. The provincial government then approved the Punjab Water Act 2019 on December 3, 2019. Under the Water Act, a Water Resources Commission (WRC) was formed under the chairmanship of the Chief Minister of Punjab. The role of the WRC is to ensure:

1. Conservation, distribution and augmentation of water resources in the Punjab.
2. Allocation of water resources for domestic, agricultural, ecological, industrial and other uses.
3. Development of wildlife and fisheries in water bodies where water is extracted from or discharged to.

A provincial water master plan is also being prepared and will be implemented after cabinet approval. A Groundwater Regulatory Authority will be established at the provincial level to maintain sustainable use of groundwater. Punjab province is becoming increasingly vulnerable to water insecurity due to multiple factors including a rapidly increasing population, economic growth and development, and climate change (PID 2018). According to the Pakistan population census of 2017, about 53% of the Pakistan's population resides in Punjab. Punjab's population, with an annual growth rate of 2.1%, has reached at 110 million, or 1.48% of the global population. Studies comparing annual water availability and population growth suggest that per capita water availability in Pakistan as well as in the Punjab has almost reached the threshold of severe water scarcity with only 1000 cubic metres per person per year (Government of Punjab 2018). Agricultural and non-agricultural water demands are likely to grow significantly which will further increase pressure on these resources. Agriculture remains under pressure, not only to produce food for the country's growing population but also because it is a source of livelihood for a majority of the population. Punjab's recent water policy (PID 2018, p. 14) estimated that water requirements are likely to increase by 18% to support increasing cropping intensity by 2025 compared to current water use (95.3 MAF), with water requirements for industry and commercial use likely to rise by 63% by 2025 due to greater urbanisation and industrialisation.

Given increasing non-agricultural water demands, it is not economically viable to divert more water to the agriculture sector. It is therefore necessary to increase agriculture water use efficiency and productivity by exploring innovative alternatives and approaches. It is also essential to study existing water management systems to identify sustainable and viable options for improved water management at the farm and system-wide levels. Keeping in view the importance of improved water use and management, this chapter provides an analysis of existing irrigation water use and management practices and approaches in the province using Cai et al.'s

(2001) sustainability framework. The analysis helps provide a way forward for improved water management that fosters greater resilience to climate change and more sustainable food security for Punjab and Pakistan.

8.1.1 Punjab's Water Resources in the National Context

The irrigation system of the Punjab consists of 13 barrages, 12 link canals and 23 major canal commands (Ahmed et al. 2018). These canal commands are stretched over 36,862 km and irrigate an area of 9.70 Mha (Akhtar 2004). The province is entitled to a total of 56 MAF of surface water flows as per the Water Apportionment Accord of 1991 (IRSA 1991). Out of the apportioned 56 MAF, 21 MAF (37.5%) is estimated to be lost during conveyance and distribution from rivers to canals and canals to distributaries and water courses before reaching farmers' fields. Besides conveyance efficiency, irrigation efficiency at the field level is very poor. About 14 MAF (25%) of water is lost due to evapotranspiration, deep percolation, leaching, and surface run-off, all exacerbated by poor management practices. Putting together the conveyance and distribution losses of 35 MAF (62.5%), only 21 MAF (37.5%) of surface water is available for crop water requirements compared with the required 42 MAF (see Fig. 8.1).

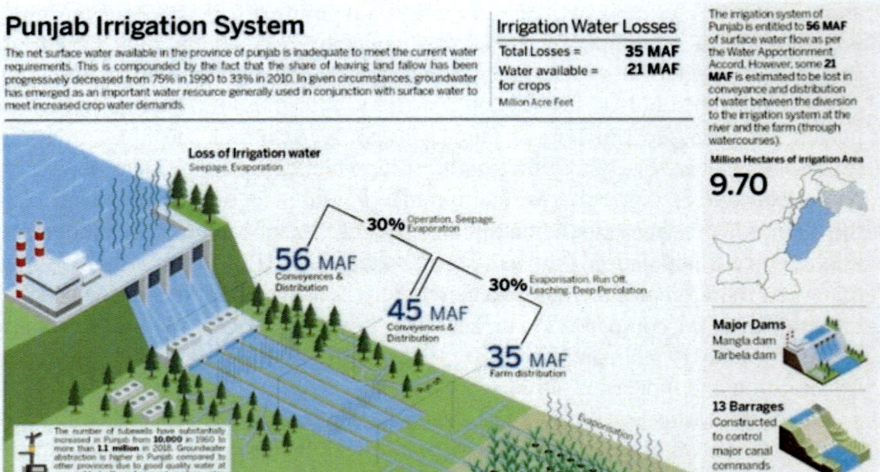

Fig. 8.1 Surface water losses and Punjab's irrigation system. (Source: Authors)

Insufficient surface water availability is compounded by increasing demands given that cropping intensities has increased from 75% in 1990 to over 170% in 2015 (Tetlay et al. 1999; Hassan and Hassan, 2017). Under these circumstances, groundwater has emerged as an important irrigation source, and is mostly used in conjunction with surface water supplies. It is estimated that around 33 MAF of groundwater is abstracted for irrigation, which is about 84% of total groundwater abstractions across the province. However, these estimates vary greatly especially given there is no effective groundwater monitoring mechanisms in the country or in Punjab. Over past decades, the proportion of groundwater in total on-farm water use has substantially increased. Currently about 60% of overall irrigation water comes through groundwater abstraction (Government of Punjab 2017). This massive increase has been propelled by unimpeded private tube-well development across the whole country, but particularly in Punjab, where the number has gone beyond 1 million in 2018–19 (Bureau of Statistics Punjab 2019). Figure 8.2 shows an exponential growth in private tube-wells in Pakistan overall and in the Punjab province. Data on groundwater abstractions indicate that groundwater abstraction is also highest in Punjab compared with other provinces.

Groundwater resources are thus rapidly depleted due to this excessive pumping and, in various parts of the province, the groundwater has become severely degraded. Inadequate records are kept of groundwater withdrawals rates, spatio-temporal recharge trends, the mixing of deep mineralised water resources with freshwater resources caused by over-abstraction, and pollution of shallow aquifers. These issues challenge the sustainability of irrigated agriculture, especially given so many farmers are increasingly reliant on poor quality groundwater for irrigation, which in turn affects their soil quality, crops yields and human health.

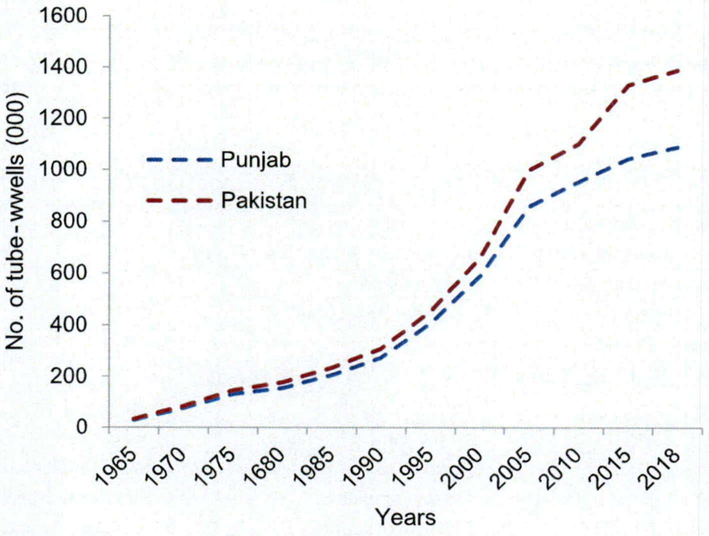

Fig. 8.2 Growth of private tube-wells in Punjab and Pakistan. (Source: FWM (2019))

8.1.2 Typology of Farm Level Irrigation Management in Punjab

Punjab's surface water distribution system consists of rivers, barrages, canals, distributaries, minors and water courses which ultimately deliver water to the farmers' fields. The canal water is distributed through a fixed weekly rotation system known as *Warabandi*. The *Warabandi* canal water rationing system aims to distribute available water supplies equitably in proportion to farm size (Bandaragoda and ur Rehman 1995; Jurriëns et al. 1996; Rinaudo et al. 1997). *Warabandi* is a supply-based system. PID supplies what is available rather than as a response to specific water requirements or demands. The system is inflexible and cannot meet the differing temporal and spatial water requirements at farm level for different crops across the different seasons (Kahlown et al. 2007; Zardari and Cordery 2009). The rigidity of this system has caused an increased dependence on groundwater for irrigation in the Indus Basin (Zardari and Cordery 2009). Land within the command area of a canal is entitled to water from that canal. Hence, water rights are de facto tied to land rights (Mustafa et al. 2013). Farmers pay a nominal annual fee called "*abiana*" to pay for PID's services. The current rate of *abiana* in Punjab for *Kharif* and *Rabi* seasons is PKR 85 and PKR 50 per acre respectively.

Irrigation water allocated to the farmer is supplied through a designed watercourse/outlet and beneficiary farmers are responsible for the cleaning and maintenance of that watercourse/outlet. The On-Farm Water Management (OFWM) program of the Punjab Government's Agriculture Department often supports water users to maximise crop and water productivity by ensuring efficient conveyance and application through a range of improved water management interventions, including watercourse improvement/ lining, laser land levelling and provision of water-saving technologies such as drip irrigation (Government of Punjab 2017).

Irrigation methods used by Punjab's farmers can be divided into two main categories i.e., surface flow and pressurised irrigation (Table 8.1). Surface flow

Table 8.1 Predominant irrigation and sowing methods used in Punjab for different crops

A. Irrigation Methods		
Surface irrigation system	**Pressurised irrigation system**	
Basin/border irrigation	Drip irrigation	
Furrow irrigation	Sprinkler irrigation	
Uncontrolled flooding		
B. Sowing and irrigation methods with respect to different crops		
Major crops	Irrigation method	Sowing method
Wheat	Basin/ border irrigation	Broadcasting
Maize	Furrow beds and ridges	Raised-bed plantation/ridges
Cotton	Furrow irrigation	Drilling/raised-bed plantation
Sugarcane	Furrow irrigation	Raised-bed plantation
Rice	Flooding	Transplantation

irrigation includes basin/ border irrigation, furrow irrigation, and uncontrolled flooding, whereas pressurised irrigation includes drip and sprinkler irrigation technologies (Latif et al. 2016). Surface irrigation is the most widely used, but the particular method adopted depends on the crop type and sowing method. For example, furrow irrigation is preferred by farmers for crops sown in rows, i.e. cotton, maize and sugarcane, whereas rice is grown through transplantation in a paddy field that is flooded to maintain a pool of water (approx. 100 mm). Table 8.1 summarises the most commonly used irrigation and sowing methods in Punjab.

Pressurised irrigation systems, often referred to as High-Efficiency Irrigation Systems (HEIS) in Pakistan, are being introduced and promoted to enhance water-use efficiency of irrigated agriculture (Government of Punjab 2017). HEIS has the potential to improve water and land productivity, and has been successfully adopted and developed around the world, including in India, where it is mostly used for sugarcane, banana, grapes and cotton crops, revealing substantial water and energy saving improvements, thus increasing crop yield and farm income (Bell et al. 2020). Despite similar climatic and ecological conditions in Pakistan adjacent to India, adoption of pressurised irrigation systems remains almost non-existent in Pakistani Punjab (2% of the total cropped area), where its out-scaling is not supported by subsidies and other incentives (Table 8.2).

One of the key barriers in adopting innovative irrigation technologies is that most Pakistani farmers have small landholdings, and this is particularly so in Punjab. Over 80% of Punjab farmers own less than 2 hectares of land and have limited resources and financial capacity to adopt innovative technologies (Abid et al. 2015). Elahi et al. (2018) also reported that small landholdings limit farmers' access to agricultural advisory and financial services that are generally biased towards larger or more influential farmers. Another important barrier in the uptake of HEIS is its high capital and operational costs, ranging from Rs. 100,000 to Rs. 180,000 per acre, and too expensive for typical small landholders (Government of Punjab 2017). Lack of awareness on the use and advantages of various pressurised irrigation systems also inhibit out-scaling of different HEISs.

Table 8.2 Cropped area irrigated by different irrigation methods in Pakistan

Cropped area and irrigation methods	Cropped area (000 hectares)	Percentage
Total cropped area	22,900	100.00%
Surface irrigation	19,270	84.15%
Under pressurised irrigation (drip and sprinkler)	400	1.75%
Others (rain-fed or another mode of irrigation)	3230	14.10%

Source: FAO (2008); Usman et al. (2016)

8.2 Sustainability Analysis of Irrigation Water Management in Punjab

As the complexity of the issues related to water resource management has increased, there have been extensive debate in the literature on defining sustainability in water resource management (Juwana et al. 2010; Loucks 2000; Ashley et al. 2004; Giuppponi et al. 2006; Mays 2006; Policy Research Initiative 2007) and developing effective tools to measure the sustainability of a system given current and projected future use patterns.

A discussion on water sustainability requires decision makers and stakeholders to consider present and future impacts of their water-related programmes and practices. Considering technical aspects only is no longer sufficient as the complexity and uncertainty of water resources calls for integrating economic, environmental and social criteria (Loucks and Gladwell 1999). Past studies (e.g. Savenije and Van der Zaag 2002, Mays 2006) suggest adopting an integrated or holistic approach to manage water-related issues covering the scope (social and developmental issues), scale (local to national) and governance (stakeholder interactions) domains of integrated water resource management.

In the past few decades, there have been increased attention on how sustainability of different systems can be measured, including of water resource management (Juwana et al. 2010), with different approaches suggested (e.g. Nardo et al. 2005; Cai et al. 2001, 2002). One approach is to develop assessment tools based on sustainability indicators. Nardo et al. (2005) defines an indicator as "a measure, either qualitative or quantitative, of facts or conditions of particular issue(s)" that if observed regularly, enable changes over time to be analysed. They further emphasise the need for a multi-dimensional approach and to identify sub-indicators that further define or explain key indicators. Cai et al. (2001) emphasise the need for selecting manageable indicators following broad guidelines and principles that can not only detect the problems as they arise but also provide an early warning system for decision makers. A number of indicator-based sustainability assessment frameworks have been developed and used (Chaves and Alipaz 2007; Policy Research Initiative 2007; Sullivan and Meigh 2007; Juwana et al., 2010). In the context of arid or semi-arid basins where irrigation is the dominant water use, Cai et al. (2001) proposes using water supply system reliability, reversibility, and vulnerability, environmental system integrity, equity in water sharing, and economic acceptability indicators to assess the sustainability of irrigation water management in arid to semi-arid basins.

Given the arid to semi-arid context for the Indus Basin where water is predominantly used use for irrigation, this study has adopted and modified Cai et al.'s (2001) framework to suit the current situation and policies of water management in Pakistan. We divided sustainability indicators into a set of three, as shown in Table 8.3, being: (1) water supply system risk and vulnerability; (2) environment system integrity; and (3) equity and economic acceptability.

Table 8.3 Sustainability indicators

Key indicators	Sub-indicators
Risk and vulnerability	Reliability
	Reversibility
	Vulnerability
Environmental system integrity	The health of floodplain ecosystem
	Water quality
	Adaptive capacity/ resilience to climate change
Equity and economic acceptability	Self-reliance
	Water use efficiency
	Equity in water sharing

8.2.1 Risk and Vulnerability

Punjab's irrigation system is at risk due to its inherent stochastic variability and lack of information on risk assessments, planning and mitigation (Young et al. 2019). Sustainable water resources management requires a stable but flexible supply system and should have enough resilience capacity to cope with extreme climatic conditions. Risk has been identified as a key factor to consider in sustainable water resource management. To assess the performance of current irrigation management in terms of risk and vulnerability, we used reliability and reversibility as criteria.

The irrigation management system in the Punjab is highly vulnerable to fluctuating surface water availability and rapidly falling groundwater tables. Historical data about Indus River flows is already showing a declining trend which is likely to be impacted by impending changes in climate which will affect glacial storage and precipitation patterns (Chaudhry 2017). Climate change is also likely to disturb the river flow patterns further in the future. Projections suggest increase in river flows due to glacier retreat and fewer but heavier bouts of precipitation for the next few decades, followed by an increasing drought due to reduced river flows (PID 2018; Rees and Collins 2004). However, the current irrigation system and the supporting infrastructure is incapable of harvesting the benefits of increased water supplies both at the national level and the Punjab province due to its limited storage capacity. Further, future reduction in water flows due to glacial retreat would have serious implications for already stressed ecosystems of the Indus River Basin and food security that is heavily dependent on water availability from its rivers.

Another indicator of water shortage in the Indus Basin is per capita water availability, which was estimated at 5000 m^3 in 1974 and has decreased to 1017 m^3 in 2015. The same is the case with Punjab, where per capita water availability was estimated at 1019 m^3 in the year 2017 (UNDP 2017). According to the Falkenmark water scarcity threshold, if the per capita water availability is at 1000 m^3, the system suffers high vulnerability and the country can be considered as water stressed (UNDP 2017; Cai et al. 2001). This indicator show that the Indus water system could be vulnerable to increasing pressure from rapidly growing population.

8.2.1.1 Reliability

Punjab province is arid to semi-arid region with irrigated agriculture that is primarily reliant on surface water from the Indus River and its tributaries. The Indus Basin is vulnerable to changes either in climate or transboundary conflicts at both national and inter-provincial levels. Over 80% of water allocated for irrigated agriculture is used for only four major crops, i.e. sugarcane, rice, maize, and wheat. Most farmers growing these crops use traditional surface irrigation practices (flood or basin irrigation) with low application efficiency that hampers overall water productivity.

Historical data shows a negative trend for Indus water inflows due to various factors, including climate change and the development of mega infrastructures by India on its western tributaries (SBP 2017). Figure 8.3 shows that over the past century the annual river inflow in the Indus River peaked at 226 BCM and troughed at 121 BCM, and that annual river flows shows a gradual decline.

Water scarcity due to competing demands for surface water, depletion of groundwater resources, and changes in climate through increasing temperature and high fluctuations in rainfall distribution is a significant threat to the agro-based economy of Punjab. Any shortages in surface water, in the short term at least, increase dependence of irrigated agriculture on groundwater resources and will lead to increased costs of production and further overuse of the groundwater resource. In the longer term, the higher pumping costs may lead to investments in improvements in surface irrigation and HEIS, but this may prove discriminatory against poorer farmers on smaller landholdings. A robust economic challenge to the agriculture sector will come from inter-sectoral transfers, as cities and industrial sectors grow, and the service and industrial sectors claim more economic wealth per unit of water. A recent water policy document by the Government of Pakistan suggests that water demands for irrigated agriculture will rise by 18% by 2025, compared with a 65% increase in water requirements for the non-agricultural sector – showing a rising inter-sectoral conflict over water use (PID 2018).

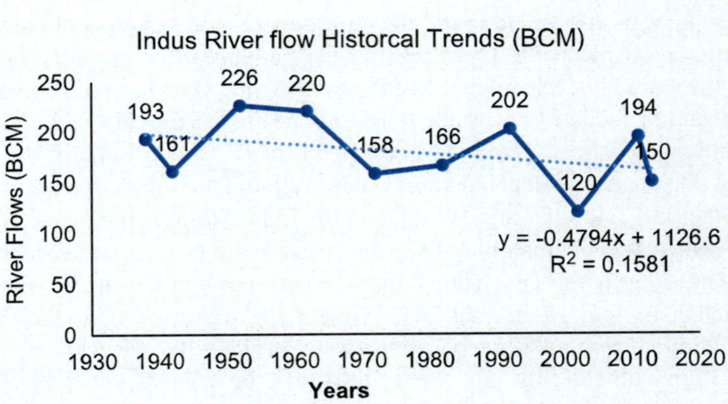

Fig. 8.3 Historical trends of Indus River flows. (Plotted from WAPDA data)

Groundwater abstraction varies across the province. Much empirical evidence has already highlighted that overexploitation of groundwater resources not only decreases groundwater levels but also limits the reliability of irrigation supplies (Watto and Mugera 2016). While availability of water supplies is subject to various factors, inter-sectoral demands are adding further impetus to the problematic reliability of the water resource situation in Pakistan.

8.2.1.2 Reversibility

Reversibility is the probability that the system may recover from a failure to some acceptable state within a specified interval of time (Cai et al. 2001). In the case of Punjab, the situation is very complicated as groundwater has become an essential source of irrigation due to increasing crop water requirements and irregular surface water supplies. An analysis of historical data of areas under different irrigation methods is presented in Fig. 8.4, and shows that the use of groundwater and conjunctive use of both surface and groundwater has increased, creating stresses on groundwater reserves. Recent studies conducted in parts of Pakistan show that due to this overexploitation, the rate of abstraction of groundwater has surpassed the recharge rate in such a way that reversing it has become impossible or highly tricky (Watto and Mugera 2015).

8.2.2 Environmental System Integrity

A guiding criterion for sustainable irrigation water management is to minimise negative environmental impacts of irrigation (Cai et al. 2001). For the Punjab context, an assessment of this environmental integrity criterion focuses on the health of floodplain ecosystem, water quality, and resilience to climate change.

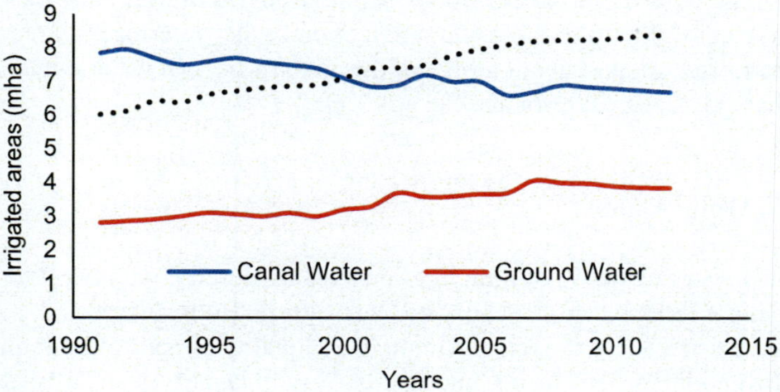

Fig. 8.4 Overtime share of groundwater in agriculture in Punjab. (Plotted using data from Bureau of Statistics Punjab 2012)

8.2.2.1 The Health of Flood Plain Ecosystem

Environmental resources and ecosystems in the province are under great threat from high levels of water withdrawal, increasing water pollution, growth of urban settlements and intensive agriculture. Loss of biodiversity, declining fish stock in the river and degradation of important ecosystems in the downstream Indus are some of the key consequences of these stresses (Irfan et al. 2019).

In particular, unsustainable water use practices have caused land degradation, affecting land quality. For instance, current surface irrigation methods with little regard to drainage are often causes of waterlogging and leaching of chemicals into the soil affecting the quality of soil and groundwater (Qureshi et al. 2008).

Another limitation of surface irrigation is uneven distribution within the field, resulting in under and over-irrigation of the crop. This is due to poorly graded fields and a mismatch between field characteristics such as infiltration, grade, field sizes, and stream sizes. The target depth of application in a surface irrigation event in Punjab is typically 100 mm. This shallow depth of application is difficult to apply uniformly, especially for pre-sowing, when the freshly ploughed and tilled soil results in high intake rates (Akbar et al. 2001). The use of open-ended basin/border strip irrigation decreases the efficiency of nitrogen and farmyard manure due to nitrogen and nutrient run-off that causes lower yields and non-point source pollution. Paddy fields often create a hard crust pan in the soils that depreciates soil properties like several pores that then hampers root growth (Fahong et al. 2004). Control of diseases and pests is also tricky with surface irrigation due to extended periods of standing water in fields or saturated soils.

There have been significant investments by the government in water drainage during the last two decades, though waterlogging still affects a large proportion of land in Punjab. Soil salinity and sodicity also constrain farmers and affect agricultural production in the region. These problems are further worsened by use of poor quality groundwater for irrigation. In fresh groundwater areas, excessive abstraction of groundwater through tube-wells has led to substantial reduction in the quality and quantity of groundwater (Watto and Mugera 2015). Waterlogging in the Indus Basin was high in 1990s due to heavy floods, while droughts during 2000–2010 resulted in a lowering of the water table and a reduction in the waterlogged area. However, recent estimates suggest about 6 Mha of the country are still waterlogged and saline (Qureshi et al. 2010; Qureshi and Sarwar 2009).

8.2.2.2 Water Quality

Pakistan's water crisis is not limited to increasing scarcity, with poor water quality also posing a serious threat for irrigated agriculture. Both surface and groundwater sources are affected by the issue. Although the quality of natural flows in the Indus Basin Irrigation System meets international water quality standards, it has been adversely affected by human factors. Surface water quality in the upper Indus Basin is excellent on certain measures, but the quality progressively decreases downstream due to the dumping of untreated agricultural effluents, human waste and industrial pollutants into the river, canals and drains. Loading of nitrogen, phosphorous, pesticide, organic, and mercury deposition shows alarming levels throughout the river (Iqbal 2013). Unchecked and uncontrolled flooding, basin and border irrigation methods have not only impacted soil but also water quality. Surface water runoffs with pesticides and fertiliser residues. According to an estimate, around 55 km^3 of wastewater is dispersed annually from farms, cities, industries, and households, with 90% of these effluents coming from the agricultural sector (Qureshi et al. 2010).

Pressures on water quantity interact with quality. Decreasing water quality can ultimately decrease the available water quantities for certain uses. Likewise, diminishing water quantities will increase concentration of any pollutants present, further eroding water quality. Thus increasing stress on water quantity and quality by intensive agriculture, urban agglomerations, and industrial concentrations will ultimately burden the Indus Basin through substantial pollution. Various studies (e.g. Azizullah et al. 2011; Ehsan et al. 2020; Waqas et al. 2017) reported the presence of heavy metals in the untreated water drained into freshwater resources and its associated negative impacts on human and animal health. According to another estimate, about 50–60 million people in Pakistan, mostly in Punjab, are at risk of arsenic poisoning from geologically contaminated groundwater (Podgorski et al. 2017). Additionally, Indus Basin aquifers are closely connected to surface waters, allowing pollutants and contaminants from surface water to easily leach into groundwater. Overexploitation of groundwater has also been accompanied by the increasing salinity of aquifers.

8.2.2.3 Resilience to Climate Change

Despite the consensus among researchers on climate change impacts on water inflows, there remains uncertainty around its exact impacts on water inflows to the Indus Basin. Historical trends confirm increasing temperature and rainfall over the last century over the entire region. All recent studies and projections suggest significant changes in climate and subsequent impacts on natural resources particularly water resources (Park 2013). Studies suggest an initial increase in Indus River inflows due to glacial retreat in combination with changes in rainfall patterns followed by substantial decline in water inflows due to limited supply of water from glacier melting – a major source of water inflows to Indus. The Indus Basin Irrigation System and related sectors are not able to bear both situations. The increase flow of water will exacerbate the serious problems of flooding and drainage in downstream Indus Basin areas in the next few decades. The water infrastructures in Pakistan that can hold water only for 30 days will not be able to hold the future increasing amount of water and may result in significant economic losses and displacement of millions of people (Young et al. 2019). Similarly, limited water availability due to decline in Indus water flows will badly affect the agricultural production in the country, which is totally dependent on Indus basin for surface and groundwater supplies. A business-as-usual attitude makes the irrigation water management ill-prepared for future changes in water availability due to climate change.

8.2.3 *Economic Acceptability and Equity*

8.2.3.1 Self-Reliance

Current water management practices are not autonomous and are dependent on numerous subsidies provided to farmers in the form of subsidised inputs, technology, and low irrigation water tariffs. The current water tariffs in Punjab do not place value on water or the actual use of water (Young et al. 2019; Planning Commission 2012). For example, water is charged on a flat rate basin in Punjab which is PKR 170 and 100 per acres for *Kharif* and *Rabi* seasons respectively since 2019, without accounting for actual consumption by various crops. Rice consumes 60% more water compared to cotton but the *abiana* rate for both crops is the same.

With such low prices, water is considered as a free commodity by users and its social and economic value is not commensurate with the costs associated with its delivery, operation and maintenance. Currently, an estimated USD 44 million is the annual subsidy for irrigation (Young et al. 2019). The provision of subsidies distorts the market and water tariffs are low compared to the operation and maintenance (O&M) cost of the irrigation system and sends no signal to encourage water use efficiency. In addition to the very low prices, the recovery of water charges (*abiana*) is also very low in the province. Figure 8.5 shows the trend over time of water charge collections and O&M expenses in Punjab. The trends show that the recovery of water charges has reduced from USD 13 million in 2010 to only USD 7 million in 2016, showing a 46% reduction since 2010, whereas the O&M expenses has increased by 32% during the same time. About 70% of the O&M budget is disbursed on salaries (Planning Commission 2012). There is a little political incentive to correct the gap between water tariffs and O&M costs or prevent a decrease in real terms of water tariffs. Agriculture support prices of certain commodities also encourage overproduction and inefficient use of resources, particularly of water.

8.2.3.2 Water Use Efficiency

Water availability is already becoming a significant issue for the sustainability of current agricultural water management in Punjab. The agricultural sector cannot grow any further using water at this rate and must reduce water use to compensate for the increased demand for water in other sectors, but there is no transparent mechanism to do so. The sectoral demand for water has increased due to urbanisation and rapid population growth. In order to meet the increasing water demands, water resources will need to be reallocated from agriculture to non-agriculture

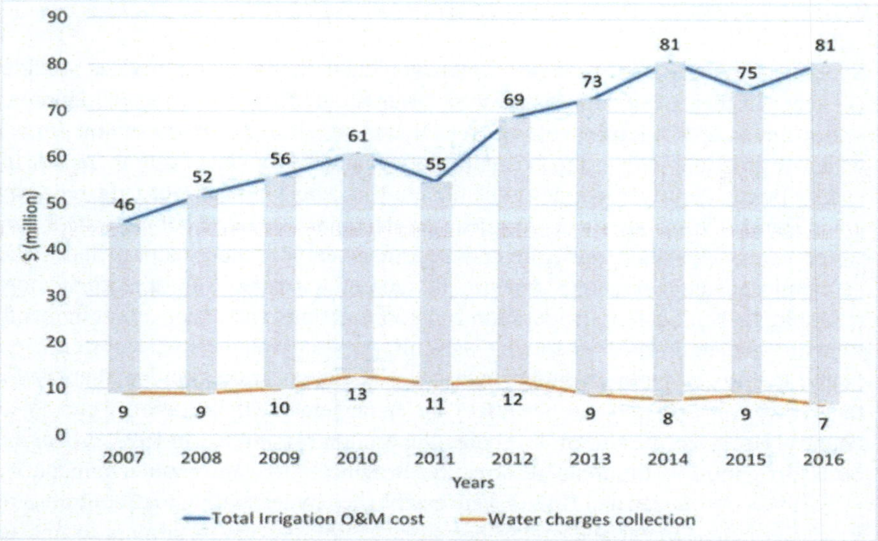

Fig. 8.5 Operation and maintenance cost and *water charges* collection trends in Punjab. (Source: Authors)

sectors. However, current irrigation management system and farm level practices show low little preparedness for this paradigm shift in future water allocations.

Despite the fact that agricultural lands in Punjab are naturally fertile and have a favourable climate that allows cultivation throughout the year, land and water productivity in Punjab remains low due to traditional on-farm water management methods. For instance, Bakhsh et al. (2015) reported the average water productivity of wheat crop under conventional irrigation methods (0.98 kg/m^3) was less than half the water productivity under drip irrigation (2.26 kg/m^3). Young et al. (2019) report that four major crops i.e. rice, wheat, cotton and sugarcane use about 80% of the water allocated for agriculture, yet contribute only 5% of the Gross Domestic Production in Pakistan – showing the economic inefficiency of current irrigation practices in Pakistan. One of the key disadvantages of using conventional irrigation methods is that it requires relatively large discharges. According to one estimate, the crop uses only half of the water applied in the field. Bakhsh et al. (2015) also show that irrigation efficiency of conventional surface irrigation methods ranges between 59–62% compared to 74–98% for pressurised irrigation systems. However, the use of pressurised irrigation system is only limited to 1% of total Punjab land and may need significant effort and investment to introduce it to small landholders, who accounts for 80% of the farming community in the province.

8.2.3.3 Equity

Socially, the current irrigation water management is fragile as the services provided are few and their availability is skewed towards elite farmers and feudal landlords (Fatima et al. 2016). It has been observed that most tail-end farmers rely on groundwater abstractions which are becoming economically unviable due to increasing energy prices. Equity continues to challenge water management for irrigated farmers in Punjab. Water shortages for areas irrigated has increased competition among water users. Upstream and influential farmers exploit their natural and socio-economic advantages at the expense of tail-end farmers, who are often poorer (Ebrahim 2019). Lack of cooperation and coordination is a big constraint in revolutionising the traditional inefficient agricultural water management system. It is feared that the widening conflicts over water resources at the provincial as well as the local level and may further affect the management of irrigation water in the country. However, the current level of collaboration is minimal. In 1997, as a part of broad irrigation institutional reform, the Provincial Irrigation and Drainage Act (1997) was introduced and PIDAs were established by legislation to decentralise the irrigation management and giving decision-making powers to farmers to manage and distribute water at command areas in cooperation with the provincial Irrigation Departments. In Punjab, the Punjab Irrigation and Drainage Authority (PIDA) was established in 1997 with the introduction of elected Area Water boards at three main canal circles i.e. Fordwah Easter Sadqia Canal (Bahwalnagar), Lower Chennab Canal (Faisalabad) and DeraJaat Canal Circle (Dera Ghazi Khan) and elected Farmer Organisations (FOs) at distributary canals and Khalpanchayat (KPs) at the watercourse scale. However, the absence of a comprehensive and conducive legal framework, lack of cooperation and ownership of institutional reforms by the Punjab Irrigation Department (PID), and rent-seeking behaviour by irrigation officers have eventually undermined this major irrigation reform process in Punjab (Young et al. 2019). With the subsequent Punjab Khal Panchayat Act 2019 being passed, all Area Water Boards and Farmer Organisations (FOs) in Punjab have been abolished along with PIDA, and PID has taken back control of all canals and watercourses (Dawn Newspaper 2019). Mustafa et al. (2017) assert that failure in implementing irrigation reforms reflects the deeper structural problems persisting within the Pakistani water bureaucracy and confirm the reluctance of PID officials to facilitate participatory institutional reforms in the irrigation sector.

As water resources become stretched ever more thinly across space, time and sectors, the risk of conflict between stakeholders increases. There is a deep-seated mistrust among the provinces on the allocation of Indus water and mechanisms to reshape common water resources. At the other end of the scale, farmers and water institutions have an imbalanced legal relationship, with water institutions disproportionately empowered by colonial legislation. Similarly, while disputes between farmers are often resolved amicably through social structures – they can also turn into long protracted legal battles, or in extreme cases, individuals taking the law into their own hands. For instance, Zulfiqar et al. (2017) reported that water thefts and disputes over water allocation often turned into violence that killed many farmers.

An increase in incidents of rift between water users in parts of Punjab and resulted violence (Dawn Newspaper 2016; Hayat 2007) show desperation among farmers and lack of trust in water managers and law enforcement agencies to take action against water misusers (The Express Tribune 2019; The Nation Newspaper 2019).

8.3 Conclusion and Recommendations

This chapter has provided a sustainability analysis of irrigation water management in Punjab using indicators of sustainability related to risk and vulnerability, environmental system integrity and economic acceptability.

The analysis suggests that irrigation management system in the province is highly unsustainable and faces significant environmental challenges. The irrigation system in the province faces uncertainty in the availability of surface water due to declining Indus inflows, whereas water demands from agricultural and non-agricultural sectors are rising. In addition, uncertainty in surface water availability has increased pressure on groundwater resources, resulting in the exponential growth of tube-wells and depletion of groundwater resources. Water scarcity is increasing in the province making the irrigation system more vulnerable.

Intensive irrigation and overexploitation of groundwater resources has resulted in negative impacts on the health of floodplain ecosystem and biodiversity in the lower Indus Basin. Unsustainable water use and agricultural practices have caused land degradation, affecting land quality with little regard to drainage. Leaching of chemicals into the soil has affected the quality of soil and groundwater. The system's resilience to cope with climatic changes is very low given its inflexible supply-driven characteristics with high dependence of local livelihoods on irrigated agriculture.

The analysis has also highlighted that the current pricing regime in the province is economically not viable due to its services being highly subsidised. On one hand, the water charges are very low and do not account the actual cost of water, while on the other hand the recovery of water charges is quite low. The gap between collection of water charges and O&M expenses has reached USD 76 million. Water and land productivity in the province remains very low. The water use for irrigation in the province is quite high compared to other irrigated regions in the world. Equity in water distribution continues to challenge water management for irrigated farmers in Punjab. The scarcity of water has increased competition among water users. Upstream and influential farmers exploit their natural and socio-economic advantages at the expense of other farmers.

Cooperation within farming communities and institutions is very weak and the level of collaboration between various institutions in the province is minimal. The participatory irrigation management system with farmer institutions has failed to deliver the dividends due to lack of cooperation and commitment from co-farmers and various institutions. Similarly, conflicts over water resources are rising at the provincial as well as the local level and may interrupt the effective supply of water for Punjab.

To improve the sustainability of irrigation water management in Punjab requires a comprehensive and inclusive approach that considers the interest of all major stakeholders. As an initial step, to reduce the stress on water resources, agricultural systems in the province need to improve current water use practices through adopting innovative farm and water management practices. Punjab's Agriculture Department has already taken some steps through its On-Farm Water Management (OFWM) wing by introducing water saving technologies and improved farm management practices. However, the transition from conventional surface methods to modernised irrigation systems, such as pressurised irrigation systems, is progressing slowly. Significant investments are required to overcome financial and technological constraints. Focus needs to turn to developing strategies to improve water efficiency within conventional irrigation methods and its dissemination among farming communities.

The capacity of farmers also needs to be developed to effectively enhance their decision-making to select the most appropriate options among available surface irrigation methods. For example, furrow could replace basin and uncontrolled flooding irrigation methods. Furrow irrigation not only reduces crop stress, waterlogging and salinity, but also has a higher field application efficiency compared to flood or basin irrigation methods. Studies have shown successes with furrow irrigation. It is easy to establish, and does not require large capital investment or expensive infrastructure upgrades as compared with pressurised irrigation. Furrow irrigation reduces other input costs in addition to water, e.g. seed costs decrease due to a defined number of rows (Singh et al. 2002). Furrow irrigation system also helps to eradicate the problem of hardpan in the soil surface, and it decreases crop lodging and increases fertiliser efficiency. Water use in surface irrigation can increase by employing conservation tillage that not only increases water infiltration but also reduces run-off and improves soil moisture storage. Conservation tillage also maintains soil fertility levels with micro-nutrients in crop production. Soil-water conservation measures incorporating crop residue and adequate land preparation for crop sowing also help to improve water use in surface irrigation.

There is also an opportunity to improve water use efficiency by introducing some form of volume-based management, which could have a positive impact on farmers' choice of irrigation methods. Studies report that there is a misconception that an increase in water tariffs may affect small farmers disproportionately. However, small farmers can be protected against increases in water tariffs through social tariffs or through a fund made available through the increased water tariffs that target subsidies to improve system efficiency, thus enhancing equity in the distribution of water. Bell et al. (2016) found that an increase in water charges not only enhances agricultural production but also improves efficiency and equity in the distribution of water among farmers. However, removing institutional excess is required to ensure the increased revenue from water tariffs contributes to improving irrigation services rather than propping up bloated irrigation bureaucracies.

A surface irrigation system provides an opportunity to improve irrigation practices through better policy measures. The Government of Punjab could enhance its policy of raising the profile of effective and efficient use of water through social appreciation and financial incentives for farmers who reduce water use. The government has already, since 2004, introduced yield competitions among farmers, with farmers rewarded with social standing and cash prizes. Similar schemes could be introduced to encourage efficient water use as the current tariff structure (low and applied by land area) provides no incentive for farmers to minimise water use.

Social awareness may contribute to increased efficiency of surface irrigation by encouraging advanced land management practices such as laser land levelling, soil conservation tillage, better land preparation, and raised-bed plantation. Village demonstration plots may promote uptake, where the use of technological innovations is disproportionately low. Another effective tool for convincing farmers may be the use of local notables (religious, political etc.) to provide leadership and act as agents of change to encourage farmers to conserve water through improved land management practices.

The current situation of groundwater over-exploitation in Punjab need to be controlled through implementing proper legislation and regulations on the use of groundwater and controlling the number of tube-wells in vulnerable regions. The focus of irrigation management also needs to change in response to increasing demand from non-agricultural sectors. PID is already working on some initiatives to transform itself from a single irrigation-focused department to a broader water resource management department taking care of water uses of non-agricultural sectors and the environment.

References

Abid M, Scheffran J, Schneider UA, Ashfaq MJ (2015) Farmers' perceptions of and adaptation strategies to climate change and their determinants: the case of Punjab province, Pakistan. Earth Syst Dynam 6(1):225–243

Ahmed W, Rais MN, Bano R et al (2018) Water sharing, governance, and management among the provinces in Pakistan using evidence-based decision support system. In: World environmental and water resources congress 2018: watershed management, irrigation and drainage, and water resources planning and management. American Society of Civil Engineers, Reston, VA, pp 220–233

Akbar G, Yasin M, Ahmed S et al (2001) Comparative performance of sprinkler and surface irrigation methods. Sarhad J Agri 17(4):601–608

Akhtar N (2004) The use of irrigation systems for sustainable fish production in Pakistan. In: Petr T (ed) Fisheries in irrigation systems of arid Asia, FAO fisheries technical paper no. 430. FAO, Rome, pp 17–39

Ashley RM, Blackwood D, Butler D, Jowitt P (2004) Sustainable water services. IWA Publishing, London

Azizullah A, Khattak MN, Richter P, Häder DP (2011) Water pollution in Pakistan and its impact on public health: a review. Environ Int 37(2):479–497

Bakhsh A, Ashfaq M, Ali A et al (2015) Economic evaluation of different irrigation systems for wheat production in Rechna Doab, Pakistan. Pak J Agric Sci 52(3):821–828

Bandaragoda DJ, ur Rehman S (1995) Warabandi in Pakistan's canal irrigation system: widening gap between theory and practice. International Irrigation Management Institute, Colombo
Bell AR, Ward PS, Shah MA (2016) Increased water charges improve efficiency and equity in an irrigation system. Ecology and Society 1(3):21
Bell AR, Ward PS, Ashfaq M, Davies S (2020) Valuation and aspirations for drip irrigation in Punjab, Pakistan. J Water Resour Plan Manag 146(6):04020035
Bureau of Statistics Punjab (2012) Punjab development statistics 2012. Bureau of Statistics Punjab, Government of Punjab, Lahore
Bureau of Statistics Punjab (2019) Punjab development statistics. Bureau of Statistics Punjab, Government of Punjab, Pakistan. http://www.bos.gop.pk/system/files/PDS-2019.pdf
Cai X, McKinney DC, Rosegrant MW (2001) Sustainability analysis for irrigation water management: concepts, methodology, and application to the Aral Sea Region. Environment and production technology division, discussion paper no. 86, International Food Policy Research Institute, Washington, DC
Cai X, McKinney DC, Lasdon LS (2002) A framework for sustainability analysis in water resources management and application to the Syr Darya Basin. Water Resour Res 38(6):21–31
Chaudhry QU (2017) Climate change profile of Pakistan. Asian Development Bank, Pakistan
Chaves HM, Alipaz S (2007) An integrated indicator based on basin hydrology, environment, life, and policy: the watershed sustainability index. Water Res Manage 21(5):883–895
Dawn Newspaper (2016, May 2) Two killed over water dispute. https://www.dawn.com/news/1255765
Dawn Newspaper (2019, May 25) PIDA abolished: Irrigation dept takes control of canals, watercourses. https://www.dawn.com/news/1484382
Ebrahim Z (2019) Irrigation in Pakistan: water theft drains Indus canals dry. The Third Pole January 4. https://www.thethirdpole.net/2019/01/04/pakistan-irrigation/
Ehsan N, Shan A, Riaz S et al (2020) Health risk assessment due to exposure of arsenic contamination in drinking water of district Shiekhupura, Punjab, Pakistan. Hum Ecol Risk Assess Int J 26(1):162–176
Elahi E, Abid M, Zhang L et al (2018) Agricultural advisory and financial services; farm level access, outreach and impact in a mixed cropping district of Punjab, Pakistan. Land Use Policy 71:249–260
Fahong W, Xuqing W, Sayre K (2004) Comparison of conventional, flood irrigated, flat planting with furrow irrigated, raised bed planting for winter wheat in China. Field Crop Res 87(1):35–42
FAO (2008) FAO water database
Fatima F, Jacoby HG, Mansuri G (2016) Decentralizing corruption? Irrigation reform in Pakistan's Indus Basin. The World Bank. http://pubdocs.worldbank.org/en/764111466186379941/DecentralizingCorruption-May16.pdf. Accessed 10 Aug 2020
Giuppponi C, Jakeman AJ, Karrssenberg D, Hare MP (2006) Sustainable management of water resources. Edward Elgar Publishing Limited, Massachusetts
FWM (2019). Federal Water Management Cell. Ministry of National Food Security and Research, Islamabad
Government of Punjab (2017) Punjab irrigated-agriculture productivity improvement project-revised(PIPIP-revised), directorate general agriculture (water management), Government of Punjab, Pakistan, Lahore. http://ofwm.agripunjab.gov.pk/system/files/PIPIP-Revised.pdf. Accessed 13 May 2020
Hassan G, Hassan F (2017) Sustainable use of groundwater for irrigated agriculture: a case study of Punjab, Pakistan. Eur Water 57:475–480
Hayat A (2007) Irrigation sector development in Punjab (Pakistan): case study of district Sargodha. Unpublished Masters dissertation, The Tema Institute, Department of Water and Environmental Studies, Linköping University
Iqbal AR (2013) Environmental issues of Indus River basin: an analysis. ISSRA Papers, 5(1):89–112. https://ndu.edu.pk/issra/ISSRA_paper_2013_IssueI.php

Irfan M, Qadir A, Ali H et al. (2019) Vulnerability of environmental resources in Indus Basin after the development of irrigation system. In irrigation-water productivity and operation, sustainability and climate change. IntechOpen

IRSA (1991) Apportionment of the water of the Indus River system between the provinces of Pakistan, Indus water accord 1991, Indus River System Authority (IRSA), Government of Pakistan. http://www.pakirsa.gov.pk/WAA.aspx. Accessed 20 May 2018

Jurriëns M, Mollinga PP, Wester P (1996) Scarcity by design: protective irrigation in India and Pakistan. International Institute for Land Reclamation and Improvement Netherlands, Wageningen Agricultural University

Juwana I, Perera BJ, Muttil N (2010) A water sustainability index for West Java. Part 1: Developing the conceptual framework. Water Sci Technol 62(7):1629–1640

Kahlown MA, Kemper WD (2007) Factors affecting success and failure of trickle irrigation systems in Balochistan, Pakistan. Irrigation Science, 26(1):71–79

Kahlown MA, Raoof A, Zubair M, Kemper WD (2007) Water use efficiency and economic feasibility of growing rice and wheat with sprinkler irrigation in the Indus Basin of Pakistan. Agric Water Manag 87:292–298

Latif M, Haider SS, Rashid MU (2016) Adoption of high efficiency irrigation systems to overcome scarcity of irrigation water in Pakistan. Proc Pak Acad Sci Life Environ Sci 53(4):243–252

Loucks DP (2000) Sustainable water resources management. Water Int 25(1):3–11

Loucks DP, Gladwell JS (1999) Sustainability criteria for water resource systems. Cambridge University Press, Cambridge

Mays LW (2006) Water resources sustainability. McGraw-Hill Professional, New York

Ministry of Finance (2019) Economic survey of Pakistan 2018–19. Finance Division, Ministry of Finance, Revenue and Economic Affairs, Government of Pakistan. http://www.finance.gov.pk/survey_1819.html

Ministry of Water Resources (2018) National water policy. Ministry of Water Resources, Government of Pakistan, Islamabad

Mustafa D, Akhter M, Nasrallah N (2013) Understanding Pakistan's water-security nexus. United States Institute of Peace, Washington, DC

Mustafa D, Gioli G, Karner M, Khan I (2017) Contested waters: subnational scale of water and conflict in Pakistan. United States Institute of Peace, Washington, DC

Nardo M, Saisana M, Saltelli A et al (2005) Handbook on constructing composite indicators: methodology and user guide, OECD statistics working paper. Ispra, Italy

Park YW (2013) The environment and climate change: outlook of Pakistan. United Nations Environment Programme (UNEP):107

PID (2018) Punjab water policy. Punjab Irrigation Department (PID), Government of Punjab, Lahore

Planning Commission (2012) Canal water pricing for irrigation in Pakistan, assessment, issues and options, Planning Commission, Government of Pakistan

Podgorski JE, Eqani SA, Khanam T et al (2017) Extensive arsenic contamination in high-pH unconfined aquifers in the Indus Valley. Sci Adv 3(8):e1700935. https://doi.org/10.1126/sciadv.1700935

Policy Research Initiative (2007) Canadian water sustainability index. Policy Research Institute on Ageing & Ethnicity

Qureshi AS, Sarwar A (2009) Managing salinity in the Indus Basin of Pakistan. Int J River Basin Manage 7(2):111–117

Qureshi AS, McCornick PG, Qadir M, Aslam Z (2008) Managing salinity and waterlogging in the Indus Basin of Pakistan. Agric Water Manag 95(1):1–10

Qureshi AS, Gill MA, Sarwar A (2010) Sustainable groundwater management in Pakistan: challenges and opportunities. J Int Comm Irrig Drain 59(2):107–116

Rees G, Collins D (2004) SAGARMATHA: snow and glacier aspects of water resources mnagement in the Himalayas. DFID project R7980—an assessment of the potential impacts of deglaciation on the water resources of the Himalaya, Centre for Ecology and Hydrology, Oxfordshire

Rinaudo JD, Strosser P, Rieu T (1997) Linking water market functioning, access to water resources and farm production strategies: example from Pakistan. Irrig Drain Syst 11(3):261–280

Savenije HH, Van Der Zaag P (2002) Water as an economic good and demand management: paradigms with pitfalls. Water Int 27(1):98–104

SBP (2017) Chapter 7, water sustainability in Pakistan – key issues and challenges, in State Bank of Pakistan's annual report 2016–17. State Bank of Pakistan

Singh SA, Yadav RK, Singh H, (2002) Furrow irrigated raised bed planting system: a resource conservation technology for increasing wheat productivity in rice-wheat sequence. In Malik RK et al. (eds.), Herbicide resistance management and zero tillage in rice-wheat cropping system, workshop proceedings, Hisar, India, 4–6 March 2002, CCS HAU Hissar, India, pp. 198–200

Sullivan CA, Meigh J (2007) Integration of the biophysical and social sciences using an indicator approach: addressing water problems at different scales. Water Resour Manag 21(1):111–128

Tetlay K, Byerlee D, Ahmad Z (1999) Role of tractors, tubewells and plant breeding in increasing cropping intensity in Pakistan's Punjab. Agric Econ 4(1):13–25

The Express Tribune (2019, June 2) Crackdown against water theft continues in Punjab. https://tribune.com.pk/story/1985058/crackdown-water-theft-continues-punjab

The Nation Newspaper (2019, January 2) Pressed by SC, irrigation dept wakes up to water theft. https://nation.com.pk/02-Jan-2019/pressed-by-sc-irrigation-dept-wakes-up-to-water-theft

UNDP (2017) The vulnerability of Pakistan's water sector to the impacts of climate change: identification of gaps and recommendations for action. A report by United National Development Programme, Pakistan, Islamabad

Usman KM, Muhammad T, Majid M et al (2016) Drip irrigation in Pakistan: status, challenges and future prospects. Russ J Agri Socio Econ Sci 56(8)

Waqas H, Shan A, Khan YG et al (2017) Human health risk assessment of arsenic in groundwater aquifers of Lahore, Pakistan. Hum Ecol Risk Assess Int J 23(4):836–850

Watto MA, Mugera AW (2015) Econometric estimation of groundwater irrigation efficiency of cotton cultivation farms in Pakistan. J Hydrol Reg Stud 4:193–211

Watto MA, Mugera AW (2016) Groundwater depletion in the Indus Plains of Pakistan: imperatives, repercussions and management issues. Int J River Basin Manage 14(4):447–458

Young W, Anwar A, Bhatti T et al. (2019) Pakistan: getting more from water. World Bank, Washington, DC. https://openknowledge.worldbank.org/handle/10986/31160

Zardari NH, Cordery I (2009) Water productivity in a rigid irrigation delivery system. Water Resour Manag 23(6):1025–1040. https://doi.org/10.1007/s11269-008-9312-2

Zulfiqar F, Shahzad S, Abid M, Ullah R (2017) Managing water commons and the extent of their climate resilience in Punjab, Pakistan. Conference Paper, Indiana University, Digital library of commons.

Chapter 9
Examining Irrigated Agriculture in Pakistan with a Water-Energy-Food Nexus Approach

James L. Wescoat Jr., Waqas Ahmed, Steven Burian, Jehangir F. Punthakey, and Ayesha Shahid

Abstract This chapter explores the applicability of the Water-Energy-Food nexus framework for the Indus Basin of Pakistan. It introduces the importance of linking water, energy, and food resources, and current sectoral gaps in Pakistan. To better understand these gaps and linkages, the chapter shows that the water-energy-food nexus is not an entirely new idea or practice in the Indus Basin. Although the terminology is new, integration of various sorts was common up to the mid-nineteenth century. The modern sectoral breaks arose with the development of large-scale colonial canal irrigation, administered under its own spatial jurisdictions, and further institutional gaps arose with the growth of electric power and fossil fuel programs. The chapter then lays out a framework for understanding the linkages among these sectors as a nexus that entails a broad range of perspectives, levels, and scales. This framework is used to characterise the problems associated with sectoral fragmentation, and then to identify ways forward in analytic, governance, and societal terms.

J. L. Wescoat Jr. (✉)
School of Architecture + Planning, MIT, Cambridge, MA, USA
e-mail: wescoat@mit.edu

W. Ahmed
US Pakistan Center for Advanced Studies in Water, Mehran University of Engineering and Technology, Jamshoro, Sindh, Pakistan
e-mail: wapathan.uspcasw@faculty.muet.edu.pk

S. Burian
Department of Civil and Environmental Engineering, University of Utah, Salt Lake City, UT, USA
e-mail: steve.burian@utah.edu

J. F. Punthakey
Institute for Land, Water and Society, Charles Sturt University, Albury, NSW, Australia
e-mail: eco@ecoseal.com

A. Shahid
Independent Researcher, Rawalpindi, Pakistan

M. A. Watto et al. (eds.), *Water Resources of Pakistan*, World Water Resources 9,
https://doi.org/10.1007/978-3-030-65679-9_9

Keywords Water-Energy-Food nexus · Nexus perspectives · Nexus levels · Scales of analysis

9.1 Introduction

Increasing population, economic growth, and development have put immense pressures on Pakistan's agriculture sector, both to provide for increasing domestic food demands and for export agriculture. To meet the growing demand for food and fibre, the agricultural sector has to increase production by 4%, which could require an increase in water supplies to the agriculture sector by 10% (Hassan et al. 2019). Exacerbating this challenge, Pakistan's water resources have to meet increasing industrial, domestic, and environmental demands, some of which may take priority over agriculture demand in the years ahead. The energy sector in Pakistan likewise faces increasing pressure to fulfil multiple uses, including those related to water and food. These include hydropower production, groundwater pumping, and power plant cooling. It is important to jointly consider the energy value in calories and the water footprint in consumptive use for production of food for humans and animals, for biofuels, etc. Given these sectoral linkages, the aim of this chapter is to assess the utility – the strengths and limitations – of addressing agricultural water demand using a water-energy-food nexus framework.

In the Water-Energy-Food (WEF) nexus framework, inter-linkages among water, energy, and food systems affect each other in complex ways. In Pakistan, increasing population, economic growth, and societal change are driving competition among all aspects of the WEF nexus, e.g., among agricultural, domestic, hydropower, industrial, recreational, and environmental uses of water resources. These challenges are leading to methods for finding synergies that reduce loss, increase efficiency, and discover new resources (see the systematic review of nexus research methods by Albrecht et al. 2018). A common methodological starting point takes one of the resources as an entry point and identifies its linkages with the other two nexus sectors (as illustrated in Fig. 9.1, which takes water as the entry resource, and analyses linkages with energy and food across multiple spatial and temporal scales). In this chapter, we begin with water for food production and extend to energy and other uses, which are also common in Integrated Water Resources Management studies (Nepal et al. 2019). Many nexus studies in Pakistan and elsewhere have begun with water-energy linkages (e.g., Siddiqi and Wescoat Jr 2013), while some begin with climate change and extreme events (Diamond 2005). As will be shown below, there is not a universal framework to address the nexus: it varies with the perspectives, levels, and sectors under consideration, as well as the regions, times, and scales of analysis. However, the underlying nexus concept is consistent: it focuses on interrelationships, synergies, and trade-offs among sectors that are often managed separately, and seeks to develop and deploy integrative performance measures.

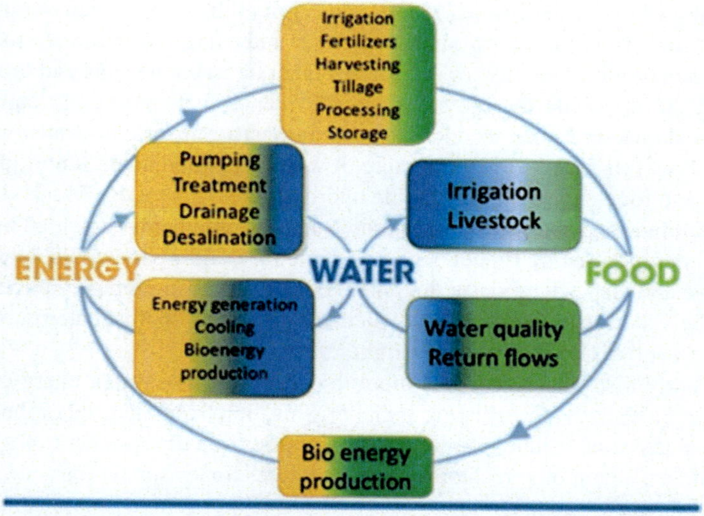

Fig. 9.1 Key linkages in the WEF nexus. (Adapted from Ferroukhi et al. 2015)

This chapter has four aims. The first is to show that the WEF nexus has deep historical geographic roots in the Indus Basin region. While the nexus terminology is new, some of the relationships among sectors are not. To demonstrate this point, the chapter begins with an historical geographic perspective on integrated water, energy, and food systems in the Indus Basin, showing that they have often operated jointly, albeit with varying levels of conscious management. The second aim is to introduce modern ideas about the WEF nexus, building upon the Keskinen et al. (2016) framework of perspectives, levels, and scales in transboundary Asian river basins. Examples from the agricultural water sector in Pakistan are presented for each aspect of the framework. The third aim is to review current sectoral fragmentation in Pakistan and thus identify the main challenges of a nexus approach. The fourth aim is to show how the nexus approach can be more fully used to advance the irrigated agriculture sector in Pakistan.

9.2 An Historical Geographic Perspective on the WEF Nexus in the Indus Basin

There is nothing new about water, energy, and food linkages in the Indus Basin, or elsewhere. Such linkages have operated, more or less effectively, long before the establishment of Pakistan, and it is worth reviewing their historical geography before proceeding to modern concepts of the WEF nexus. All organisms and organisations expend and store energy to acquire water and food that they then transform into energy for metabolism, movement, reproduction, and so on (Meadows and

Meadows 1999). All of the early pastoral, agricultural, and urban settlements of Indus Valley civilisation drew on natural water and energy sources for food production, which provided energy for ever more elaborate water systems and food storage (Wright 2009). While there is no explicit evidence that these early communities consciously viewed their economic activities as a "nexus" of water, food, and energy, it was that, and something more. It was understood, then as now, that water, energy, and food depended on variable hydro-climatic processes. Monsoon fluctuations required food and water surplus and storage. Settlements expended human and animal energy to secure timber, soil, and stone for building materials, with secondary reuse and disposal of waste materials. Harappan communities implicitly understood the water-energy-food nexus when creating their sophisticated wells, baths, granaries, and drainage systems (Wright 2009).

The early historical nexus was more than that, because water, energy, and food relations are also vital for shaping social organisation from the level of the family to the social division of labour in larger communities, the development of specialised crafts, management of agro-biodiversity, strategies for seasonal migration, and conflict management (e.g., Zimmerer and de Haan 2019). These complex social relations are addressed in modern research in cultural ecology and political ecology (e.g., Perreault et al. 2015), and it is noteworthy that few of these ecological aspects of complex social organisation have been addressed in much detail in modern water-energy-food nexus research to date.

The focus here is on agricultural water use in the Indus Basin, which involved two main types of nexus. The first relied on large natural inundation canals, many of them former river channels, which created favourable sites for floodplain cultivation and natural animal watering sites. Their floodwaters were distributed and drained by potential energy, and the biomass that thrived on their banks enhanced caloric energy accumulation by animals and humans. At a smaller scale, communities in upland areas constructed myriad shallow seasonal dug wells, some of which were deepened and reinforced with brick masonry. Still others had geared water lifting devices driven by animals that supported sedentary cultivation (Crooke 1989). These wells did not require much energy to build or maintain, but they did require considerable human and animal energy for daily water collection, food production, and migration from annually depleted locations to more favourable sites.

These early precolonial nexus patterns changed in the mid-nineteenth century with the massive expansion of permanent canal irrigation, on one hand, and agricultural revenue collection, on the other (Ali 1988; Gilmartin 2020; Shahid et al. 2019). Canals and revenue extraction in Punjab, Multan, and Thatta dated back at least to the Mughal period, as recorded in the *Ain-i Akbari* (Institutes of Akbar, c. 1595). But the canals were few in number and scale until the British took experience gained in the Yamuna Basin and expanded it across Punjab with a massive administrative apparatus to build and operate the system of canal commands separately from other colonial administrative departments. For agricultural revenue purposes, by comparison, provinces (*subahs*) were divided into districts (*parganahs* and *tehsils*), which were divided into revenue villages (*mouzas*). By the end of the nineteenth century, district administrative policies shifted from a primary emphasis on crop revenue extraction to multiple sectors of economic development, driven by colonial political goals (Ali 1988; Gilmartin 2020). Each province of the Indus was governed from a distant presidency, Punjab from Delhi, and Sindh from Bombay, which aggravated irrigation conflicts across provincial and princely state boundaries long before independence in 1947.

Precolonial forms of the WEF nexus were thus broken during the nineteenth century in ways that have yet to be fully bridged. Surface irrigation canals cut across revenue districts in "canal commands" that had an entirely separate spatial logic and administrative apparatus from the districts that govern agriculture and related land resources. Provincial irrigation departments have a different hierarchy of administrative units than those of the agricultural and rural development sectors. The energy sector followed yet another trajectory. Solar insolation was, and is, the most important form of energy for the agricultural sector, though that is sometimes omitted in modern nexus research. Biofuels (grass, wood, and dung) have likewise been used for millennia for heating and cooling. A significant proportion of crops were devoted to fodder to sustain animal draft power for water lifting and other farm labour. In addition, most perennial irrigation canals flow by gravity (i.e., kinetic energy) aided by the potential energy of low impoundments behind barrages. Canal commands follow those alignments. Some canals employed lift structures to raise water to a canal head that then flowed by gravity along the higher *bar* lands of the *doabs* (Bedi 2003).

The nineteenth century introduced coal for steam engines, rail transportation of agricultural goods, and later for electricity generation. The second half of the twentieth century witnessed development of large dams at Mangla and Tarbela that enabled longer periods of perennial irrigation, as well as hydropower generation to supplement thermal energy production. The electric energy generation (GENCOs) and distribution regions (DISCOs) that developed during the twentieth century had yet another set of spatial service areas and regulatory policies. Thus, by the early twentieth century, the water, energy, and food sectors were deeply divided. Nexus relationships linked these sectors at the farm and community levels despite the institutional gaps that constrained and distorted them. Some of the best examples may be found in villages of Gilgit-Baltistan where late-twentieth micro-hydro facilities were linked with irrigation and domestic water and sanitation improvements. By contrast, tube-well proliferation powered by fossil fuels in Balochistan led to depletion of historic *karez* water distribution systems.

Sectoral fragmentation worsened during the mid-twentieth century with the technological revolution in groundwater pumping, which began with federal waterlogging and salinity wells and rapidly extended to millions of unregulated private tube-wells that now provide the bulk of irrigation water for crop production and diminish the significance and maintenance of irrigation department canals. Limited subsidies and frequent power outages led most tube-well operators to rely on diesel fuel. Constraints on the development of new large dams have further increased reliance on thermal power production.

The water, energy, and agricultural sectors have thus fragmented with the advent of each new technology, program, and problem at the federal, provincial, local, and private enterprise levels. In Pakistan's federal system of government, water is a provincial subject, which limits the federal role in integration. The inherently integrated water-energy-food nexus of pre-colonial times became thoroughly fragmented from the late-nineteenth through the early-twenty-first century. The water-energy-food nexus has endured in various ways at the farm level – it had to – but that was in spite of fragmented governance and growth of sector stakeholders.

It is in this institutionally fragmented context that purposeful attempts at "integration" have been put forward. One of the earliest efforts was in "valley projects," like the Sutlej Valley Project of the 1930s, which sought to link water and agricultural development (see chapter by Wescoat et al. in this volume). "Integrated river basin development" followed in the 1940s and 50s. It explicitly linked hydropower, land, and water development, but it was only partially adopted in the post-partition context (Lieftinck et al. 1968; Michel 1967). Partial integration of surface water management was followed by the tube-well revolution in the 1970s, which led to largely unsuccessful calls for "conjunctive management" of surface water and groundwater uses (van Steenbergen 2020). These efforts were followed by calls for "Integrated Water Resource Management" (IWRM), promoted and financed by international water organisations (Lenton and Muller 2009). Most recently, the Government of Pakistan declared that, "The National Water Policy is based on the concept of Integrated Water Resources Management…" (Ministry of Water Resources 2018). This declaration specified 33 aspects of IWRM, which are forward-looking and reasonable, albeit difficult to achieve. IWRM has been criticised as both over- and under-ambitious (Giordano and Shah 2014, and other articles in the special issue of the 2014 *International Journal of Water Resource Management* devoted to IWRM).

The National Water Policy also mentions water-food-and-energy security and the water-energy nexus, and it situates them within the IWRM paradigm. It thus renders water-energy-food relationships as a subset of IWRM, rather than the reverse. As Giordano and Shah (2014) point out, IWRM puts water at the centre of an (overly) expansive water management approach. The nexus paradigm, by contrast, usually treats the three sectors as equally important and encourages, "multi-tiered institutional arrangements" and mechanisms for joint decision-making (Benson et al. 2015). In the IWRM policy framework, the nexus paradigm becomes most relevant at the farm level where it is often measured with metrics of water use efficiency and sustainability, e.g., "more crop per drop" (Ministry of Water Resources 2018). Government authorities have developed some inter-sector coordination processes, but bureaucratic resistance and related politics work against the fulfilment vis-à-vis rhetoric of this goal (Arfan et al. 2020; Shahid et al. 2019).

As underscored above, the Indus has been an important laboratory for basin-wide agricultural water planning for more than 70 years, beginning with regional soil and water land evaluation studies in the 1950s and evolving to the new National Water Policy (Wescoat et al. 2000). The 1960 Indus Waters Treaty led to massive investment in agricultural water development by the World Bank and a consortium of donor countries, guided in part by new tools of river basin optimisation planning (Maass et al. 1962). A three-volume study by Lieftinck et al. (1968) explicitly linked the water and power sectors in its scope of investment planning. However, later versions of the World Bank's Indus Basin Model Revised (IBMR) focused on agro-economic optimisation of water infrastructure investment treating hydropower alternatives as assumptions, scenarios, or constraints (Ahmad and Kutcher 1992; Duloy and O'Mara 1984; WSIPS 1990). At the turn of the century, these models were further adapted to assess climate change (Wescoat and Leichenko 1992; Yu et al. 2013). The national energy sector is modelled separately (International Resources Group 2010), as are some water subsectors that require shorter time steps like flooding and snowmelt (Wescoat and Leichenko 1992).

In summary, the Water-Energy-Food Nexus can be regarded as the latest attempt to address the fragmented situation in natural resources management that developed in the mid-nineteenth century, and to advance the prospects for integrated environmental management in Pakistan. The past decade has witnessed increasingly critical assessments of water governance problems, which help explain some of the root causes of water, energy, and food problems (Akhter 2015; Mustafa 2013; Naqvi 2013). Each sector has conducted its own attempts at reform and integration, with limited coordination. Electricity load shedding in summer months is less attributed to scarcity than to fiscal failings of revenue collection, payment practices, and circular debt. These institutional problems make it prudent to be sceptical as well as receptive to purely analytical approaches in the Indus Basin. That said, increasingly sophisticated frameworks and analyses of water, energy, food relationships are underway, which shed light on scientific and technical linkages, experiments, and reforms. The next section of the paper introduces a modern WEF nexus framework and discusses its potential relevance in Pakistan.

9.3 The Modern WEF Nexus Framework: Perspectives, Levels and Scales

The modern water, energy, food (WEF) nexus has been conceived in a variety of ways that are reviewed in this section. Figure 9.2 depicts a nexus framework that encompasses multiple perspectives, levels, and scales for identifying synergies, co-benefits, and trade-offs (Keskinen et al. 2016; Liu et al. 2018). In this section, we briefly discuss these perspectives, levels and scales, and consider their potential for addressing WEF problems more holistically in Pakistan.

9.3.1 Perspectives

Keskinen et al. (2016) suggest that the nexus can be approached from three perspectives which are: (i) analytical; (ii) governance; (iii) and discourse. We build upon these three and add three more which are: (iv) resource recovery; (v) environmental livelihoods; and (vi) economic optimisation. To elaborate:

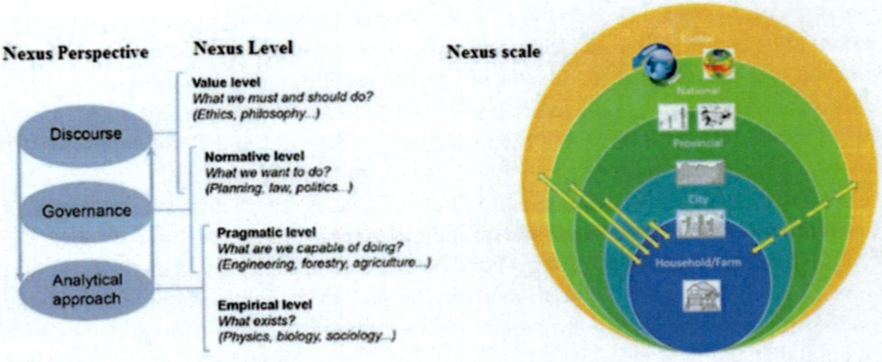

Fig. 9.2 WEF nexus perspectives, levels, and scales. (Adapted from Keskinen et al. 2016)

9.3.1.1 System Perspective

The WEF nexus has often been approached through quantitative analyses of the basin as a system of interconnected resource stocks and flows, in which the outputs of a system are balanced by its inputs and change in storage (Maass et al. 1962). Systems analysis has been the pre-eminent perspective employed in river basin modelling and management in Pakistan from the Indus Waters Treaty period to the present (Duloy and O'Mara 1984; Lieftinck et al. 1968; Yu et al. 2013). It includes simulation and optimisation modelling.

9.3.1.2 Governance Perspective

Linkages between sectors and their respective stakeholders are increasingly addressed in terms of institutional processes, politics, and power dynamics that affect policy implementation, effectiveness, and coherence (Benson et al. 2015). Governance research in Pakistan has increased in ways that help interpret these political and policy issues, experiments, and failings (Akhter 2015; Mustafa 2013).

9.3.1.3 Emerging Discourse Perspective

The nexus is also a mode of communication among interconnected sectors. Discourse analysis sheds light on how issues are framed, and redefined to establish new relationships among the WEF sectors. Analysis of policy documents like the National Water Policy can offer insights into the evolving language and logic of resource management (Arfan et al. 2020).

9.3.1.4 Resource Recovery Perspective

Scott et al. (2015) view the nexus as a means for increasing "resource recovery" by establishing new linkages across sectors. Their argument is based in part on the view proposed by Kurian and Dietz (2013) that adaptive management involves the creative transformation of social-ecological system uncertainties. Scott (2013) has also drawn attention to the risk of Jevons Paradox in efficiency-driven approaches, where increased efficiency can lead to increased overall consumption (e.g., groundwater depletion). Gilmartin (2003) has criticised the emphasis on waste and efficiency as a specifically colonial perspective on the political economy of irrigation in South Asia (though it may be noted that it was also prevalent in early-twentieth century conservation movements in the US). The resource recovery perspective begins with one resource sector, known as the "entry sector," while the other two sectors are then studied to identify opportunities for recovering and recycling resources that might otherwise be lost in conventional conservation practice (cf. Yang et al. 2016).

9.3.1.5 Environmental Livelihood Security Perspective

Biggs et al. (2015) approach the WEF nexus from an Environmental Livelihood Security (ELS) perspective. They ascribe the security focus to the World Economic Forum, and build upon critical perspectives on livelihood assessment (e.g., Scoones 2009). The livelihoods perspective has special significance for run-of-the-river schemes in upstream areas of the Indus Basin where small-scale irrigation and micro-hydro systems are closely adapted to variability in snow and ice hydrology. However, it is also relevant for the downstream livelihoods of Indus Delta fisher people imperilled by diversions upstream, wastewater discharge, and overfishing.

9.3.1.6 Economic Optimisation Perspective

Water, energy, and food systems have often been studied with multi-objective models for optimising net economic benefits. This perspective is closely related to the systems perspective, though the latter often employs simulation models, as well as optimisation. Optimisation modelling has a long history in the Indus Basin Irrigation System (from Lieftinck et al. 1968 to Yang et al. 2016). Interestingly, the difference between simulation and optimisation results in climate change modelling may be interpreted as a metric of adaptation (i.e., the best response the models can make to simulated first-order impacts). Optimisation modelling in the Indus Basin of Pakistan has placed increasing emphasis on exogenous economic variables, international trade, and subsidy effects in recent studies, though integration with national energy models has yet to occur (Dorosh et al. 2006; Yu et al. 2013).

One of the long-term aims of Water-Energy-Food Nexus research in Pakistan is to link and try to synthesise these multiple perspectives, though that has not been attempted to date in the Indus Basin or elsewhere. Several nexus studies have employed partially coupled models complemented by commissioned studies that draw together quantitative and qualitative analyses (e.g., Wescoat and Leichenko 1992; Yu et al. 2013).

9.3.2 *Levels of Analysis*

There are several levels of nexus analysis, which can be generalised in four main categories: the value level, normative level, pragmatic level, and empirical level (Keskinen et al. 2016):

9.3.2.1 Value Level

Nexus analysis has important underlying ethical and philosophical values that include integration, holism, equity, security, and so on. Natural resources management is primarily concerned with instrumental values, i.e., in which "resources," by definition, are means to various social ends. Environmental management and social policy may also make claims about intrinsic values (e.g., human rights, spiritual beliefs, deep ecology). There is increasing research on water ethics, agricultural ethics, and food ethics, less so energy ethics, but they have not been closely linked to one another. Pakistan, and all countries, have diverse, sometimes divergent, and conflicting value systems with respect to water, energy and food, which make it difficult to conceive of unified water-energy-food nexus values (Mustafa 2013).

9.3.2.2 Normative Level

Norms in this framework refer to laws, policies, politics, and planning processes that we collectively call institutions. Institutional norms stipulate what ought to happen, and in principle they are or should be closely related to the societal values noted above (Wescoat 2013). Institutions include unwritten "rules in use" as well as black-letter law and policy (Ostrom 1992). Institutional fragmentation across the water, energy, and food sectors thus calls for a normative level of institutional analysis and reform.

9.3.2.3 Pragmatic Level

This level is related to the practice of resource management, guided by core concepts, strategies, methods, and adjustments based on outcomes. Pragmatism is a philosophy of action in which resource users learn by doing, experimentation, and risk mitigation in ways that may alter institutional norms and values (Wescoat

1992). Briscoe et al. (2005) introduced the idea of "principled pragmatism" to Indus Basin planning, which is an over-simplified version of this level that addresses the most practicable reforms first. The pragmatist level has considerable potential for theoretical and methodological development in Pakistan.

9.3.2.4 Empirical Level

Observation, measurement, and analysis of resource patterns, processes, and linkages constitute an empirical level of analysis. Pakistan is relatively data rich in all three sectors of the nexus. It has had detailed research on data quality issues, and it is constantly developing new tools for measurement, analysis, and modelling (Wescoat et al. 2018). Akhter (2017) has questioned the politics of over-emphasising empirical analysis, while others have called for finer-grained data-driven management (Ahmad et al. 2013).

Although more fully developed than other levels, the empirical level is still underdeveloped with respect to all three pillars of the water-energy-food nexus. For example, water resources are primarily treated in terms of liquid supplies and rarely in terms of key water quality parameters relevant for irrigated agriculture (e.g., soil moisture and salinity) or hydropower (e.g., water temperature and turbidity). Snow and ice hydrology and their associated energy fluxes are included in Upper Indus snowmelt modelling, but those daily models are not yet directly linked with monthly IBMR modelling. Surface water–groundwater fluxes are estimated with gross mass balance calculations, though that may change with new satellite remote sensing data (e.g., Soil Moisture Active Passive (SMAP) satellite imagery at the regional scale, and drone survey data at the farm scale). Field research by the International Water Management Institute (IWMI) on the Hakra distributaries has gone furthest in integrated empirical irrigation studies (Awan et al. 2016; Shah et al. 2016).

Similarly, energy usually denotes the generation, distribution, and use of electricity and different fuel types in nexus research. Other important forms of energy are recognised but less often incorporated in nexus modelling. For example, solar and radiant energy are incorporated in crop models but less so in water-driven nexus models. Potential energy is central to hydropower planning and groundwater pumping, but is rarely quantified in nexus analysis (though see Siddiqi et al. 2012). Latent heat exchange is considered in snow and ice hydrology and evapotranspiration models, but rarely as energy fluxes in nexus modelling.

Food is often treated as gross production, hectarage, and/or yields for on-farm consumption and market sale. Water and energy footprints are increasingly incorporated in nexus research. However, food processing, storage, and cold chain require additional water and energy analyses that are rarely estimated in nexus research. Likewise, caloric energy and nutritional intake for human and animal metabolism and development warrant explicit treatment in nexus research (e.g., Zaidi et al. 2018).

It is interesting to consider the relationships among these four levels of analysis in contemporary international environmental policy. The United Nations Sustainable Development Goals (SDGs) address water, energy, and food security – with interventions at each level of analysis – but they do so separately for each sector. For example, the water sector is divided into a Water, Sanitation and Hygiene (WASH) subsector and marine ecosystem subsector, but not an irrigation subsector. One SDG focuses on "Affordable and Clean Energy," and another focuses on "Zero Hunger." Several SDGs are integrative across sectors and levels of analysis, e.g., climate action, responsible growth, and gender equity, but not the water-energy-food nexus.

These SDGs can only be achieved through long-term change in social values across sectors, which in turn require interventions at the normative level, i.e., legislation, policies and governance reform. Interventions for one resource should not compromise those for other resources, requiring innovative implementation and adjustment at the pragmatic level, such as incentives for water and energy technologies that promote resource recovery. The interdependent sectors of water, energy, and food, are arguably better managed with an integrated nexus approach than through 17 separate SDGs.

9.3.3 *Scales of Analysis*

Nexus research necessarily operates at multiple scales, but what does that mean? Scale can refer to the magnitude or frequency of physical phenomena, to the relative size of places, and/or to the hierarchy of places in social, economic, and political terms. Scale is thus a complex and politically contested subject (Mustafa 2007; Norman et al. 2012). All of its meanings and associated issues have relevance in the Indus Basin Irrigation System of Pakistan.

9.3.3.1 Greater Pakistan Scale

Indus Basin water supplies reportedly support 90% of staple crop and 100% of cash crop production in Pakistan (Ringler and Anwar 2013). Although many studies cite such figures, it is not clear how they address large areas of the Upper Indus, distant areas of Balochistan, and rain-fed (*barani*) areas of Punjab. The full scale of "Pakistan" in all of its territorial, hydrological, and socio-economic extensions is rarely addressed comprehensively. Remote areas of Pakistan may use less water and produce less food crops, but they have nexus linkages that are extremely important at local and regional scales. Water used for food production is also needed for hydropower energy production, municipal and industrial use, and ecosystem services, which create trade-offs that can benefit from a nexus approach, both to coordinate uses, recover resources, and find synergies. To push this argument further, all of Pakistan's resources – water, energy, food, climate, and more – have transboundary extensions and linkages, both physically in monsoon rainfall, runoff, and groundwater recharge; and economically in local and international food and energy markets. Climate change research in Pakistan also has macro-regional and global linkages.

9.3.3.2 Indus Basin Irrigation System Scale

With these larger scale perspectives in mind, we now scale down to the Indus Basin Irrigation System (IBIS) of Pakistan, reportedly the largest contiguously irrigated area in the world. Yang et al. (2016) studied the WEF nexus for agricultural production in the IBIS region. They used the Indus Basin Model Revised (IBMR) to focus on two conceptualisations of the nexus: (i) water-energy use for agriculture; and (ii) food production-hydro power generation trade-offs. They also studied the system under changing climate conditions, alteration of water allocation policies, and new infrastructure investments.

The IBMR model is subdivided into agro-climatic zones (ACZs) as the primary scale of water use and agricultural production, the results of which are aggregated to the provincial and basin scales. IBMR is an economic optimisation model in which water, energy and food produce economic values that the model maximises, subject to various physical and cost constraints. It should be noted that this scale of analysis and planning gained importance following the 1960 Indus Waters Treaty, which led to massive national and international irrigation infrastructure investment under the auspices of the Pakistan Water and Power Development Authority (WAPDA), culminating in the Water Sector Investment Planning Study (WSIPS 1990). While the IBMR has been used in many studies at the federal level, it is not clear how decision-makers have used its results. It remains analytically important but perhaps of diminishing significance for planning purposes following passage of the 18th Amendment to the Constitution of Pakistan in 2010, which devolved sector planning to the provincial scale.

9.3.3.3 Provincial Scale

Siddiqi and Wescoat (2013) studied the WEF nexus at the provincial scale, which has increased in significance following the 18th Amendment. They conceptualised the system as one of water and energy resource use for crop production purposes. This is an important aspect of much nexus research, in which food operates as the dependent variable, while water and energy are independent variables, i.e., in which the three pillars of the nexus are categorically different. In basins like the Indus, water is often treated as the primary independent variable for food production, while energy, capital, labour, and other technologies are both dependent (e.g., hydropower) and independent variables (e.g., pumping costs) associated with the water sector. Ghani et al. (2019) recently conducted a life cycle analysis of the water-energy-food nexus for bioethanol production in Punjab that incorporates water, carbon, and ecological footprints. Aggregating results to the provincial scale is important because water and food decisions are made by provincial decision makers. Basin-scale results are important for assessing impacts of climate change, water allocation polices, and infrastructure investments at the inter-provincial and federal scales.

Siddiqi and Wescoat (2013) conclude that while the provincial scale is highly relevant for policy formulation and public sector investment, resource analysis requires more rigorous baseline analysis at the farm scale as well. The IBMR was built upon a survey of farm production functions in the agro-climatic zones. The farm survey was recently updated by the World Bank's Water Sector Capacity Building and Advisory Services Project (WCAP) in the past decade though there are sampling challenges in a basin as large and varied as the Indus (Young et al. 2019). Recent farm-level irrigation productivity programs administered by provincial agricultural departments have focused on precision land levelling, micro-irrigation, and watercourse improvement with results that continuously change and improve farm production functions (World Bank 2017).

These impressive studies may miss an important point regarding nexus integration at the provincial scale by the Planning and Development Boards and Departments of Finance, not because they use nexus models, but rather because they compile sector analyses and proposals in order to allocate resources across sectors on an annual and medium-term basis (interviews with Mr. Muhammad Ahmed Bodla and Muhammad Jehanzeb Khan of the Punjab Planning and Development Board 2019). A detailed study of how those decisions are made would shed light on the provincial WEF nexus in practice.

9.3.3.4 Sub-Provincial Scales

Sub-provincial scales of nexus analysis pose a number of interesting, and challenging, issues. To date, these scales have included: (1) agro-climatic zones (ACZs used in the IBMR); (2) watersheds; (3) canal commands and distributaries; (4) divisions, districts, tehsils, and mouzas; and (5) electricity generation companies (4 GENCOs) and distribution company service areas (11 DISCOs). The differences between these sub-provincial planning regions pose some of the greatest challenges for nexus analysis and planning. For example, ACZs provide some integration of water and agricultural regions, but they are not official planning regions. Watershed planning is also unofficial, and has waxed and waned in the basin, with greatest applicability in mountainous (*kohi*) areas of Gilgit-Baltistan and hilly (*pahari*) areas of Balochistan and Punjab that are less well integrated with the IBIS. Provincial irrigation departments are subdivided into canal command, secondary, distributary, and watercourse offices. Perhaps the outstanding example of integrated local irrigation research is that of IWMI in the Hakra Branch of Fordwah Eastern Sadiqiya Canal Division in southern Punjab. This work has involved over three decades of modelling water deliveries, on-farm water management, and irrigation productivity (Awan et al. 2016; Shah et al. 2016).

A related field of research has focused gaps between sub-provincial scales of irrigation and agricultural administration. Historically, provincial agriculture and irrigation departments evolved in ways that created a 'great divide' at all scales, from the smallest *mogah* level at water course outlets to canal headworks and barrages. Irrigation departments are responsible for ensuring delivery of allocated water shares through their canal infrastructure, while agriculture is responsible for supporting farmers on how to best use water and other inputs for crop production at the farm level (Gill and Mushtaq 1998). There are massive applied scientific literatures on farm productivity in Pakistan, with linkages to credit, price, cost, and technologies (as well as energy and water inputs), which could extend the frontiers of nexus research. In an effort to address the separation between these two departments, On-Farm Water Management wings were created within provincial agriculture departments to encourage adoption and use of water conservation technologies by farmers. For example, the Punjab On-Farm Water Management (OFWM) wing subsidised water conservation technologies like sprinkler irrigation systems, drip irrigation and laser land levelling, some of which have transitioned to commercial development. Notwithstanding these advances, the OFWM wings in Punjab and Sindh continue to function with the historical understanding that watercourses and farms fall under the jurisdiction of the agriculture departments while all irrigation infrastructure above the watercourse is in the domain of the irrigation departments, that is, the 'great divide' continues to exist between the departments as it has since colonial times.

Efforts have also been made to improve irrigation operations and recovery of *abiana* (water charges) through participatory irrigation management (PIM), which established farmer organisations (FOs) and area water boards (AWBs) in Punjab and Sindh (Ahmad et al. 2020). These efforts were mostly limited to operational functions rather than longer-term strategic planning. Energy usage and management in agriculture, primarily for groundwater pumping, remains more of an individual farmer affair, with no effective provincial groundwater policies to date. While operational management occurs at all of the sub-provincial levels from divisions to districts, tehsils, and mouzas (villages), medium- and long-term planning remains limited to the provincial and federal levels (Shahid et al. 2019).

In summary, the water-energy-food nexus may be most effectively studied at three scales: first, at the farm scale where resources are collectively mobilised. Farm-scale results can then be jointly aggregated to the village, tehsil, and district hierarchy, on one hand; and to the distributary, secondary, and canal command irrigation hierarchy, on the other. Second, are the informal institutions emphasised by Elinor Ostrom (1992), which operate at all scales, in which day-to-day communications link irrigation, agricultural, and energy officers at the provincial and sub-provincial scales. The third promising scale for nexus research involves the federal and provincial departments of finance, budget, and planning and development where fiscal integration occurs to some extent.

9.3.4 *Summary*

To date, WEF nexus research in Pakistan has emphasised perspectives on the nexus as a system subject to economic optimisation but guided by discursive, governance, and political processes. Systems research has focused on optimising the net economic benefits of agricultural production, and on increasing resource efficiency and recovery. The resource recovery and sustainable livelihoods perspectives have unrealised potential.

Recent research has expanded its scope to encompass the normative and pragmatic levels of resource management, but it has yet to explore in depth the underlying values associated with resource use, conflict, and trade-offs. In this review, we drew attention to the shift in scale from IBIS basin level studies of water, food, and energy analysis that were prominent in the half-century following the 1960 Indus Waters Treaty, to the increasing importance of provincial and sub-provincial scales of analysis following the 18th Amendment to the Constitution in 2010. This shift underscores the importance of the governance perspective at all levels and scales of nexus research. While a number of governance experiments have been attempted, the split between irrigation and agricultural institutions continues to be hard-coded in ways that constrain WEF nexus policies. Even in India where all water agencies were brought under a single Jal Shakti Ministry in 2019, sectoral fragmentation persists on a pragmatic level.

9.4 WEF Nexus Challenges in Pakistan's Irrigated Agriculture Sector

This section returns to some of the irrigated agriculture challenges in Pakistan introduced above, and it indicates how a WEF nexus framework and approach could help address them. While aggregate national level estimates vary, the agriculture sector reportedly contributes 20–25% of Pakistan's Gross Domestic Product, and provides livelihoods to about 40% of the labour force (Ministry of Water Resources 2018). Total arable land in Pakistan is about 22 Mha out of which 17 Mha is irrigated, and 5 Mha land is rain-fed (Qureshi 2005). Thus, the irrigated agriculture sector is by far the largest water consumer in the country, with an estimated 90% of consumptive water used for irrigation purposes. Similar figures are reported for other arid and semi-arid regions of the world. Irrigated agriculture also consumes significant amounts of energy for pumping, processing, storage, and transport. Increasing the productivity of irrigation offers significant opportunities to secure more food, use water and energy more judiciously, and thereby meet challenges of improving food security, livelihoods, and poverty alleviation in Pakistan. Central to these aims is improving water distribution equity at the canal command and farm levels to increase food security. Minimising energy use for groundwater pumping, fertiliser production, weeding, and harvesting are also important.

Equity of water, energy, and food resource allocation and consumption is crucial. Resource inequities contribute to poverty and suffering, in which over 20% of Pakistan's population is food insecure, and 44% of its children under 5 are stunted due to poor maternal and early childhood nutrition (Planning Commission 2011). Below, we consider related challenges in the irrigated agriculture sector that may be alleviated if addressed with a water, energy and food nexus approach.

9.4.1 Low Crop Yields and Water Productivity

The inter-related problems of low crop yields and water productivity limit the attainment of sustainable livelihoods in Pakistan. Table 9.1 shows average yields for the four major crops in Pakistan compared with those of other countries. Wheat yield in Pakistan is lowest among all the countries listed, and half of the world's average yield. Similarly, rice and sugarcane yields are lower than the world averages. Cotton yield is a little higher than the world average but it is lower than that of Egypt, which has similar climatic conditions. Low yields are driven in part by low water productivity. Pakistan's water productivity for cereals is only one-third that of India (0.13 kg/m^3 in Pakistan, compared to the 0.39 kg/m^3 in India). Water productivity for wheat production in Pakistan (0.5 kg/m^3) is only half of that in India (1.0 kg/m^3). What can a nexus approach contribute?

Pakistan has had limited adoption of nexus innovations that involve low water use–high yield methods, such as the System of Rice Intensification (SRI) (SRI-Rice 2020). SRI methods are a good example of the WEF resource recovery perspective discussed above (Scott et al. 2015). Contrary to expectations, lower water and agricultural chemical inputs, when combined with scientific bed preparation, mulching and weeding practices, result in higher crop yields. On first glance it appears paradoxical that lower inputs can give higher yields, but the modified soil cultivation and labour inputs more than substitute for conventional irrigation inputs. SRI methods are now being extended to other crops, including wheat, sugarcane, small grains, and horticultural crops both in Pakistan and other regions (ibid.; Asif Sharif pers. comm. 2020; Pedaver Facebook page).

9.4.2 Low Staple and Export Crop Diversification

The agriculture sector consumes the largest share of total renewable water resources and a significant share of energy resources, which are used to produce a small number of staple crops. Wheat and rice contribute the largest share of caloric intake in the country. Wheat is a winter (*rabi*) season crop with relatively low irrigation requirements compared to rice. Rice is consumed domestically, and a significant portion is exported to earn foreign exchange reserves. Cotton is exported both in

Table 9.1 Average crop yields for countries [ton/ha] (Qureshi 2005)

Country	Wheat	Rice	Sugarcane	Cotton
Pakistan	2.5	3.07	47.8	1.92
India	4.9	2.97	69.0	0.85
Egypt	5.9	8.49	107.4	2.26
China	3.8	6.34	–	2.87
USA	6.5	6.69	64.2	1.79
World	4.8	3.83	64.7	1.62

bales and as cloth textile products. While there is value addition in the cotton crop, rice is mainly exported in its raw form. This raises concerns about whether return on exports reflects the full economic value of water and energy resources consumed, and whether the land and water used for surplus rice production could be used for other staple and cash crops to address Pakistan's nutrition needs. Water consumed in growing export crops is traded in terms of virtual water that is exported to other countries (see Konar et al. 2013). Production and trade of these crops could be regulated to reduce overall water stress (estimated at the high stress level of 1100 m^3/cap/year), promote sustainable use of water, and produce a more diverse mix of crops for domestic nutrition. However, at present, these nexus relationships are obscured by price distortions, limited regulation, and weak analytic links between resource use, crop diversification, and nutrition benefits.

9.4.3 Energy Use for Groundwater Pumping

Canal irrigation operates for the most part through gravity-fed potential energy, along with solar energy for plant growth. Electricity and fossil fuels (high speed and light diesel oil) are direct energy inputs in the irrigated agriculture sector (Siddiqi and Wescoat 2013). These inputs support agricultural mechanisation, cold storage, transport of crops to market, and groundwater pumping. It is estimated that the energy intensity for crop production in Punjab increased from 1.03 MJ/kg in 1994–95 to 1.86 MJ/kg in 2009–10; and that this increase was primarily due to increased energy used for pumping groundwater (ibid.). Energy use will continue to increase both in areas of declining freshwater groundwater levels, and low-lying areas of increased waterlogging and salinity, especially if government provides energy price subsidies, as has occurred in Punjab, India (Shah 2008). Higher energy pumping costs would significantly increase the price of staples, especially winter crops such as wheat, which depend on groundwater due to limited canal supplies in that season (Tahir and Habib 2001). A nexus strategy for these issues would link increased use of alternative energy sources, such as solar-powered tube-wells, high efficiency low water use micro-irrigation systems, and low input/high yield agricultural technologies such as SRI. Additionally, a nexus approach would take advantage of the changing role of canal irrigation, which is becoming a low-energy gravity-fed groundwater recharge system as well as a crop water delivery system in Pakistan. In the future, conjunctive management of surface water and groundwater may be as much about energy pumping cost management as it is about water supply management. From a WEF nexus perspective, if Pakistan's canal distribution system is purposely adapted to serve as a groundwater recharge mechanism, in contrast with its current accidental role, pumping lifts could be sustained at low levels. Irrigation energy costs could also be managed by recycling and reuse of municipal and agricultural waste streams. Additional innovations would be needed for low cost pumping, precision water delivery, plant water uptake, and soil salinity management, which may be the greatest long-term scientific, technological, and economic challenge. These nexus adjustments could usher in a new irrigated agriculture regime, and further reduce the need for large dams. The combination of glaciers and aquifers would serve as the main water storage and potential energy management zones.

9.4.4 Dated Surface Water Allocation Systems

There are four main tiers of surface water allocation in the Indus Basin, and each of them has a bearing on the WEF nexus: (i) distribution between India and Pakistan governed by the 1960 Indus Waters Treaty; (ii) distribution among provinces, governed by 1991 Water Apportionment Accord; (iii) distribution among canal commands within provinces, governed by historical allocations; and iv) distribution within command areas, governed by a rotational system of turns known as *warabandi*. The 1960 Indus Waters Treaty between India and Pakistan has proven remarkably resilient over the past 60 years, but it faces increasing political challenges from India and limited scientific or strategic attention within Pakistan. New questions are being raised about the unspecified water shares of Afghanistan on the Kabul River, and on competing hydropower development plans on the Jhelum and Chenab in the Jammu and Kashmir region. The former is being addressed through creative bilateral workshops, and the later through international arbitration and mediation processes specified in the Indus Waters Treaty. International cooperation has occurred in the fields of snow and ice hydrology and flood warning systems (e.g., ICIMOD 2020). Proposals for regional South Asian water cooperation have been linked to multi-track diplomacy for energy, food, and trade (World Bank 2020), but these are constrained by the current geopolitical context.

Each of these allocative institutions has proven remarkably resilient on a time scale of decades, but faces new stresses. Within the Indus Basin of Pakistan, surface water allocation has a substantial impact on water and energy use for food production. Climate change research has not examined the treaty-based depletion of the eastern rivers in detail, or the geopolitical uncertainties of other rivers. Yang et al. (2014) studied the impact of climate change on water allocation in the Indus Basin at the basin scale and the intra-provincial scale, taking ACZs as the primary scale for agricultural and economic analysis. Results suggested that while water allocation based on the 1991 Water Apportionment Accord would not hinder efficient allocation of water in the future, refining water allocations between the canal command areas within the provinces could bring higher economic benefits. Yang et al. (2016) also showed that climate change would tend to increase energy consumption, which could be mitigated by keeping surface water allocation policies flexible, and thus reducing energy needs.

Water allocation at the distributary level is more complex. Mustafa (2013) has drawn attention to the joint vulnerability of tail-end canal water users to water scarcity and flood hazards associated with actual vis-à-vis decreed allocations. IWMI's research in the Hakra Branch of the Fordwah Eastern Sadiqiya canal command has documented *waribandi* delivery dynamics, anomalies, and linkages in great detail, along with the role of tube-well pumping in risk reduction (Shah et al. 2016). New low-cost ultrasound flow measurement and telemetric data transmission technologies developed by LUMS Centre for Water Informatics and Technology are leading to further measurement, monitoring, and analytic advances in several locations of Punjab and more recently Sindh (Ahmad et al. 2013; Muhammad et al. 2016). Integration of these new data with on-farm energy, crop, and food budgets could help update surface water allocation systems, and launch the next generation of nexus production functions, research, and management (Vinca et al. 2020).

9.4.5 Lack of a Regulatory Groundwater Policy Framework

Surface water rights have a long and detailed, if contested, history, but groundwater has just the opposite problem. Groundwater extraction in Pakistan is largely unregulated. It has increased dramatically in the past 40 years, especially, in the Punjab and Balochistan provinces, with over a million private tube-wells in the former province alone. This development has enabled increased cropping intensity and improvement in livelihoods, but it has come at substantial costs. In parts of the major *doabs* in Punjab, for example, groundwater depletion is occurring, and in some areas farmers are forced to use marginal quality groundwater due to over-pumping, aggravated by inequities in canal water deliveries. Increasing use of groundwater pumping where salinity levels are high poses cumulative risks for irrigated cropping systems. Groundwater processes are estimated from monitoring well records (rather than measurement of tube-well withdrawals) and remote sensing of irrigated soil moisture flux. Groundwater management is further constrained by limited institutional capacity for implementing regulations. The absence of groundwater management means that groundwater pumping rates and costs are uncertain, and their contributions to agricultural productivity are also uncertain outside of specific research sites. For instance, Punthakey et al. (2015) estimate the sustainable yield for Rechna Doab as 10 ± 1 BCM. However, there is no institutional mechanism at present to monitor these withdrawals or their impacts. Policy mechanisms that enable community-based aquifer management and co-ownership of aquifer shares (e.g., in a system of correlative rights) could help the transition to sustainable groundwater management. Using energy prices and food policies to help achieve these ends would render this a nexus approach.

9.4.6 Adaptation of a Climate-and-Society Smart Irrigated Agriculture

Water-energy-food nexus issues in Pakistan are all subject to climate variability and change. Those farmers located in the mid portions of the *doabs* in Punjab who are using marginal quality groundwater are adapting by using groundwater for three irrigations and canal water for the last irrigation to flush salts out of the root zone. Will that adaptive strategy remain feasible in conditions of climate change? Unregulated pumping of marginal quality groundwater results in soil salinity and accumulation of salts in the plant root zone. This can cause secondary salinisation and land degradation and reduce agricultural production. These irrigation problems are amplified by increasing average temperatures, and their management is disrupted by extreme hydro-meteorological events (Yu et al. 2013; Yang et al. 2014).

In the eastern *doabs* of Punjab, farmers are adopting laser land levelling, raised bed cultivation and other climate-smart agriculture options that use less water for higher yields (see Pedaver Facebook page). These are no-regrets innovations that pay off under current as well as forecasted climate conditions. Imran et al. (2018) found a significant difference in the costs of production and benefits (farm income) between adopters of Climate Smart Agriculture (CSA) versus conventional cotton farmers. Faced with decreasing profit margins and higher input costs, a new movement of 'Nature-Inspired Farming' called Paradoxical Agriculture is also taking off (LUMS Centre for Water Informatics and Technology 2020). Using principles of regenerative agriculture, which relies upon minimum or no tillage, extensive use of mulching, cover crops, soil moisture conservation, raised bed farming, and investment in developing soil organic matter, Paradoxical Agriculture seeks to optimise the WEF nexus at farm level. A key reason for the methodology's spread has been its claims of resilience to disturbance conditions (e.g. pests, weeds, changing weather patterns), as well as propagation of crop intensification technologies and reduced use of inputs like weedicide and fertiliser, thereby reducing input costs and increasing profit margins for farmers. A number of trials are underway to study the success of this methodology in achieving these objectives and optimising nexus management at the farm level (see Pedaver Facebook page for examples). Future efforts need to explain the nexus resource recovery mechanisms in these emerging lines of climate smart agriculture in Pakistan.

9.4.7 Governing the WEF Nexus

Given the substantial number of WEF nexus issues at all levels from the farm level to district, province, basin, and country, there is a need for a multi-scale governance approach. As noted above, the provincial planning and development boards already play an important role in rationalising water, energy and agriculture sector programs and budgeting. The current IWRM-based approach to water resource management, by comparison, does not do justice to agriculture and energy components of the nexus. Within each sector, governance issues vary. For example, canal irrigation is managed on a supply-based paradigm of rotational deliveries that is suboptimal for agriculture water management, food production, and food security. Groundwater use, by comparison, is demand-driven for those who can afford it, but with no public regulation at all. Each aspect of agricultural prices, marketing, inputs, and processing has different public and private governance structures, as does each major component of the energy sector. Departmental mechanisms for coordination among agencies focus on day-to-day operational conflicts, e.g. addressing issues of water theft, rather than long-term strategies and planning. These mechanisms are considered weak and ineffective by stakeholders across the board.

Attempts to manage the nexus currently happen in a top-down manner, with irrigation departments setting parameters within which food and energy considerations are addressed (Shahid et al. 2019). Depending on the amount of water available to each farmer, he/she has to choose methods of irrigation at the farm level, make appropriate crop choices, and determine how much energy to invest in groundwater abstraction. As noted above, nexus integration in practice occurs to some degree at the farm level. Governance of the water-energy-food nexus requires a rebalancing of the water, energy, and food components of irrigated agriculture, more rigorous use of farm-level experience, and a more level playing field for experts, planners and managers in each sector to develop holistic solutions.

Attempts at participatory irrigation management (PIM) and area water boards (AWBs) moved some distance in this direction by including farmers in the operation of irrigation systems instead of limiting their role to the farm level. However, consensus across the board has been that without support from the irrigation department's engineering expertise, and clear financial incentives and benefits, participatory irrigation management experiments have not been successful (Ahmad et al. 2020). Irrigation infrastructure in the Indus Basin System has hard-coded inequalities in its very design that developed by privileging certain local elites to act as the hinge between the communities and the state (Gilmartin 1994, 2020). Devolving operational powers to farmer organisations (FOs) often replicates local power structures and reinforces existing inequalities that put tail-end canal irrigators at even greater disadvantage. Keeping this risk in historical perspective, it may be noted that as far back as 1998 Canal Councils were proposed at the tehsil level where representatives of the agriculture department, irrigation department, and farmers could participate on an equal basis to improve agricultural productivity (Gill and Mushtaq 1998).

In an effort to reconstitute participatory irrigation management through infrastructure development, the Government of Punjab devised Irrigation Management Units (IMUs) as part of the Lower Bari Doab Canal Improvement Project (Asian Development Bank 2006). These went a step further than the previous AWBs and FOs by having multiple functions, including operations and maintenance (O&M), engineering guidance, conjunctive management of groundwater, on-farm water management strategies, and agricultural services (Asian Development Bank 2006). In addition to improving irrigation infrastructure, the project sought to set up demonstration farms, improve measurement of groundwater usage, and disseminate water conservation technologies in ways that address all aspects of the irrigated agriculture nexus. While the program made progress in its engineering and infrastructure development components, its 2018 completion report explained that the, "institutional structure of the arrangement at appraisal changed from project financed IMUs to the PIDA, PID, and PMU" (i.e. Project Management Unit) (Asian Development Bank 2018, p. 6). It appears that this change in the design of the IMU from being a multi-sectoral body with representation from PID and OFWM to a simpler version where the irrigation department provided direct support to the AWBs diluted the nexus governance potential of these participatory farmer bodies.

While the fate of the IMUs is yet to be determined, there is a longstanding need for multi-stakeholder governance at the meso-scale to strengthen horizontal links between the irrigation and agriculture departments (incorporating energy resources), and vertical links between the micro (farm-level) and macro (provincial-level) scales, a need identified by all stakeholders to achieve holistic problem solving of the water-energy-food nexus in irrigated agriculture of Pakistan. Linking districts, distributaries, and DISCOs at the meso-scale is a major priority. With detailed data now available from irrigation departments on canal flows and groundwater levels; from the agriculture departments on soil fertility, yields, inputs, and prices; and from energy departments on access, use, and outages, it is possible to develop integrated governance strategies as far down as the tehsil and distributary levels. Boundary spanning organisations like the IMUs that combine cross-sectoral indicators (like resource recovery instead of efficiency) will be key to overcoming the "great divide" among the irrigation, agriculture, and energy departments and enabling a move towards nexus-based governance.

9.5 Distributaries, Districts and DISCOs: A Way Forward for Implementing WEF Nexus Thinking in the Irrigated Agriculture Sector

Experience indicates that implementation of WEF nexus ideas and methods will not be straightforward. It will require that current problems be more deeply understood in historical, theoretical, political, and practical terms. Irrigation nexus problems also need to be mapped with appropriate scales, levels, and perspectives to envision more productive and equitable outcomes. In this section, we address the challenges presented in this chapter in terms of nexus perspectives, levels and scales relevant for Pakistan.

Problems of low crop yield, diversification, and water productivity can be approached through a combination of the resource recovery, systems analysis, and governance perspectives discussed above. Analysis of water and energy inputs, when analysed from multiple perspectives can help farmers and policy makers decide on crop choices and methods for improving returns, minimising resource costs, and distributing benefits equitably. That approach requires explicit assessment of the ethical values associated with water, energy, and food – individually and collectively – as part of the governance perspective, normative regulations, and pragmatic strategies for implementation. This type of implementation framework has been explored, to some extent, in the emerging discourse perspective, but it has not been addressed comprehensively to date (Briscoe et al. 2005; Kamal et al. 2012; Mustafa 2013; Ministry of Water Resources 2018; Stewart et al. 2018; Yang et al. 2014; Young et al. 2019; Yu et al. 2013). The current discourse focuses primarily on water-energy nexus relations and IWRM, which are becoming reasonably well-established, and now need to be extended to food and climate (cf. Siddiqi and Anidon 2011).

It is not difficult to envision a new analytic approach to these nexus issues. One strategy would incorporate more energy variables, trade-off functions, and distributive constraints into existing irrigation optimisation models like the IBMR. A related step would be to spatially link the canal commands with ACZ and district subregions in the IBMR model. This step would integrate agricultural and energy data available at the district and tehsil scales with hydrologic and irrigation data in the irrigation model (e.g., as in Kirby and Mobin-ud-Din 2016; Stewart et al. 2018). A third, pragmatic, strategy involves coupled modelling, where the outputs from one model (e.g., the national energy model) become inputs for the IBMR model, and vice versa. Coupled modelling approaches could also include linkages between daily time step models for flood hazards, snowmelt, and energy demand models, with monthly time step models used for water resources planning and climate adaptation. Another valuable option involves commissioning qualitative studies of nexus issues that are pragmatically weighed with quantitative model outputs (e.g., Wescoat and Leichenko 1992).

These nexus modelling strategies are exciting, but can only go so far. They depend on advances at the empirical level, some of which may seem "paradoxical" (LUMS Centre for Water Informatics and Technology 2020; Pedaver Facebook page). For example, the greatest empirical need for the nexus approach incorporates on-farm innovations that lie outside current model calibrations. The IBMR model was based on detailed farm surveys of variables and processes that are now obsolete (e.g., draft animal labour). The World Bank WCAP project sought to update those surveys, but the sampling challenges in a basin as large and diverse as the Indus are formidable. How can they incorporate new experiments in crop intensification such as SRI (that are extending to many other crops including high value horticultural crops)? Empirically these practices seem paradoxical because they suggest that lower inputs yield higher outputs (Asif Sharif, pers. comm. 2020). However, a more comprehensive analysis would reveal that these innovative practices involve: (1) inputs that are not presently incorporated in crop models (e.g., raised bed preparation, mulching, and other cultivation practices); (2) resource recovery (e.g., of plant available minerals through soil biota activity, soil moisture conservation through lateral furrow irrigation and mulching, etc.); (3) more detailed energy fluxes (e.g., solar insolation, soil temperature regulation, and their phenological effects); and (4) crop yield mechanisms yet to be discovered. This empirical knowledge may require a complete overhaul of Indus Basin nexus models.

Scientific and analytical innovations are exciting, but their complexity is matched by that of nexus water governance challenges. From a pragmatic perspective, even day-to-day coordination between water, energy, agricultural, and administrative departments require work that few seem willing to expend beyond the minimum required or commanded. On a normative level, institutional fragmentation within the water sector, as well as with other sector agencies, has been described around the world for at least a century. Nominally, agencies would appear to agree, and they do come together for special studies when required. But in practice, the gaps between agencies continue to exceed their linkages. Institutional analysts focus on the lack of incentives and rewards for horizontal and vertical coordination (compared with the risks). Some problems may stem from the top-down hierarchical culture of sector agencies that involve competition for resources, influence, and power among political appointees and civil servants at the top, and strategic obedience of those below (see the diverse studies of these phenomena by Freeman et al. 1989; Ostrom 1992; Shahid et al. 2019; Uphoff 1992). Uphoff (1992) described irrigation reforms that required a decade of experimental intervention. Ostrom (1992) stressed that no amount of formal regulations and policy documents will redirect the *informal* rules that drive institutional behaviour, which she termed "rules in use".

In addition to governance and institutions, greater attention must be given to social and environmental values that shape water, energy and food systems. What are these values? They are not articulated at a deep level in IWRM studies, the National Water Policy, or WEF nexus research. Values are a major nexus research frontier, in part because they identify principles and justifications for social change, and social change is what is needed in the Indus Basin of Pakistan, and in every river basin around the world. In this paper, we have traced the roots of WEF nexus fragmentation back to colonial irrigation and agricultural revenue policies, policies associated with social manipulation, conflict, environmental degradation, and disasters for at least 100 years in the provinces of Punjab and Sindh (Akhter 2015; Gilmartin 2020; Mustafa 2013; Naqvi 2013). Sectoral and federal fragmentation persisted in the post-independence period in Pakistan, as they have in India, the US, and elsewhere. Even in India, where all water agencies were recently brought under the umbrella of a single Jal Shakti Ministry, institutional fragmentation in Ostrom's terms persists, vertically as well as horizontally. Sectoral integration is thus not just an administrative and normative challenge. It is a cultural problem that probably requires decadal and generational processes of social change.

Generational change, in turn, depends upon new modes of social learning of unprecedented scale and commitment. What changes might be possible if currently required courses in "Pakistan Studies" focused on water, energy and food values for the Indus River basin? Students would learn about ideals and methods for advancing each of the perspectives, levels, scales of the WEF nexus for the benefit of all people, organisms, and habitats in Pakistan, and beyond.

References

Ahmad M, Kutcher GP (1992) Irrigation planning with environmental considerations: a case study of Pakistan's Indus Basin. World Bank Technical Paper No. 166. The World Bank, Washington, DC

Ahmad Z, Asad EU, Muhammad A et al (2013) Development of a low-power smart water meter for discharges in Indus Basin irrigation networks. In: Shaikh FK, Chowdhry BS, Ammari HM et al (eds) Wireless sensor networks for developing countries. Springer, Berlin/Heidelberg

Ahmad B et al (2020) Impact of institutional features on the overall performance assessment of participatory irrigation management: farmers' response from Pakistan. Water 12:497. https://doi.org/10.3390/w12020497

Akhter M (2015) Infrastructure nation: state space, hegemony, and hydraulic regionalism in Pakistan. Antipode 47(4):849–870. https://doi.org/10.1111/anti.12152

Akhter M (2017) Desiring the data state in the Indus Basin. Trans Inst Br Geogr 42:377–389. https://doi.org/10.1111/tran.12169

Albrecht TR, Crootof A, Scott CA (2018) The water-energy-food nexus: a systematic review of methods for nexus assessment. Environ Res Lett 13(4):043002. https://doi.org/10.1088/1748-9326/aaa9c6

Ali I (1988) The Punjab under imperialism, 1885–1947. Princeton University Press, Princeton

Arfan M, Ansari K, Ullah A et al (2020) Agenda setting in water and IWRM: discourse analysis of water policy debate in Pakistan. Water 12(6):1656. https://doi.org/10.3390/w12061656

Asian Development Bank (2006) Punjab irrigated agriculture investment program. https://www.adb.org/projects/37231-013/main. Accessed 4 Aug 2020

Asian Development Bank (2018) Completion report: Punjab irrigated agriculture investment program. https://www.adb.org/projects/documents/pak-37231-013-pcr. Accessed 4 Aug 2020

Awan UK, Anwar A, Ahmad W, Hafeez M (2016) A methodology to estimate equity of canal water and groundwater use at different spatial and temporal scales: a geo-informatics approach. Environ Earth Sci 75:409. https://doi.org/10.1007/s12665-015-4976-4

Bedi BPL (2003) Harvest from the desert: the life and work of Sir Ganga Ram. National College of Arts, Lahore

Benson D, Gain A, Rouillard J (2015) Water governance in a comparative perspective: from IWRM to a 'nexus' approach? Water Alter 8(1):756–773

Biggs EM, Bruce B, Boruff B et al (2015) Sustainable development and the water–energy–food nexus: a perspective on livelihoods. Environ Sci Policy 54:389–397. https://doi.org/10.1016/j.envsci.2015.08.002

Briscoe J, Qamar U, Contijoch M et al (2005) Pakistan's water economy: running dry. World Bank, Washington, DC

Crooke W (1989 reprint) A glossary of north Indian peasant life. Oxford University Press, New Delhi

Diamond J (2005) Collapse: how societies choose to fail or succeed. Penguin, London

Dorosh P, Niazi MK, Nazli, H (2006). A social accounting matrix for Pakistan, 2001-02: methodology and results. PIDE Working Paper No. 2006:9, Pakistan Institute of Development Economics (PIDE), Islamabad

Duloy JH, O'Mara GT (1984) Issues of efficiency and interdependence in water resource investments: lessons from the Indus Basin of Pakistan. World Bank Staff Working Paper 665, World Bank, Washington, DC

Ferroukhi R, Nagpal D, Lopez-Peña A et al (2015) Renewable energy in the water, energy and food nexus. IRENA, Abu Dhabi

Freeman D, Bhandarkar V, Sinn E et al (1989) Local organizations for social development: concepts and cases of irrigation organization. Westview Press, Boulder

Ghani HU, Silalertruksa T, Gheewala SH (2019) Water-energy-food nexus of bioethanol in Pakistan: a life cycle approach evaluating footprint indicators and energy performance. Sci Total Environ 687:867–876. https://doi.org/10.1016/j.scitotenv.2019.05.465

Gill MA, Mushtaq K (1998) Managing irrigation for environmentally sustainable agriculture in Pakistan: case study of Agriculture Department of Punjab province. IIMI Report No. C-!0, International Irrigation Management Institute, Lahore

Gilmartin D (1994) Scientific empire and imperial science: colonialism and irrigation technology in the Indus Basin. J Asian Stud 53(4):1127–1149. https://doi.org/10.2307/2059236

Gilmartin D (2003) Water and waste: nature, productivity and colonialism in the Indus Basin. Econ Polit Wkly 38(48):5057–5065

Gilmartin D (2020) Blood and water: the Indus River Basin in modern history. University of California Press, Berkeley

Giordano M, Shah T (2014) From IWRM back to integrated water resources management. Int J Water Res Dev 30(3):364–376. https://doi.org/10.1080/07900627.2013.851521

Hassan et al (2019) Future water demand modelling using water evaluation and planning: a case study of the Indus Basin in Pakistan. Sustain Water Resour Manag 5:1903–1915. https://doi.org/10.1007/s40899-019-00343-0

ICIMOD (2020) Indus Basin Initiative. International Centre for Integrated Mountain Development (ICIMOD), Kathmandu, Nepal. https://www.icimod.org/initiative/indus-basin-initiative/. Accessed 27 July 2020

Imran MA, Ali A, Ashfaq M et al (2018) Impact of Climate Smart Agriculture (CSA) practices on cotton production and livelihood of farmers in Punjab, Pakistan. Sustainability 10(6):2101. https://doi.org/10.3390/su10062101

International Resources Group (2010) Pakistan Integrated Energy Model (Pak-IEM). Model Design Report, Volume 1. Report prepared for Asian Development Bank, Islamabad, Pakistan. https://www.decisionwaregroup.com/assets/pakiem_model-design-report_final-update.pdf. Accessed 25 Jul 2020

Kamal S, Amir P, Mohtadullah K (2012) Development of Integrated River Basin Management (IRBM) for Indus Basin: challenges and opportunities. WWF-Pakistan, Lahore

Keskinen M, Guillaume JHA, Kattelus M et al (2016) The water-energy-food nexus and the transboundary context: insights from large Asian rivers. Water 8(5):193. https://doi.org/10.3390/w8050193

Kirby M, Mobin-ud-Din A (2016) Time series (1980–2012) crop areas and production in the districts and canal commands of Pakistan. A report prepared under the Sustainable Development Investment Portfolio of the Department of Foreign Affairs and Trade, CSIRO Land and Water, Australia

Konar M, Hussein Z, Hanaski N et al (2013) Virtual water trade flows and savings under climate change. Hydrol Earth Syst Sci 17:3219–3234. https://doi.org/10.5194/hess-17-3219-2013

Kurian M, Dietz T (2013) Leadership on the commons: Wealth distribution, co-provision and service delivery. J Dev Stud 49(11):1532–1547

Lenton R, Muller M (2009) Integrated water resources management in practice: better water management for development. Routledge, New York

Lieftinck P, Sadove AR, Creyke TC (1968) Water and power resources of West Pakistan. Johns Hopkins University Press, Baltimore

Liu J, Hull V, Godfray HCG et al (2018) Nexus approaches to global sustainable development. Nat Sustain 1(9):466–476. https://doi.org/10.1038/s41893-018-0135-8

LUMS Centre for Water Informatics and Technology (2020) Workshop on Nature Inspired Sustainable Agriculture. LUMS Centre for Water Informatics and Technology (WIT) https://wit.lums.edu.pk/news/general-news/wit-holds-workshop-nature-inspired-sustainable-agriculture. Accessed 27 July 2020

Maass A, Hufschmidt MM, Dorfman R et al (eds) (1962) Design of water resource systems. Harvard University Press, Cambridge

Meadows A, Meadows P (eds) (1999) The Indus River: biodiversity, resources, humankind. Oxford University Press, Karachi

Michel AA (1967) The Indus Rivers: a study of the effects of partition. Yale University Press, New Haven

Ministry of Water Resources (2018) National water policy and water charter. Ministry of Water Resources, Government of Pakistan. https://mowr.gov.pk/index.php/national-water-policy-2018/. Accessed 10 Aug 2020

Muhammad A, Haider B, Ahmad Z (2016) IoT enabled analysis of irrigation rosters in the Indus Basin Irrigation System. Procedia Engineering 154:229–235. https://doi.org/10.1016/j.proeng.2016.07.457

Mustafa D (2007) Social construction of hydropolitics: the geographical scales of water and security in the Indus Basin. Geogr Rev 97(4):484–501. https://doi.org/10.1111/j.1931-0846.2007.tb00408.x

Mustafa D (2013) Water resource management in a vulnerable world: the hydro-hazardscape of climate change. I.B. Tauris, London

Naqvi SA (2013) Indus waters and social change: the evolution and transition of agrarian society in Pakistan. Oxford University Press, Karachi

Nepal S, Neupane N, Belbase D et al (2019) Achieving water security in Nepal through unravelling the water-energy-agriculture nexus. Int J Water Res Dev. https://doi.org/10.1080/07900627.2019.1694867

Norman ES, Bakker K, Cook C (2012) Introduction to the themed section: water governance and the politics of scale. Water Alter 5(1):52–61

Ostrom E (1992) Crafting institutions for self-governing irrigation systems. ICS Press, San Francisco

Perreault T, Bridge G, McCarthy J (2015) The Routledge handbook of political ecology. Routledge, New York

Planning Commission (2011) National nutrition survey. Planning Commission, Government of Pakistan, Pakistan Institute of Development Economics, Islamabad

Punthakey JF, Khan MR, Riaz M et al. (2015) Optimising canal and groundwater management to assist water user associations in maximizing crop production and managing salinisation in Pakistan: Assessment of groundwater resources for Rechna Doab, Pakistan. Final Report to ACIAR, Ecoseal

Qureshi AS (2005) Climate change and water resources management in Pakistan. In: Mirza MMQ, Ahmad QK (eds) Climate change and water resources in South Asia. Taylor & Francis, London, pp 197–230

Ringler C, Anwar A (2013) Water for food security: challenges for Pakistan. Water Int 38(5):505–514. https://doi.org/10.1080/02508060.2013.832122

Scoones I (2009) Livelihoods perspectives and rural development. Journal Peasant Studies 36(1):171–296. https://doi.org/10.1080/03066150902820503

Scott CA (2013) Resource-use efficiency, once paradigm now paradox: the socio-hydrology of waste. Paper presented at the AAAS 2013 Annual Meeting, Boston, 15 Feb 2013. https://aaas.confex.com/aaas/2013/webprogram/Paper9286.html. Accessed 4 Aug 2020

Scott CA, Kurian M, Wescoat JL Jr (2015) The water-energy-food nexus: enhancing adaptive capacity to complex global challenges. In: Kurian M, Ardakanian R (eds) Governing the nexus: water, soil and waste resources considering global change. Springer, Cham, pp 15–38. https://doi.org/10.1007/978-3-319-05747-7_2

Shah T (2008) Taming the anarchy: groundwater governance in South Asia. Routledge, New Delhi

Shah MAA, Anwar AA, Bell AR, Zia UH (2016) Equity in a tertiary canal of the Indus Basin Irrigation System (IBIS). Agric Water Manag 178:201–214. https://doi.org/10.1016/j.agwat.2016.09.018

Shahid A, Siddiqi A, Wescoat JL Jr (2019) Reimagining the planning of irrigation and agriculture in Punjab, Pakistan. In: Khan S, Adams T (eds) Indus River Basin: water security and sustainability. Academic, New York, pp 89–415

Siddiqi A, Anidon L (2011) The water–energy nexus in Middle East and North Africa. Ener Policy 39:4529–4540

Siddiqi A, Wescoat JL Jr (2013) Energy use in large-scale irrigated agriculture in the Punjab province of Pakistan. Water Int 38(5):571–586. https://doi.org/10.1080/02508060.2013.828671

Siddiqi A, Wescoat JL Jr, Humair S, Afridi K (2012) An empirical analysis of the hydropower portfolio in Pakistan. Energy Policy 50(C):228–241. https://doi.org/10.1016/j.enpol.2012.06.063

SRI-Rice (2020) SRI International Network and Resources Center (SRI-Rice). Cornell University. http://sri.ciifad.cornell.edu/. Accessed 12 July 2020

Stewart JP, Podger GM, Ahmad MD et al. (2018) Indus River System Model (IRSM) – a planning tool to explore water management options in Pakistan: model conceptualisation, configuration and calibration. South Asia Sustainable Development Investment Portfolio (SDIP) Project Technical report, CSIRO, Australia

Tahir Z, Habib Z (2001) Land and water productivity: trends across Punjab canal commands, Pakistan, IWMI Working Paper No. 14. International Water Management Institute (IWMI), Colombo. https://doi.org/10.3910/2009.155

Uphoff N (1992) Learning from Gal Oya: possibilities for participatory development and post-Newtonian social science. Cornell University Press, New York

van Steenbergen F (2020) Groundwater and surface water in the mega-irrigation systems of Pakistan: the case for conjunctive management. Water Knowledge Note. World Bank, Washington, DC. http://documents.worldbank.org/curated/en/174121579803266899/Groundwater-and-Surface-Water-in-the-Mega-Irrigation-Systems-of-Pakistan-The-Case-for-Conjunctive-Management

Vinca A, Parkinson S, Byers E et al (2020) The NExus Solutions Tool (NEST) v1.0: an open platform for optimizing multi-scale energy-water-land system transformations. Geosci Model Dev 13:1095–1121. https://doi.org/10.5194/gmd-13-1095-2020

Wescoat JL Jr (1992) Common themes in the work of Gilbert White and John Dewey: a pragmatic appraisal. Ann Assoc Am Geogr 82:587–607. https://doi.org/10.1111/j.1467-8306.1992.tb01719.x

Wescoat JL Jr (2013) Reconstructing the duty of water: a study of emergent norms in socio-hydrology. Hydrol Earth Syst Sci 17:4759–4768. http://www.hydrol-earth-syst-sci.net/17/4759/2013/hess-17-4759-2013.pdf

Wescoat JL Jr, Leichenko RM (1992) Complex river basin management in a changing global climate: Indus River Basin case study in Pakistan, a national modelling assessment. Collaborative Paper No. 5, Center for Advanced Decision Support for Water and Environmental Systems (CADSWES), Boulder

Wescoat JL Jr, Halvorson S, Mustafa D (2000) Water management in the Indus Basin of Pakistan: a half-century perspective. Int J Water Res Dev 16:391–406

Wescoat JL, Siddiqi A, Muhammad A (2018) Socio-hydrology of channel flows in complex river basins: rivers, canals, and distributaries in Punjab, Pakistan. Water Resour Res 54(1):464–479. https://doi.org/10.1002/2017WR021486

World Bank (2017) Pakistan – Punjab irrigated agriculture productivity improvement program project: additional financing. World Bank, Washington, DC. http://documents.worldbank.org/curated/en/448911512270028837/Pakistan-Punjab-Irrigated-Agriculture-Productivity-Improvement-Program-Project-additional-financing

World Bank (2020) South Asia Water Initiative (SAWI). https://wwwworldbankorg/en/programs/sawi Accessed 27 July 2020

Wright R (2009) The ancient Indus: urbanism, economy and society. Cambridge University Press, Cambridge

WSIPS (1990) Water sector investment planning study. 5 vols. WAPDA and World Bank, World Bank Publications

Yang YCE, Brown C, Yu W et al (2014) Water governance and adaptation to climate change in the Indus River Basin. J Hydrol 519(PC):2527–2537. https://doi.org/10.1016/j.jhydrol.2014.08.055

Yang YE, Ringler C, Brown C, Mondal MAH (2016) Modeling the agricultural water–energy–food nexus in the Indus River Basin, Pakistan. J Water Resour Plan Manag 142(12):04016062

Young W, Anwar A, Bhatti T et al. (2019) Pakistan: getting more from water. World Bank, Washington, DC. https://openknowledge.worldbank.org/handle/10986/31160

Yu W, Yang YC, Savitsky A et al (2013) The Indus Basin of Pakistan: The impacts of climate risks on water and agriculture, Directions in development series. World Bank, Washington DC

Zaidi S, Bhutta Z, Hussain SS, Rasanathan K (2018) Multisector governance for nutrition and early childhood development: overlapping agendas and differing progress in Pakistan. BMJ Glob Health 3(Suppl 4):e000678. https://doi.org/10.1136/bmjgh-2017-000678

Zimmerer K, de Haan S (eds) (2019) Agrobiodiversity: integrating knowledge for a sustainable future. MIT Press, Cambridge

Chapter 10
Groundwater Crisis: A Crisis of Governance?

Fazilda Nabeel

Abstract This chapter delves into Pakistan's groundwater situation by engaging in a historical case study of the province of Punjab from the colonial to the contemporary period. Drawing on archival research, document analysis and interviews conducted as a part of the author's doctoral research, Punjab's groundwater crisis is analysed to make two key arguments. First, despite Punjab's endowment of surface water, the province is heavily dependent on groundwater for meeting its agricultural, domestic and industrial water demand. Second, in contrast to a conventional narrative that characterises Pakistan's groundwater crisis as one arising from the absence of groundwater governance, this chapter argues that the crisis was brought about by the *very nature of groundwater governance* in this region over time. Through a careful historical analysis of the techniques of groundwater governance in colonial, post-colonial and contemporary Punjab, the chapter highlights the strategic importance of groundwater to achieve Pakistan's development goals, and in turn how this has shaped the nation's approach towards groundwater use and management over time. The paper concludes that the nature of groundwater governance in Pakistan over time has been focused on groundwater exploitation for agricultural expansion and economic development, with little emphasis on sustainable management of the resource.

Keywords Groundwater governance · Governance challenge · Depletion and contamination · Management approaches · Groundwater use

F. Nabeel (✉)
Department for International Development, University of Sussex, Brighton, UK
e-mail: f.nabeel@sussex.ac.uk

M. A. Watto et al. (eds.), *Water Resources of Pakistan*, World Water Resources 9,
https://doi.org/10.1007/978-3-030-65679-9_10

10.1 Pakistan's Groundwater Crisis in the Global Context

Groundwater accounts for about 97% of the available global non-frozen freshwater resources (IGRAC 2015), contributing over 40% of global food production, and 50% of global drinking water requirements (IGRAC 2015). Global groundwater consumption is continuously growing with a 1–3% increase in annual use. Between 1960 and 2000, groundwater use more than doubled, from 312 km^3 to 734 km^3 per year (Fienen and Arshad 2016). Pakistan's dependency on groundwater follows the global trend, increasing from 8% of total irrigation water use in 1960 to 40% in 1996, and reaching 60% in 2006 (Briscoe et al. 2005). While these numbers are averages for the whole of Pakistan, groundwater dependency for irrigation varies considerably across the country based on surface water availability and groundwater quality. In areas where surface water supplies are intermittent or unreliable, groundwater forms up to 80% of irrigation water requirements.

Pakistan has become the fourth largest abstractor of groundwater globally, primarily withdrawing from the Indus Basin aquifer, the second most overstressed groundwater basin in the world (Margat and van der Gun 2013; Buis and Wilson 2015). Groundwater in Pakistan is primarily used for agriculture (94%) with 6% for domestic use and the industrial sector. It provides a buffer against variability in the monsoon rainfall, which occurs for a short period in summer. In 2016, there were 1,385,000 tube-wells in Pakistan, predominantly located in the Punjab province (Ministry of Finance 2019). Existing studies highlight rapidly deteriorating groundwater in some areas of Pakistan, both because the annual rate of abstraction exceeds recharge and because of the rampant contamination of aquifer systems (Wescoat et al. 2000; MacDonald et al. 2016; Raza et al. 2017). To further add to the groundwater crisis, climate change and variability in rainfall affect the recharge of water stored in aquifers, while also leading to an increased demand for groundwater (Taylor et al. 2013). The improvement in well-drilling techniques and tube-well technology has further improved the accessibility of groundwater as a resource. In addition, rapidly proliferating solar pumps have increased the independence of groundwater users, not having to rely on intermittent electricity supplies from the state grid or recurring costs of fuel for diesel-powered tube-wells.

Contamination of groundwater systems from agricultural, industrial and urban sources of pollution creates environmental hazards and risks to life. A principal source of groundwater contamination in Pakistan occurs through secondary salinisation resulting from irrigation. The extent of salinity varies greatly between provinces. Almost half of the farmland in Sindh and Balochistan is affected, while 10% of the farmland in Punjab and KPK is affected by salinity (Zulfiqar and Thapa 2017). Excessive flood irrigation also results in water from the root zone of crops leaching into the groundwater and contaminating underlying aquifers with fertilisers, pesticides and salts. Untreated industrial effluent from water-polluting industries such as textile, leather and several others contains heavy metals such as cadmium, cobalt, copper, mercury, nickel, lead, tin and zinc, which also leaches into the groundwater. Another critical aspect of groundwater pollution in Pakistan is arsenic contamination, which is estimated to affect about 50 million people in the Indus Basin who are at risk of poisoning from the high level of arsenic in the soil (Ravenscroft et al. 2009; Podgorski et al. 2017).

10.2 Growing Dependency on Groundwater: A Case of Pakistan's Punjab Province

Punjab presents an intriguing case to analyse the problem of groundwater governance in Pakistan, not least because the province has the majority of the country's tube-wells. Punjab is also the most populous province, with 53% of the country's population (Punjab Bureau of Statistics 2019), and accounts for 60% of Pakistan's agricultural exports. The province is heavily dependent on groundwater to support its agrarian and industrial economies. In 2014–15, more than three quarters (76.8%) of the total irrigated area in Punjab was dependent on groundwater for irrigation, whether exclusively by wells and tube-wells, or in combination with water from canals (Fig. 10.1).

Despite being the land of five rivers, Punjab, as Pakistan's breadbasket, is the most dependent on groundwater for irrigation compared to other provinces (Young et al. 2019, and see Fig. 10.2). Groundwater is the main irrigation source for Punjab's major agricultural crops, accounting for between two-thirds and three-quarters of total irrigated area for wheat, rice, cotton and sugarcane when taking into account exclusive and conjunctive use of tube-wells (Punjab Bureau of Statistics 2017). In the year 2015–16, an overwhelming 78.4% of the total irrigated areas under wheat was irrigated partially or solely with groundwater sources. Similar statistics for rice,

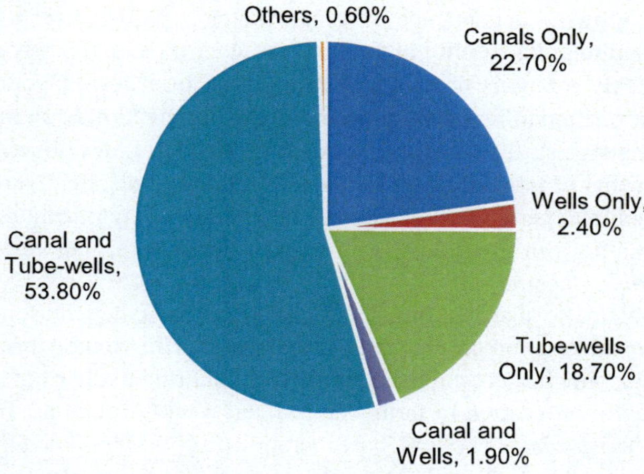

Fig. 10.1 Source of irrigation water in Punjab, Pakistan. (Source: Punjab Bureau of Statistics 2014–2015)

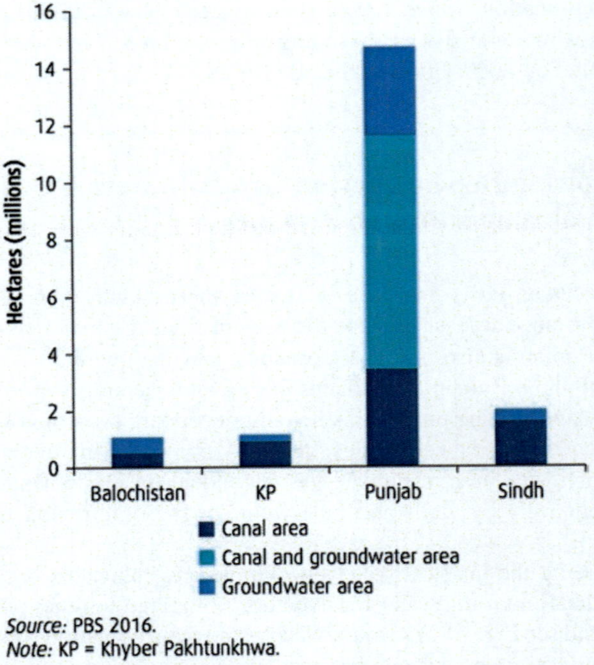

Fig. 10.2 Punjab's groundwater use in comparison to other provinces. (Source: Young et al. 2019, p. 16)

sugarcane and cotton show an overwhelming 88.7%, 69% and 68.6% of total irrigated area was dependent on groundwater irrigation (Punjab Bureau of Statistics 2017).

Punjab's growing dependence on groundwater is partly in response to overall variability and unpredictability in total inflows over time, and partly due to inflexibility in weekly rotations of its distribution at the farm level (Watto and Mugera 2016). Most of Punjab lies within the Indus Basin Irrigation System, which is the largest contiguous surface irrigation system in the world, developed through massive investments in large-scale surface infrastructure under British rule and in the early post-colonial period. Recent studies have shown a significant decline in average annual inflows in the Indus Basin in both the eastern and western tributaries (Cheema 2012; Cheema and Pawar 2015).

The problematic distribution of surface water is a key reason for Punjab's increased dependency on groundwater. Canal water is distributed through the *warabandi* system, which was established under colonial rule to distribute available surface supplies in proportion to farm size in fixed weekly rotations. In practice, the rigidity of the *warabandi* system, bureaucratic corruption and inequitable distribution between head and tail-ends of canals means farmers are increasingly reliant on groundwater pumped through tube-wells for timely access to irrigation water at crucial times in the cropping season (Jacoby and Mansuri 2018; Zardari and Cordery 2009). Since the *warabandi* system was established, cropping intensities have more than doubled from 70 to more than 150% (Shah 2007). Some areas of Punjab's sugarcane-wheat zone support crop intensities of 234% due to groundwater use (Mirza and Latif 2012). This was in particular due to the introduction of high yielding seeds and fertilisers during the Green Revolution of the 1960s.

The 1960 signing of the Indus Waters Treaty also induced a gradual transition to groundwater use. The Treaty was an agreement on the division of the Indus River system following partition between India and Pakistan. Under the Treaty, the three western tributaries of the Indus (Indus, Jhelum, Chenab) were given to Pakistan and the three eastern tributaries (Ravi, Bias and Sutlej) to India. However, the Treaty also gave rights to India for power generation and non-consumptive use on the three western tributaries before they entered Pakistani territory. In the last few decades, intensive hydropower development upstream by India on the western tributaries has substantially reduced downstream flow to Pakistan, thus making users more dependent on groundwater (Mustafa 2010).

Punjab also has a sizeable industrial sector, with more than 48,000 plants that together contribute 24% of the province's GDP (Planning and Development Department 2015). These include water intensive and water-polluting industries such as textiles, leather, pulp, chemicals and light manufactured goods. Pakistan's textile industry contributes 54% of the country's overall exports, with a water footprint of 12,251 million cubic metres of water (Linstead et al. 2015), with a sizeable proportion of the industry located within Punjab. Groundwater is also the main source of drinking and domestic water use for 87% of Punjab's population, with many districts (such as Bhakkar, Gujranwala, Gujrat, Hafizabad, Jhang, Kasir, Mandi Bahauddin, Nankana Sahib, Narowal, Okara, Sheikhupura and Sialkot) entirely dependent on groundwater as their sole source of drinking water (Tahir et al. 2010).

The groundwater crisis in Punjab manifests itself in the form of persistent decline in the depth of the aquifer due to over-abstraction, as well as groundwater pollution due to seepage from agricultural and industrial effluents. According to a recent study, groundwater levels across most of Punjab are falling by 16–55 centimetres per year, with large urban centres such as Lahore experiencing higher depletion rates of 40–150 centimetres per year (Basharat et al. 2015). In parts of Punjab with historically good quality groundwater such as the Rechna Doab, agricultural and industrial activity as well as urban growth has meant that groundwater levels are falling by two to three metres annually. Falling groundwater levels due to over abstraction necessitate deeper drilling to pump from the underground aquifers, thus marginalising poorer farmers due to increased costs of drilling and pumping (Nabeel 2020).

In addition to declining quantity, there is degradation of quality of groundwater resources. Punjab has many of the hotspots where people are at risk of arsenic poisoning (Podgorski et al. 2017), which is especially alarming given groundwater is often the main source and sometimes the only source of drinking water. Pesticide seepage and industrial pollution has contaminated shallow groundwater aquifers

Fig. 10.3 A factory in Sheikhupura that has built unlined wells for the disposal of untreated effluent

with hazardous chemicals and metals. Industries routinely discharge untreated effluent, often via open drains, while those located away from the drainage network find other 'creative' ways for effluent disposal. During the fieldwork that informed this chapter, the author found unlined wells or '*khuwas*' being used in Sheikhupura to discharge untreated effluent, which slowly percolated down into the underlying shallow aquifer (see Fig. 10.3). This resulted in severe contamination of groundwater used in adjoining areas, making the water unfit for drinking and even for irrigation. Poor farmers at the tail-end of canals bear a disproportionate burden of polluted groundwater aquifers, as compared to wealthier farmers who often live near the head of canals where plentiful surface water seepage continues to replenish groundwater reserves. This is worrying because farmers at the tail-ends are more dependent on groundwater for their irrigation requirements, as opposed to farmers at the head who have ready access to surface water (Nabeel 2020).

10.3 A Crisis of Governance? Insights from the History of Groundwater Governance in Punjab, Pakistan

Existing studies have identified that groundwater crises result from governance arrangements that are weak (e.g. Faysse et al. 2014) or lacking (e.g. van Steenbergen 1995). These include analyses of the situation in Punjab and Pakistan generally, which have pointed to the absence of or weaknesses in legal, institutional and regulatory policies governing groundwater use, and the need to strengthen these to achieve resource sustainability (Qureshi et al. 2010; Khair et al. 2015; Watto and Mugera 2016). A recent World Bank report also takes a similar view about water governance: "The governance challenges relate to inadequate legal frameworks for water at federal and provincial levels, and the incompleteness of policy frameworks and the inadequacy of policy implementation" (Young et al. 2019, p. xviii). This chapter builds on existing studies, but presents a refined view. Instead of attributing the groundwater crisis due to the absence of, or weak, governance of the resource, the crisis is attributed to the very techniques of groundwater governance occurring in Punjab over time.

The chapter uses a historical analysis of the techniques and mechanisms of groundwater governance from the colonial to the contemporary period to highlight the role of the state in creating Punjab's groundwater crisis. The chapter finds that through each historical period, groundwater is needed for the state to fulfil its developmental goals, and in turn the significance of groundwater shapes the state's approach towards its governance, ultimately leading to the contemporary groundwater crisis. The analysis begins with an overview of groundwater governance in the colonial period (Sect. 10.3.1), before highlighting the role of the state in initiating large scale public groundwater development in the post-colonial period (Sect. 10.3.2), and finally reviewing the mechanisms of governance that drive contemporary governance of groundwater (Sect. 10.3.3). The chapter ends with reflections on significance of groundwater to Pakistan's economy as well as the complex nature of historical groundwater governance that has led to the contemporary crisis (Sect. 10.4).

10.3.1 Colonial Approach to Groundwater Governance

The use of dug wells for irrigation and domestic purposes has been recorded in the Indus Basin as early as the fourteenth century, when access to subsoil water was made possible through the use of the Persian wheel. Babur, the founder of the Mughal Dynasty, made reference to Persian wheels being used in 1519 for irrigation of profitable crops such as rice and sugarcane in Bhera and Khoshab, areas which are part of present day Punjab in Pakistan (Siddiqui 1986). However, the small-scale nature and yield of such irrigation, powered by animals, meant that well irrigation was for the most part managed by individual cultivators. Upon British annexation of Punjab in 1849, the colonial state continued to encourage this form of irrigation through (subsidised) *tuccavi* loans for well construction, a practice that had also been used earlier by pre-colonial rulers.

In the early colonial period, dug wells remained the dominant form of irrigation across the north western provinces of British India, particularly in Punjab and Uttar Pradesh. In the year 1885, about 16% of the total cultivated area in Punjab was irrigated by wells (*chahi*) – whether as the sole source of irrigation or in conjunction with other sources. By comparison, the area irrigated by canals (*nahri*) in Punjab

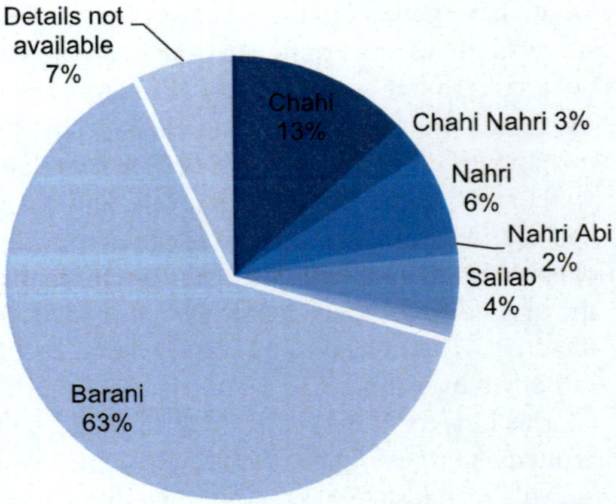

Fig. 10.4 Cultivated area in Punjab by source of irrigation in 1885. (Source: Statistical Abstract Relating to British India, 1885. Area labelled *barani*: rain-fed, *chahi*: well-irrigated, *nahri:* canal-irrigated)

constituted about 11% of the total cultivated area including area irrigated by inundation canals. As shown in Fig. 10.4, most of Punjab's agriculture was still rain-fed *(barani)* until the end of the nineteenth century.

Despite the British Government's efforts to expand the canal network in the latter half of the nineteenth century, groundwater use through wells remained an integral part of the irrigation mix in Punjab, for three key reasons: First, well irrigation was critical for areas not supplied by the canal network. As over half of Punjab's agriculture was rain-fed for most of the nineteenth century, the variability and lack of rainfall frequently led to distressed crop outcomes. The colonial government's construction of large-scale engineered canal works was ambitiously pursued to reduce such fluctuations in output. However, the only areas that benefitted from canal irrigation were those that could be commanded by gravity flow. Canal irrigation was not suitable for upland terrains, areas above canal heads and other areas thought to be inaccessible. Hence wells represented a suitable method to irrigate uplands, and to allow an increased and more flexible allocation of water and cultivation intensity (Ali 1988).

Second, wells provided supplementary irrigation in canal command areas for *Rabi* (winter) crops, when canal flows were inadequate. Well irrigation was particularly crucial for wheat, which was the "most important of all Punjabi crops" occupying up to one-third of total cropped area in the province by the turn of the twentieth century (Annual Report of the Department of Land Records and Agriculture 1908, p. 5). The strategic importance of wheat as an export crop for Britain's global imperial economy underscores the significance of well irrigation for the colonial government at that time. Wheat comprised 4 per cent of the value of exports out of the province in the year 1900, with United Kingdom being the top recipient of the produce (Statistical Abstracts Relating to British India 1899–1900).

Third, well irrigation had been declared as a form of 'protective irrigation' under the evolving colonial irrigation policy, helping to prevent famine in times of drought (Report of the Indian Famine Commission 1880). The devastating outcome of periodically occurring famines in British India, such as the ones that occurred in Madras in 1877–78, had compelled the colonial government to look towards encouraging forms of irrigation that would serve to "prevent" famines, rather than incurring expenditure on post-famine relief efforts. In British Punjab more than half of the cultivated area was rain-fed until the end of the nineteenth century, making the fate of agriculture highly dependent on rainfall distribution. Well irrigation (together with perennial canals) provided a resilient means of irrigation during times of droughts when other methods such as tanks and inundation canals would "dry up" (Danvers 1877).

It is interesting to chart the colonial approach to groundwater governance, which evolved over the course of British rule from the mid-nineteenth to mid-twentieth century. Until the early twentieth century, apart from its role to provide 'protection', groundwater was primarily a small-scale method of irrigation managed by individual cultivators, at their own expense, in areas not supplied by canals or for use as supplementary irrigation during winter months. Mechanisation and tube-well technology at the turn of the twentieth century showed promise of increased yield of groundwater for irrigation. The freedom from dependency on animal power to extract groundwater allowed by tube-wells meant that groundwater was no longer to be limited as a small-scale means of irrigation. This enabled the colonial government to consider large-scale sponsorship of public tube-well schemes in Punjab, not just for agriculture but also for municipal water supply. It represented a paradigm shift in the colonial government's approach towards groundwater, enabling the government to charge tube-well users for groundwater, in a manner similar to irrigation dues paid by users of canal irrigation.

From the beginning of the twentieth century, the colonial state used a two-pronged approach to groundwater development: (a) activities supporting farmers to enhance yields of existing dug wells by mechanically deepening them; and (b) exploring technological and economic feasibilities of private and public tube-well irrigation schemes. The colonial state offered expertise on both fronts through the Punjab Department of Agriculture's Engineering Section in the first half of the twentieth century. The colonial state also incentivised uptake of tube-wells by private cultivators by encouraging them to cultivate government wasteland through installation of tube-wells (Ali 1988). Tube-well installation subsequently became a pre-condition for granting of tenure – lessees were required to construct tube-wells as a part of the lease obligations.

Despite the colonial state's efforts to promote groundwater exploitation and uptake for irrigation, groundwater irrigation did not achieve a breakthrough by the end of the colonial rule. Part of the reason lay in the broader political economy situation during British rule – the two World Wars, budgetary constraints, and high prices of obtaining imported materials for tube-well boring and installation stymied the progress in groundwater development. Also, because cultivators had to bear (subsidised) costs of installation of wells and tube-wells, the high prices and lack of availability of materials during war years deterred the demand for groundwater irrigation. Another important set of factors hindering groundwater development despite the colonial state's intent concerned the difficulty in achieving a breakthrough in generating hydroelectric power for running tube-wells. The availability of low cost hydroelectric power was critical to improving the economic feasibility of tube-wells as opposed to higher cost diesel-powered pumps. In addition, technical issues with tube-well technology and high maintenance expenses deterred farmers from installing tube-wells for groundwater irrigation (Punjab Department of Agriculture 1938).

Notwithstanding the limited success of tube-well irrigation during colonial rule, the colonial state's policies towards groundwater development have left lasting impacts on the way the resource is governed in the region. In particular, the legal and administrative changes in the governance for groundwater brought about during the colonial period completely re-configured the groundwater-society dynamic in Punjab, sowing the seeds for the explosive growth in post-colonial use of groundwater, and in many ways contributing to the current crisis. In the pre-colonial period, the rights to sub-soil water were largely customary, arising from local social relations, kinship and "tradition", and usually determined by the share of investment in jointly owned wells within communities. This system of customary rules governing groundwater rights, along with social and economic relations at the village level were studied by the colonial administration and carefully recorded in village administration papers called "*shart wajib-ul-arz*".[1] The shares in investment of jointly constructed wells determined precisely when each shareholder's bullocks had to be yoked and unyoked at certain times of the day to allow the cattle for another shareholder to draw water for irrigation (Gilmartin 2015). In 1882, groundwater rights were formally tied to private property rights over land after passing of the Indian Easements Act based on common law principles of England (Gilmartin 2015, p. 109). According to the Easements Act, owners of land have the right to "collect and dispose within his own limits of all water under the land which does not pass in a defined channel and all water on its surface which does not pass in a defined channel" (Indian Easements Act 1882, Section 7). The change in legal entitlement to water meant that those who held titles to the land could sell their property rights to land and water without consideration of others in the community (Singh 1991). These changes, along with parallel improvements in mechanisation improving the volume and ease of groundwater extraction, meant that groundwater was effectively rendered as a resource that could be extracted to a cultivator's desire, without consideration or dependency on others in the community.

Despite the colonial state's policies and strong intent to pursue large-scale groundwater development in Punjab, it could not be materialised by the eve of independence in 1947. As the next section will illustrate, it was only after independence and the creation of Pakistan that large-scale state-owned schemes sponsored by international agencies such as USAID and the International Bank for Reconstruction and Development were established. Though these 'public tube-wells' were initially installed with the intention to alleviate the problem of waterlogging and salinity, within a decade there was a rapid and concomitant rise in installation of private tube-wells by individual cultivators as tube-well technology became commonplace and tube-wells started to be manufactured locally. The explosive growth of

[1] The *shart wajib ul arz* regulated the management of the village commons or *shamilat* in the villages ranging from cultivation of *shamilat* by proprietors and tenants, grazing rights by proprietors and non-proprietors of the village, the use of wells, the right to plant and cut trees in the *shamilat*.

tube-well irrigation, accompanied by the availability of high-yielding variety of seeds, fertilisers and mechanisation of agriculture, led to the post-colonial 'Green Revolution'. Yet, in many ways, the post-colonial approach represented continuity in the prior colonial approach to groundwater governance, as detailed in the next section (Sect. 10.3.2).

10.3.2 Post-colonial Groundwater Development

After the partition of the Indian subcontinent and independence from British Rule in 1947, the newly established state of Pakistan sought to invest in public groundwater development to fulfil three key objectives: (a) to increase water availability for food self-sufficiency and reduction of foreign exchange spent on imported grains; (b) to relieve waterlogged and saline soils which had rendered about a quarter of West Pakistan's area uncultivable at the time; and (c) to enhance the agricultural growth rate for its predominantly rural-based economy.

Pakistan's first decade after independence was politically and economically unstable. In the early 1950's, the economy was provided with a temporary boom during the Korean War due to increased demand for jute products from East Pakistan (now Bangladesh). After the end of the Korean War, the decline in foreign exchange earnings in Pakistan and unfavourable monsoons led to a shortage of food, medicine and essential consumer goods. This was resolved with the help of foreign aid by bringing in food grains as well as aid from development programs. In the country's first five-year national development plan period, Pakistan had to pursue extensive import of food grains, consuming a significant $700 million in foreign exchange, in addition to freight charges paid on food grains received as aid.

The partition of the subcontinent in 1947 also involved a separation of both land and water resources across the Indus Basin. In the aftermath of the Indus Waters Treaty of 1960 that divided the surface waters of the Indus, the Pakistani state sought to replenish water 'lost' to India by constructing massive diversion works for water from the western rivers to the fields that the eastern rivers formerly irrigated. The Indus Basin Development Fund made this possible, which was a financial agreement between Pakistan, the World Bank and a group of capitalist states including Australia, Canada, New Zealand, UK and USA (Akhter 2015, p. 70). The Indus Basin Project was the largest integrated irrigation project in the world at the time, with funding of about $1200 million (Michel, 1967).

On one hand, Pakistan struggled with the need to secure its share of surface waters of the Indus, while on the other hand the growing problem of waterlogging and salinity presented a sizeable constraint for agricultural growth and the problem of food security. Groundwater levels and the loss of land to waterlogging and salinity had been recognised as a growing problem even under colonial rule as early as 1908 when the first comprehensive survey of water table in the Chenab colony was conducted. However, even as the colonial government sought to understand and ameliorate the problem by propagating the value of tube-wells as a vertical drainage solution, there was "little progress" on the issue of waterlogging by the eve of independence in 1946 (Gilmartin 2015, p. 237).

After independence, a series of programs were initiated by the Pakistani government to control waterlogging via vertical tube-well drainage, but with inconclusive results, perhaps because of the limited scope of these scattered projects (Johnson 1989, p. 2). The Rasul Scheme, originally proposed as a tube-well irrigation and drainage scheme by the colonial government in 1927, was finally sanctioned in 1944. The scheme had been delayed first because of jurisdictional issues of where the power supply would come from, then because of financial stringency during World War 2, and finally due to chaos created by the partition. When the project was finally functional, it was deemed a 'failure' (Michel 1967, p. 460). In addition to the Rasul scheme, the 495 tube-wells installed in the Chaj Doab were so close to the canals that, rather than relieve waterlogging, they accelerated seepage. Most of the 762 tube-wells installed in the Rechna Doab failed or declined in yield due to blockage of strainers and incrustation (when deposits of carbonates build up on well strainers).

After a series of lukewarm efforts at ameliorating waterlogging and analysing prospects for groundwater development by the newly independent Pakistani state, it was international expertise and financial assistance that came to the rescue and initiated large-scale groundwater development in Pakistan. In 1953, the Pakistani government reached an agreement with the predecessor of USAID for "technical assistance in soil and water resources investigation" initiating the "West Pakistan Groundwater Survey" nicknamed "Project 035" whose immediate objective centred on the "the provision of soils and water data essential for agricultural development, improved irrigation, salinity control, and land reclamation in West Pakistan" (Taylor Jr 1976, p. 87). As a result of Project 035, a Pakistani counterpart for managing project activities was established as the Groundwater Development Organisation (GWDO) within the West Pakistan Department of Irrigation. GWDO was later transferred to West Pakistan Water and Power Development Authority (WAPDA) in 1960 where it was reorganised in the form of Water and Soils Investigation Division (WASID).

The production of knowledge resulting from Project 035 in the form of soil data and maps, water table maps and hydrographs, exploratory well logs, water quality analysis and maps and technical reports eventually fed into the design and construction of the first Salinity Control and Reclamation Program (SCARP 1) in 1961, a pilot project in the Chenab colony funded by a low interest American loan. The SCARP program's logic was simple and something that had been advocated as a solution to waterlogging and salinity even in the colonial period – installation of public tube-wells for vertical drainage that would lower the water table, while providing additional water for irrigation. The West Pakistan Water and Power Development Authority (WAPDA), an institution created to manage the construction of surface infrastructure under the Indus Basin Plan, was also given charge of designing, monitoring and implementing the public tube-well programs drawing on the advice of many foreign consultants readily assigned to the project (Michel, 1967).

This was a time immediately after the signing of the Indus Water Treaty of 1960 when foreign donors committed huge sums of money to construct dams and barrages on the western rivers allocated to Pakistan as compensation for the loss of water from the eastern rivers allocated to India under the treaty. The post-treaty Cold War environment provided the perfect opportunity for the then President Ayub Khan of Pakistan to visit the US to lobby for more funding highlighting his "concern with waterlogging and salinisation which threatened the future of Pakistan's agriculture" (Taylor Jr 1976, p. 89). President Kennedy responded by appointing a special scientific presidential panel headed by Richard Revelle, the then Science Advisor to the US Secretary of Interior. The knowledge products from the Revelle Report led to a massive expansion of SCARP programs in the 1960s and 1970s. With up to 15,000 state owned and managed tube-wells, this was the most comprehensive response to the issue of waterlogging and salinity in Punjab ever attempted. These public tube-wells were usually installed at the head of a tertiary water channel, thus supplementing and compensating for irregular canal water supplies to village watercourses (Gilmartin 2015, p. 239).

The investment in public tube-wells through SCARP Projects was consciously planned by the state, but the parallel growth of private tube-wells mushroomed almost unnoticed until a survey was done in 1964. In 1950, the Department of Agriculture reported only 177 private tube-wells, all of which had been installed with assistance from the Department from 1939 onwards (Johnson 1989, p. 1). In subsequent decades, private tube-wells had a greater share than public ones in increasing the role of groundwater in overall irrigation mix. By 1996, there were about 300,000 private tube-wells, most of which were in Punjab. Groundwater accounted for 8% of total irrigation water in 1960, rising rapidly to 40% in 1996 and reaching 60% in 2006 (Briscoe et al. 2005). As seen in Fig. 10.5, the province of Punjab led the private tube-well "revolution" in Pakistan.

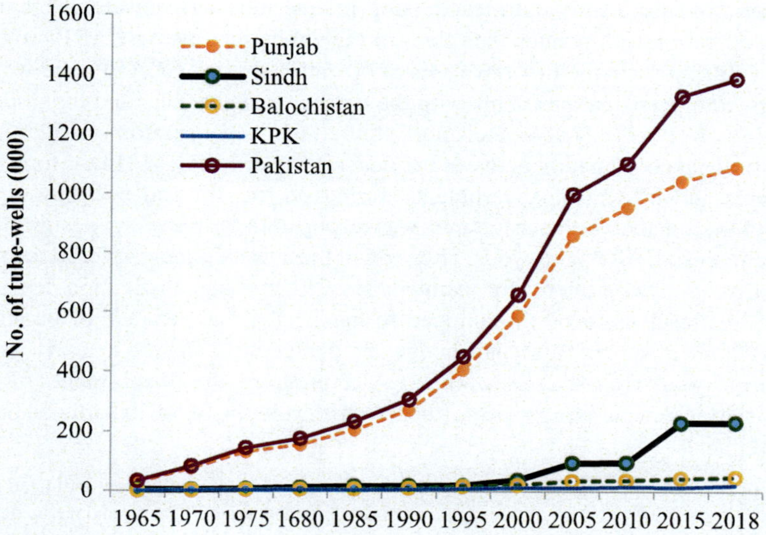

Fig. 10.5 Private tube-well growth in the post-colonial period. (Source: FWM 2019)

Most accounts of private tube-well development in Punjab document it as a silent revolution often attributing its phenomenal growth to the individual incentives of the rational farmer (Watto and Mugera 2016; Molle et al. 2003). While the improved economic feasibility of tube-wells in the post-colonial period improved their uptake, the analysis in this section suggests that the techniques of formal governance of groundwater played an active role in bringing about this breakthrough in tube-well irrigation. The Pakistani state offered subsidies for private tube-well construction, power connection, drilling, and electricity tariffs during the 1960s and 1970s as incentives to increase cropping intensity and to fulfil targets for agricultural growth (Government of Pakistan 1978). During this period, a tube-well construction subsidy was given to small and marginal farmers on tail-ends of canals and rain-fed areas, in addition to a separate subsidy on connection to the electricity grid, as well as a subsidy on drilling costs. A diesel subsidy of 20% was also introduced in 1972, benefiting farmers with diesel-operated tube-wells. By April 1979, the scheme supported 11,500 tube-wells at a total cost of Rs. 132 million (Anson 1984, 34). Agricultural tube-wells extended a subsidised electricity rate of 0.35 per kilowatt per hour, compared to non-agricultural customers who paid 66% more than the subsidised rate. Apart from direct subsidies, the Agriculture Development Bank of Pakistan (ABDP) provided institutional credit for private tube-well development. Between 1964 and 1981, the total number of tube-wells increased from 23,000 to 186,000, with an average increase of 9500 tube-wells each year. ABDP's loans contributed to an approximate 20% of the tube-well units installed each year within this period (Anson 1984, p. 135).

Perhaps farmers would have invested in private tube-wells even without the subsidies. However, state subsidies clearly indicate the state's intent to harness groundwater resources for agricultural production and economic growth. Another reason for subsidising and incentivising private tube-well development was the high operation and maintenance costs of the public tube-well SCARP program. It was evident within the first few years of operation that the SCARP tube-wells suffered from persistent problems with the design, maintenance, and high operating costs, as well as staff absenteeism of public tube-well operators. In addition, the institutional coordination between the Federal Government, WAPDA, foreign consultants and provincial governments during project life did not play out well. WAPDA, together with consultants, was responsible for planning, design and construction of SCARP tube-wells, after which these were handed over to the provincial irrigation departments for maintenance. The provincial irrigation departments found themselves to be ill-equipped to handle the maintenance of public tube-wells both from expertise and budgetary perspectives. Thus, as early as 1965, merely 5 years from the start of the SCARP program, the Government of Pakistan had started encouraging private tube-well development in its official development plan:

> Construction of private tube-wells has been accelerating rapidly and is now estimated to be running at 6500 per year, of which two thirds are diesel-operated. The rapid development of private tube-well now requires a new tactic – a strategy of public-private development. It is

> proposed to give private tube-wells every assistance in the form of imports, credit and electrical connections. With such assistance, some 40,000 new tube-wells can be expected during the plan period (Government of Pakistan 1965, p. 294).

While SCARP tube-wells had strong operating and maintenance issues during the 1970s, the government assisted private tube-well development programs were doing quite well at the time. Up to 1982–83, a subsidy of Rs. 211.6 million had been given to farmers against which 19,433 private tube-wells have been installed. In the Sixth Five Year Plan, an additional Rs 102 million were to be allocated for private tube-well subsidy for installation of 8195 additional private tube-wells, the majority of which were to be located in Punjab (Government of Pakistan 1983). In the Eighth Five Year Plan it was decided that no new public tube-well program would be initiated in fresh groundwater zones. From then on, SCARP tube-wells to alleviate waterlogging and salinity would only be installed in saline groundwater zones after the 1990s (Government of Pakistan 1983), and soon afterwards the existing SCARP tubewells were also ushered into a World Bank sponsored privatisation program. The Pakistani state's subsidies for groundwater irrigation are an important part of the explanation of the post-colonial "tube-well revolution" in Pakistan. This was accompanied by concomitant efforts at improvements in tube-well technology and rural electrification after the construction of the Tarbela dam.

The centralised development of groundwater through publicly owned and operated tube-wells, and the private groundwater boom that followed suit, considerably transformed the post-colonial groundwater-society dynamic. Perhaps the most profound effect of the state's intervention in development of centralised groundwater schemes was the demonstration effect to the farmers. Public tube-wells transformed the way society valued groundwater. Whereas private tube-well development had been repeatedly found to be economically unfeasible by adopters in the colonial period, technological improvements and the availability of the locally made pumps transformed this relationship in the post-colonial period.

Many of the legacies of colonialism inherited by the post-colonial state had of course contributed to the boom in atomistic tube-well development. As discussed in the previous section, the legal rights to groundwater defined in the colonial period held groundwater entitlement as linked to rights to private properly over land. Treating groundwater as a private resource linked to rights over land, with little or no state oversight, effectively meant that the only cost to using groundwater was the cost of pumping it from the ground. While the cost of pumping groundwater was higher than the fixed *abiana* farmers paid for their canal water allocation, they did not have to wait for their turn to get water to irrigate their crops. The lack of flexibility in surface water supplies, and the on-demand nature of private groundwater pumps, meant farmers were willing to invest, leading to the private pump boom.

10.3.3 Historical Contingency and the Contemporary Groundwater Governance Problem

Techniques and mechanisms of formal governance of groundwater in the colonial and post-colonial period have been instrumental in the making of Punjab's current groundwater crisis. While governance mechanisms towards groundwater have changed between historical periods, formal governance of groundwater has consistently focused on exploiting groundwater for agricultural expansion, as this chapter has shown. For the colonial state, tube-wells were promoted as an additional means of irrigation, especially for areas not served by canals and for important winter crops such as wheat. In the post-colonial period, the loss of productive land to waterlogging led the state to carry out large-scale public groundwater development, serving twin objectives of increasing the water budget for agricultural intensification and relieving waterlogging and soil salinity. The post-colonial state also offered sustained incentives and subsidies that eventually led to a revolution in Punjab's private tube-well uptake. This section shows how, even in the contemporary period, state policies indirectly incentivised increasing dependence on groundwater as surface water supplies become more unpredictable. Further, the section highlights how the legal entitlement to groundwater and the institutional responsibility for its management is currently muddled because of the blend of old and new institutions and laws created by successive policy regimes.

Contemporary groundwater governance in Punjab indirectly incentivises groundwater overexploitation for agricultural and industrial use. The agricultural sector has held strategic importance for Pakistan and Punjab's economy during the historic past as well as in the contemporary period. Keeping in mind the importance of the rural electorate that comprises nearly half of the total population, successive political governments have subsidised agricultural inputs such fertilisers, pesticides, tube-wells, tractors and improved seeds (Dorosh and Salam 2007; Pursell et al. 2011). In the contemporary period, the agriculture sector contributes about 21% to the country's GDP, but employs 45% of the labour force, and contributes a staggering 80% to Pakistan's export earnings (including forward linkages to agro-based industries such as textiles) (Ministry of Finance 2018). Wheat, rice, cotton and sugarcane are key crops for Pakistan's economy, accounting for two thirds of the total cropped area in Pakistan and contributing one-third to the total agricultural GDP (Ministry of Finance 2018). As has already been pointed out, between two-thirds and three-quarters of the area under these crops is irrigated by groundwater, whether exclusively or conjunctively with canal water. What is even more worrying is the very low water use efficiency of these crops, with 80% of the country's water being used to generate less than 5% of its GDP (Ministry of Finance 2018). In addition to the heavy water footprint of major agricultural crops, livestock production (dairy, poultry, sheep and goats), accounting for nearly half of the agricultural GDP, is also water intensive. State policies have incentivised the uptake of water intensive crops, and subsidised farm inputs that increase water requirements such as fertilisers, both historically as well as in the contemporary period. For instance, the fertiliser subsidy is the single largest subsidy in the total subsidies budget, with wheat and sugar subsidies falling at the second position (Government of Pakistan 2017). Rice, another water intensive crop, is Pakistan's second largest export commodity, and yields the government about $2 billion in foreign exchange annually (The Express Tribune 2017).

Favourable wheat and sugar subsidies benefit capitalist interests of ruling political elites, particularly sugar mill owners in Pakistan. Cotton growing districts in Punjab such as Rahim Yar Khan, Bahawalpur and Multan have documented a shift in cropping pattern towards growing sugarcane because of the artificially inflated support price of sugar, despite the heavy water requirement of the crop (Arif Nadeem, Former Secretary for Agriculture Punjab, pers. comm., 2017). The crop water requirement for sugarcane is almost double the requirement for cotton; hence the recent expansion in the area under sugarcane, and the concomitant decline in the area under cotton, is bad news for groundwater over abstraction (Ebrahim 2020). Pakistan's sugarcane production is highly water inefficient, having the second highest total water footprint in the world compared to the world average, with 69% of the total sugarcane growing in Punjab province (Scholten 2009). In addition to price support mechanisms for primary produce, the government supports key water intensive agro-based export sectors of textile, leather, sports goods, surgical goods and carpets through its "zero rated regime" (The News 2017). Pakistan's government has also made efforts to increase exports of highly water intensive halal meat to gulf countries and to China (Dawn Newspaper 2018). It is estimated that the total water footprint of beef can be up to six times the water footprint of rice.[2] Agricultural and economic policies place considerable emphasis on water intensive agribusiness production and export, facilitating allocation of precious groundwater resources based on users' ability to extract.

Thus, the state's techniques for groundwater governance in the contemporary period have largely drawn on a mix of state subsidies and trade policy incentives – policies that have over time made Pakistan emerge as the top groundwater exporter in the world (ahead of the United States and India) (Dalin et al. 2017). Dalin et al. (2017) show that about 11% of global non-renewable groundwater use for irrigation is embedded in international food trade, of which two-thirds are exported by Pakistan, USA and India alone. For Pakistan, the cultivation and export of crops such as rice account for a major share of the non-renewable groundwater use.

The rise of powerful agro-industrial corporate interests in contemporary Punjab's increasingly globalised economy are also shaping groundwater governance processes in the later post-colonial and contemporary period. Notwithstanding the significance of agriculture to Pakistan's economy, industry is an equally important sector with a 21% contribution to the country's GDP (Ministry of Finance 2017). While the contribution of industry to GDP is similar to that of agriculture, official estimates suggest that industrial water use is "negligible" or between 1 and 2% of

[2] Beef requires the most water, at 1847 gal./lb., followed by sheep at 1248 gal./lb. Chicken at 518 gallons of water per pound. Milk by itself uses only 122 gallons of water per pound. Rice requires an average of requiring 299 gallons of water per pound of processed rice. (www.waterfootprint.org)

total water use while agriculture consumes more than 90% (Basharat et al. 2015). These official government estimates of industrial water use (which is not currently metered or measured in Punjab), and the consequent corporate water footprint, are questionable in the light of other studies. A recent WWF Pakistan study estimates the total blue, green and grey water footprint of four key industrial sectors of Punjab (textile, leather, sugar and paper) to be 30,000 million cubic metres annually. This represents a significant water requirement when put in perspective of total Indus river flows (Linstead et al. 2015). While this study does not isolate groundwater footprint from overall water footprint, these estimates are indicative of the water intensive nature of Pakistan's agribusiness and manufacturing industries.

The legal apparatus of the state for groundwater governance is a critical part of the governance puzzle, with contemporary legislation on groundwater based on colonial law and a jumble of post-colonial amendments. As noted above, the most decisive historical change in legal entitlements to groundwater was brought about by the colonial state in the Indian Easements Act, which tied groundwater entitlement to property rights over land. The post-colonial Pakistan state issued a series of legal provisions to support groundwater management, but none of the post-independence legislation around groundwater has been fully or consistently implemented (van Steenbergen and Gohar 2005). The Punjab Soil Reclamation Act of 1952 set up a Soil Reclamation Board to control problems of waterlogging and salinity that affected fertile lands in the newly independent country. The key responsibilities of the Soil Reclamation Board also included issuing licenses to landowners to install private tube-wells for the pre-determined land reclamation areas. However, the Board was suspended after a few years, with these responsibilities shifting to the provincial irrigation departments. After the establishment of WAPDA under the Pakistan Water and Power Development Authority Act (1958), the responsibility for groundwater development and monitoring rested with WAPDA under the World Bank sponsored SCARP programs.

There is also confusion between federal, provincial, and local responsibility for groundwater governance between various levels of government. Water management falls under provincial jurisdiction in Pakistan, though it can be developed and regulated by two federal bodies - the Council of Common Interests and the Water and Power Development Authority (WAPDA). The Water and Power Development Authority Act of 1958 gives the federal body a mandate to develop water resources in Pakistan with consent from the provinces. In the provinces, the control over groundwater, pollution, and overexploitation lies with the respective provincial irrigation departments. In principle, local governments in rural and municipal areas also have the right to regulate the use of private water sources and tube-wells for drinking water (Local Government Ordinance 2001). However, this mandate of the local government impinges on the rights of property granted by the Indian Easements Act of 1882, which effectively grants unchecked private ownership of the common pool groundwater resource flowing under the land to which one has ownership rights. Further, a 2006 Amendment to the Canal and Drainage Act of 1873 allows the provincial government to intervene in case of groundwater overexploitation. In practice this provision has hardly been deployed by provincial governments, implying little recognition or effort on the part of the government to counter groundwater depletion.

The recently passed Punjab Water Act of 2019 stipulates radical changes in the way entitlements to use groundwater are administered. According to the Act, a 22 member Punjab Water Resources Commission controls surface water resources, water in lakes or reservoirs, water in drains, *as well as groundwater resources* (Punjab Water Act 2019) (http://punjablaws.gov.pk/laws/2743.html). Thus, the Act makes the Commission responsible for conserving, "controlling" and allocating effectively all the water resources in Punjab. For groundwater use, this specifically means that the Commission has the authority to issue licenses for abstraction and disposal of groundwater for agriculture, domestic, industrial and mining purposes. The implications of the recent legislation for groundwater users or the groundwater crisis are not clear yet. It is not clear how the state will set up a system of licensing and abstraction for the sustainable use of the resource under the new law, which awaits implementation.

10.4 Implication for Water Policy and Groundwater Governance in Pakistan

The chapter has analysed the historical importance of groundwater using a case study of Punjab province in Pakistan over the last 150 years to emphasise how groundwater has supported the state's developmental goals in each period. The analysis underscores groundwater's critical contribution to colonial hydrology and the post-colonial Green Revolution, and presents compelling evidence on groundwater's role in Punjab's contemporary water intensive export economy. It highlights how throughout Punjab's history groundwater has helped to fulfil critical development objectives, in turn shaping governance policies that incentivised exploitation of the resource. The chapter shows how governance techniques and mechanisms have focused on groundwater exploitation for agricultural expansion. Whether in an effort to control waterlogging by vertical drainage or as a supplementary source of irrigation, groundwater has been critical in doubling cropping intensities from the colonial to the contemporary period. State policies over time have been instrumental in transforming the groundwater-society relationship in Punjab. Large scale public tube-well development in the post-colonial period changed the way the society valued groundwater, paving the way for the explosion in private tube-well investments. At the same time, the combination of agricultural subsidies in both the post-colonial and contemporary periods mean that the dependency on tube-well irrigation is deeply entrenched. The character of state policies towards groundwater so far has been focused on groundwater exploitation to foster economic development, without groundwater sustainability concerns.

By highlighting the indispensable historical contribution of groundwater to Punjab's economy, the chapter demonstrates why there needs to be a higher policy priority on groundwater conservation. In the contemporary scenario, the water resources policy debates in Punjab (and Pakistan) are dominated disproportionately by surface water development, primarily the need to increase storage by the construction of dams on western tributaries of the Indus in a race against upstream India. Securing transboundary surface waters shared with India, and the use of the rivers as issues of 'national security' has placed an overwhelming pressure on successive governments to prioritise surface water infrastructure development, especially given the context of mounting development and population pressures. While there has been recent recognition of groundwater overexploitation, and a mention of the need for sustainable groundwater management in the country's first ever national water policy in 2018, there has been little follow up in terms of implementation. This chapter focuses on the historical analysis of the importance of groundwater for the region's development, and an emphasis on its current role as the major contributor to the agro-industrial economy, thus emphasising the need to focus on groundwater conservation. Incorporating historical analysis into groundwater policy will bring attention to the legacies of colonialism and the contribution of post-colonial policies and programs by the state embedded in specific political economy issues of each era. This chapter has highlighted the many ways in which the current groundwater crisis in Punjab is historically contingent on the colonial approach to groundwater governance. The techniques of groundwater governance used in the colonial period – particularly the legal, administrative and institutional changes – have left behind strong legacies for post-colonial and contemporary governance approaches. An appreciation of the history of groundwater development and the role of state in Punjab will also lead to an appreciation of the blend of old and new institutions created by successive governance regimes. This will bring attention to the need to clearly demarcate responsibilities for groundwater between various levels in the state – federal, provincial, local – as well between institutions at the same level.

The chapter also highlights the need for reviewing and engaging with the production of knowledge around groundwater. The dominant rhetoric in most technical and policy studies is to blame the traditional agriculture sector as "inefficient" and "wasteful". At the same time, industrial water use is deemed as negligible, and multinational companies are increasingly advertising their water use efficiency by subscribing to international standards of water efficiency such as the Alliance for Water Stewardship (AWS) and Sustainable Rice Partnership (SRP). It is worth noting that subscription to international standards of water efficiency can merely represent a "greenwashing" of operations for many businesses, while making little difference to the actual water footprint. There is a need to engage in critical knowledge production to review and measure industrial water use in order to effectively manage demand for sustainable groundwater development.

This chapter does not make the usual recommendations for groundwater policy and governance in Pakistan. It also refrains from defining conceptions of what "good" groundwater governance is, in order to appreciate its complex and messy nature. The experience of other countries dependent on groundwater illustrates that there is no magic formula that "works" for sustainable groundwater management (Mukherji and Shah 2005). Hence the analysis in this study does not emphasise a particular approach to groundwater governance over another. Rather the aim is to bring to light how Pakistan's experiences with various techniques of governance have actively contributed to the contemporary groundwater crisis. Perhaps the implementation of the Punjab Water Act 2019 and the control of groundwater by the state will be decisive for the governance trajectory of this resource. It is also possible that little will change on ground, as the reality of countries like Mexico and Spain that have re-centralised the ownership of groundwater to be controlled by the state shows. Pakistan first needs to unpack and put its policies and legislation concerning groundwater into careful implementation, before we find out the impact of these on the suitability of this precious resource.

References

Akhter M (2015) The hydropolitical cold war: the Indus Waters Treaty and state formation in Pakistan. Polit Geogr 46:65–75. https://doi.org/10.1016/j.polgeo.2014.12.002

Ali I (1988) The Punjab under imperialism, 1885–1947. Princeton University Press, Princeton. https://doi.org/10.1515/9781400859580

Anson R (1984) Public and private tube-well performance: emerging issues and options. World Bank. http://documents1.worldbank.org/curated/en/648291468287749149/pdf/multi-page.pdf

Basharat M, Sultan SJ, Malik AS (2015) Groundwater management in Indus Plain and integrated water resources management approach (IWASRI Publication No. 303). International Waterlogging and Salinity Research Institute (IWASRI), Lahore, Pakistan

Briscoe J, Qamar U, Contijoch M et al (2005) Pakistan's water economy: running dry. World Bank, Washington, DC

Buis A, Wilson J (2015) Study: third of big groundwater basins in distress. NASA, Washington, DC. https://www.nasa.gov/jpl/grace/study-third-of-big-groundwater-basins-in-distress

Bureau of Statistics Punjab (2017) Punjab development statistics 2017. Bureau of Statistics, Planning and Development Department, Government of Punjab. http://bos.gop.pk/developmentstat

Bureau of Statistics Punjab (2019) Punjab in figures 2019. Bureau of Statistics, Planning and Development Department, Government of Punjab

Dalin C, Wada Y, Kastner T, Puma MJ (2017) Groundwater depletion embedded in international food trade. Nature 543(7647):700–704. https://doi.org/10.1038/nature21403

Danvers FC (1877) A century of famines, being particulars of all famines that have visited India since the year 1770 and an inquiry as to the best means of providing against them. IOR/V/27/830/15. India Office Records and Private Papers, British Library, London

Dawn Newspaper (2018, June 4) Meat exports begin to grow again. https://www.dawn.com/news/1411794

Department of Land Records and Agriculture (1908) Annual report, Punjab Department of Land Records and Agriculture. India Office Records and Private Papers, British Library, London

Dorosh PA, Salam A (2007) Distortions to agricultural incentives in Pakistan. World Bank. http://documents.worldbank.org/curated/en/155531468313744488/pdf/560780NWP0PK0v1UBLIC10Pakistan10508.pdf

Ebrahim Z (2020) Pakistan abandoning cotton for water guzzling sugar cane. The Third Pole. April 9. https://www.thethirdpole.net/2020/04/09/pakistan-sugarcane/

Faysse N, Errahj M, Imache A et al (2014) Paving the way for social learning when governance is weak: supporting dialogue between stakeholders to face a groundwater crisis in Morocco. Soc Nat Resour 27(3):249–264. https://doi.org/10.1080/08941920.2013.847998

Fienen MN, Arshad M (2016) The international scale of the groundwater issue. In: Jakeman A et al (eds) Integrated groundwater management: concepts, approaches and challenges. Springer, Cham, pp 21–48. https://doi.org/10.1007/978-3-319-23576-9_2

FWM (2019) Federal Water Management Cell. Ministry of National Food Security and Research, Islamabad

Gilmartin D (2015) Blood and water: the Indus River Basin in modern history. University of California Press, Oakland

Government of Pakistan (1965) The third five year plan 1965–70. WAPDA Library, Lahore

Government of Pakistan (1978) The fifth five year plan 1978–83. WAPDA Library, Lahore

Government of Pakistan (1983) The sixth five year plan 1983–88. WAPDA Library, Lahore

Government of Pakistan (2017) Federal Budget 2017–18: Budget in Brief. Finance Division, Government of Pakistan, Islamabad. http://www.finance.gov.pk/budget/Budget%20in%20Brief%202017-18.pdf

IGRAC (2015) Transboundary aquifers of the World Map 2015. International Groundwater Resources Assessment Centre (IGRAC, UNESCO-IHP, Delft. https://www.un-igrac.org/resource/transboundary-aquifers-world-map-2015

Jacoby HG, Mansuri G (2018) Governing the commons? Water and power in Pakistan's Indus Basin. World Bank policy research working paper no. 8351. https://doi.org/10.1596/1813-9450-8351

Johnson RL (1989) Private tube well development in Pakistan's Punjab: review of past public programs/policies and relevant research. IIMI Country Paper Pakistan No. 1, International Irrigation Management Institute (IIMI), Colombo

Khair SM, Mushtaq S, Reardon-Smith K (2015) Groundwater governance in a water-starved country: public policy, farmers' perceptions, and drivers of tubewell adoption in Balochistan, Pakistan. Groundwater 53(4):626–637. https://doi.org/10.1111/gwat.12250

Linstead C, Sayed AH, Naqvi SA (2015) Water footprint of key industrial sectors in Punjab Pakistan. WWF-Pakistan, Lahore http://67.222.34.53/wsp/pdf/WFK_industrialSector.pdf

MacDonald AM, Bonsor HC, Ahmed KM et al (2016) Groundwater quality and depletion in the Indo-Gangetic Basin mapped from in situ observations. Nat Geosci 9(10):762–766. https://doi.org/10.1038/ngeo2791

Margat J, van der Gun J (2013) Groundwater around the world: a geographic synopsis. CRC, Boca Raton

Michel AA (1967) The Indus Rivers: a study of the effects of partition. Yale University Press, New Haven

Ministry of Finance (2017) Pakistan economic survey 2016–17: agriculture. Ministry of Finance, Government of Pakistan http://www.finance.gov.pk/survey_1718.html

Ministry of Finance (2018) Pakistan economic survey 2018–19: agriculture. Ministry of Finance, Government of Pakistan. http://www.finance.gov.pk/survey/chapters_18/02-Agriculture.pdf

Ministry of Finance (2019) Pakistan economic survey 2018–19. Ministry of Finance, Government of Pakistan. http://www.finance.gov.pk/survey/chapters_19/Economic_Survey_2018_19.pdf

Mirza GM, Latif M (2012) Assessment of current agro-economic conditions in Indus Basin of Pakistan. In Proceedings of international conference on water, energy, environment and food nexus: solutions and adaptations under changing climate, Lahore, Pakistan. 4–5 April 2012

Molle F, Shah T, Barker R (2003) The groundswell of pumps: multilevel impacts of a silent revolution. Paper presented at the ICID Asia regional workshop, Taipei, Taiwan, November 10-12

Mukherji A, Shah T (2005) Groundwater socio-ecology and governance: a review of institutions and policies in selected countries. Hydrogeol J 13(1):328–345. https://doi.org/10.1007/s10040-005-0434-9

Mustafa D (2010) Hydropolitics in Pakistan's Indus Basin. US Institute of Peace, Washington, DC

Nabeel F (2020) Political ecology of groundwater governance in Pakistan's Punjab. Unpublished thesis, University of Sussex, Brighton UK

Planning and Development Department (2015) Punjab growth strategy 2018: accelerating economic growth and improving social outcomes. Planning and Development Department, Government of Punjab, Lahore

Podgorski JE, Eqani SAMAS, Tasawar K et al (2017) Extensive arsenic contamination in high-pH unconfined aquifers in the Indus Valley. Sci Adv 3(8):e1700935. https://doi.org/10.1126/sciadv.1700935

Punjab Department of Agriculture (1938) Report on the operations of the Department of Agriculture, 1938/39–1940/41. IOR/V/24/11. India Office Records and Private Papers, British Library, London

Punjab Water Act (2019) This Act was passed by the Punjab Assembly on 20 November 2019; assented to by the Governor of the Punjab on 10 December 2019; and was published in the Punjab Gazette (Extraordinary), dated 13 December 2019, pages 2507–2529

Pursell G, Khan A, Gulzar S (2011) Pakistan's trade policies: future directions. IGC working paper 11/0361. International Growth Centre (IGC), London School of Economics and Political Science, London

Qureshi AS, McCornick PG, Sarwar A, Sharma BR (2010) Challenges and prospects of sustainable groundwater management in the Indus Basin, Pakistan. Water Resour Manag 24(8):1551–1569. https://doi.org/10.1007/s11269-009-9513-3

Ravenscroft P, Brammer H, Richards K (2009) Arsenic pollution. Wiley-Blackwell, Oxford doi: https://doi.org/10.1002/9781444308785

Raza M, Hussain F, Lee J et al (2017) Groundwater status in Pakistan: a review of contamination, health risks, and potential needs. Crit Rev Environ Sci Technol 47(18):1713–1762. https://doi.org/10.1080/10643389.2017.1400852

Report of the Indian Famine Commission (1880) Part I, Famine relief. IOR/L/E/6/12, File 551. India Office Records and Private Papers, British Library, London

Scholten W (2009) The water footprint of sugar and sugar-based ethanol. Thesis report, Department of Water Engineering and Management, Univerity of Twente, Enschede, The Netherlands. http://essay.utwente.nl/58701/1/scriptie_W_Scholten.pdf

Shah T (2007) The groundwater economy of South-Asia: an assessment of size, significance and socio-ecological impacts. In: Giordano M, Villholth KG (eds) The agricultural groundwater revolution: opportunities and threats to development. CABI, Wallingford, pp 7–36

Siddiqui IH (1986) Water works and irrigation system in India during pre-Mughal times. J Econ Soc Hist Orient 29(1):52–77. https://doi.org/10.1163/156852086X00036

Singh C (1991) Water rights and principles of water resources management. Indian Law Institute, New Delhi

Statistical Abstract Relating to British India (1885) Nos. 27–30. IOR/L/PARL/2/291. India Office Records and Private Papers, British Library, London

Statistical Abstract Relating to British India (1899–1900) Nos. 31–34. IOR/L/PARL/2/292. India Office Records and Private Papers, British Library, London

Tahir MA, Rasheed H, Imran S (2010) Water quality status in rural areas of Pakistan. PCRWR publication no. 143-2010. Pakistan Council of Research in Water Resources (PCRWR), Islamabad

Taylor GC Jr (1976) Historical review of the international water-resources program of the US geological survey 1940–70. United States Government, Washington, DC

Taylor RG, Scanlon B, Döll P et al (2013) Ground water and climate change. Nat Clim Chang 3(4):322–329. https://doi.org/10.1038/nclimate1744

The Express Tribune Rice (2017, February 1) Exporters: commerce minister promises 50% subsidy for brand promotion. https://tribune.com.pk/story/1313170/rice-exporters-commerce-minister-promises-50-subsidy-brand-promotion/

The News (2017, May 18) Govt assures exporters zero rating regime for 5 major sectors. https://www.thenews.com.pk/print/205036-Govt-assures-exporters-zero-rating-regime-for-5-major-sectors

van Steenbergen F (1995) The frontier problem in incipient groundwater management regimes in Balochistan (Pakistan). Hum Ecol 23(1):53–74. https://doi.org/10.1007/BF01190098

van Steenbergen F, Gohar S (2005) Groundwater development and management. In: Briscoe J, Qamar U (eds) Pakistan water economy running dry – background papers. Oxford University Press, Oxford

Watto MA, Mugera AW (2016) Groundwater depletion in the Indus Plains of Pakistan: imperatives, repercussions and management issues. Int J River Basin Manag 14(4):447–458. https://doi.org/10.1080/15715124.2016.1204154

Wescoat JL, Halvorson SJ, Mustafa D (2000) Water management in the Indus Basin of Pakistan: a half-century perspective. Int J Water Res Dev 16(3):391–406. https://doi.org/10.1080/713672507

Young W, Anwar A, Bhatti T et al. (2019) Pakistan: getting more from water. World Bank, Washington, DC. https://openknowledge.worldbank.org/handle/10986/31160

Zardari NH, Cordery I (2009) Water productivity in a rigid irrigation delivery system. Water Resour Manag 23(6):1025–1040. https://doi.org/10.1007/s11269-008-9312-2

Zulfiqar F, Thapa GB (2017) Agricultural sustainability assessment at provincial level in Pakistan. Land Use Policy 68:492–502. https://doi.org/10.1016/j.landusepol.2017.08.016

Chapter 11
Spatial Variability of Groundwater Storage in Pakistan

Kamal Ahmed, Muhammad Arif Watto, Shamsuddin Shahid, Nadeem Nawaz, and Najeebullah Khan

Abstract In recent years, the use of Gravity Recovery and Climate Experiment (GRACE) data has emerged as a valuable tool for investigating groundwater resources in data-scarce regions. This chapter reports on a study carried out to investigate and understand trends in the spatial variability of groundwater storage of Pakistan. We used three sets of GRACE data in this study: the data of Center for Space Research (CSR); Jet Propulsion Laboratory (JPL); and GeoforschungsZentrum Potsdam (GFZ). These datasets cover the period 2003 to 2016 at a spatial resolution of 1.0° × 1.0°. The rate of change in groundwater storage was assessed using Sen's slope and the significance of the change was estimated using a modified version of the Mann-Kendall (MK) test. The results showed that the degree of groundwater storage change was different across different regions. The decline in groundwater storage was noticeably higher in areas of intensive agriculture, with a range of −9.02 to −11.3 cm/year, and was higher in the south of the country compared to other parts. The MK test revealed the declines were significant for 35 grid cells across different parts of the country. The findings confirm the need for measures to be taken to prevent overuse of groundwater resources to avert a crisis in the near future.

Keywords Groundwater depletion · Groundwater storage variability · Groundwater storage trends · GRACE · Pakistan

K. Ahmed (✉) · N. Nawaz · N. Khan
Department of Water Resource Management, Lasbela University of Agriculture, Water and Marine Sciences, Balochistan, Pakistan

M. A. Watto
University of Agriculture Faisalabad, Sub-campus Depalpur, Okara, Pakistan
e-mail: arif.watto@uaf.edu.pk

S. Shahid
School of Civil Engineering, Faculty of Engineering, Universiti Teknologi Malaysia (UTM), Johor Bahru, Malaysia
e-mail: sshahid@utm.my

M. A. Watto et al. (eds.), *Water Resources of Pakistan*, World Water Resources 9,
https://doi.org/10.1007/978-3-030-65679-9_11

11.1 Introduction

Groundwater has become an important resource in sustaining agriculture and ensuring food security in Pakistan. Due to Pakistan's location within a predominantly semi-arid region, the agriculture sector relies heavily on irrigation sourced from both surface water and groundwater. In Pakistan, irrigation demand has rapidly increased due to agricultural intensification, and the extensive network of surface water is devoid of enough supplies to meet demand. Over the past few decades, the use of groundwater has, therefore, become an increasingly critical resource across the country due to seasonal variability of surface supply. Since the early 1960s, the share of groundwater used for irrigation across Pakistan has increased from just 8% in 1960 (Byerlee and Siddiq 1994) to over 60% in 2019 (Qureshi and Ashraf 2019). Yet, it is anticipated that reliance on groundwater will increase as demand for irrigation increases mainly due to the effects of climatic change (Qureshi and Ashraf 2019; Rasul 2016).

Currently, Pakistan is ranked fourth most groundwater using country globally, consuming about 6.6% of global groundwater used (Margat and Van der Gun 2013; Siebert et al. 2010). Pakistan irrigates about 5.2 million hectares with groundwater, which is about 4.6% of the total global groundwater-fed cropland area (Siebert et al. 2010). Consequently, the available groundwater resources are under enormous pressure from over-drafting in many regions of the country (Rodell et al. 2009; Wada et al. 2010; Watto and Mugera 2016). It is estimated that Pakistan abstracts about 61 km^3 of groundwater each year to meet the current irrigation water requirements, which is higher than the annually replenished groundwater reserves of 55 km^3, resulting in groundwater depletion (FAO 2020). Estimates indicate that groundwater abstractions between 2000 and 2009 have resulted in 3.61 km^3 of groundwater depletion in Pakistan annually (Döll et al. 2014).

The massive increase in groundwater use has been aided by unimpeded tube-well development throughout the country over the last few decades (Fig. 11.1). Perhaps the most intriguing factors of this groundwater revolution were the Indus Waters Treaty and the Green Revolution (Watto and Mugera 2016). The Indus

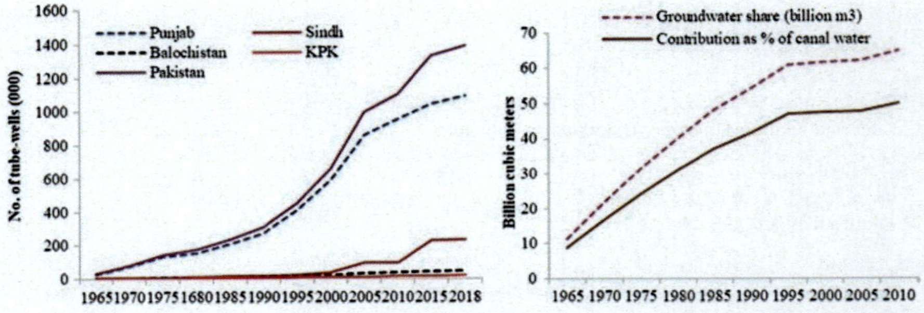

Fig. 11.1 Historical trends of conventional tube-well development and the share of groundwater use in irrigation in Pakistan. (Source: Watto and Mugera 2016; FWM 2019)

Waters Treaty deprived Pakistan of multi-million cusecs of water that used to flow into Pakistan via its eastern rivers. The Green Revolution of the 1960s helped increase cropping areas and intensities by almost twofold and were largely supported through groundwater abstractions. The government incentivised tube-well development through various support policies, such as expanding grid electricity to rural areas, subsidising electricity and diesel, and providing long-term low-interest loans (van Steenbergen 2002).

The exploitation of groundwater resources helped farmers achieve higher yields, increase cropping intensities and ultimately fetch better economic returns. These benefits further accelerated private tube-well development and contributed to increased cultivation of water-intensive crops such as rice and sugarcane (Ahmad et al. 2004; Falcon and Gotsch 1968; Kazmi et al. 2012; Meinzen-Dick 1996; Mohammad 1964; Mohammad 1965; Nulty 1972). Increased irrigation water demands accompanied by seasonal variability of surface water flows and inflexibility of the *Warabandi* (surface water allocation) system led groundwater use on a large scale (Watto and Mugera 2016). In the late 1970s, the government started promoting private tube-wells in the country which increased from about 30,000 in 1965 to over 1.3 million in 2018 (statistics as shared by the Federal Water Management Cell, Ministry of Food Security and Research, Government of Pakistan). As a result, groundwater contribution to irrigation supplies also increased by 467% since 1960 (Watto and Mugera 2016).

Such extensive and continued groundwater abstractions, especially for irrigation, have resulted in a significant lowering of groundwater tables across many parts of the country (Qureshi et al. 2010; Watto and Mugera 2016; Young et al. 2019). Rapidly falling groundwater tables not only mean groundwater pumping is becoming economically unviable but also undermine the broader environmental integrity on which sustainability of irrigated agriculture relies (Kahlown and Azam 2002; Kelleners and Chaudhry 1998; Khan et al. 2008; Kijne 1999; Qureshi et al. 2010). In Punjab, declining groundwater tables require tube-wells to be re-installed at greater depths every few years. Similarly, in Balochistan, much of the traditional Karez irrigation system no longer has water as aquifers become depleted. In some valleys of Balochistan, farmers have installed submersible electrical pumps at greater depths to chase falling groundwater tables which, for example in Kuchlagh valley in Quetta, has exhausted the aquifers due to extensive pumping of groundwater over the last three decades (van Steenbergen et al. 2015).

11.2 Geography and Climate of Pakistan

Pakistan is an agriculture-dependent country located (Fig. 11.2) in the South Asian region (Ahmed et al. 2019). The climate of the country is predominantly arid to semi-arid with a mean temperature of 35 °C in summer and below 5 °C in winter (Khan et al. 2018). Precipitation is sporadic and varies from less than 50 mm in the

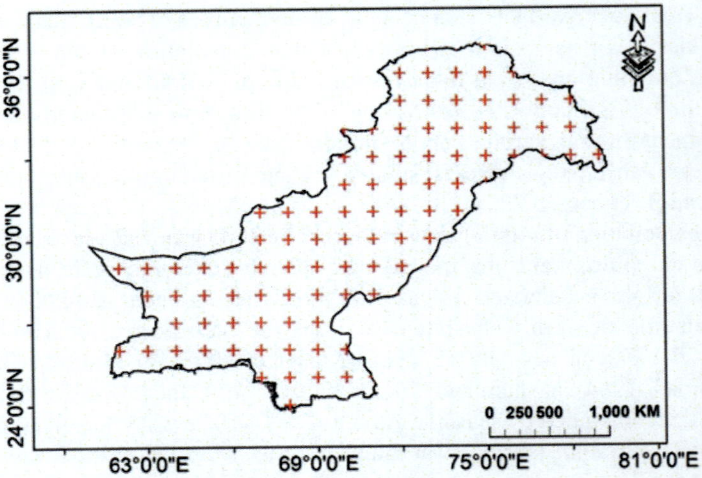

Fig. 11.2 Map of Pakistan: Plus signs are the grid cells where GRACE CSR, GFZ, and JPL data were extracted

southwest to 400 in the north (Ahmed et al. 2017). Pakistan's climate can be classified into four seasons namely: monsoon, between June and September; post-monsoon, October and November; winter, between December and March; and pre-monsoon, April and May.

The country receives the majority of rainfall (58.5%) during the monsoon. However, rainfall is not sufficient for agricultural production in most of the country. A large portion of agriculture is undertaken by using Indus River water (Iqbal et al. 2019), even though most rivers in the country are non-perennial. Less rainfall and high temperatures make water bodies dry up during non-monsoon months. Therefore, the majority of the farmers depend on groundwater for agriculture (Ahmed et al. 2015; van Steenbergen et al. 2015).

11.2.1 Groundwater Resources of Pakistan

As the world's fourth-largest groundwater withdrawing country (after China, India, and the USA), Pakistan relies on groundwater to provide 60% of its irrigation water supply, 90% of its domestic and almost 100% of its industrial water supply (Qureshi and Ashraf 2019). The Indus Basin is a flat unconsolidated region of alluvial deposits, covering 16 million hectares, underlain by an extensive, contiguous and unconfined aquifer (Qureshi et al. 2008). Across the Indus Basin, natural variations in alluvial deposits, climate, and the pattern of irrigation channels influence the groundwater typology. In Balochistan, groundwater is mainly confined in consolidated sedimentary landscapes (Young et al. 2019).

Pakistan's renewable groundwater resources are replenished through multiple ways, including recharge from rainfall, river flows, floods, canal seepage, and irrigation return flows. However, an exact recharge from these sources and ultimately an overall assessment of the country's renewable groundwater resource is difficult to make, with estimates varying. Briscoe et al. (2005) estimated that Pakistan's annual renewable groundwater resources are 63 km^3, FAO (2016) estimated internally generated groundwater resources as being 55 km^3, while WSTF (2012)'s estimates were relatively higher at 74 km^3.

Similarly, estimates by different studies on groundwater recharge across Pakistan from also differ. For instance, van Steenbergen and Gohar (2005) estimated that direct rainfall contributes 21%, seepage from canals and distributaries 45%, irrigation return flows 26%, river recharge 6%, and other sources 2% to the total groundwater recharge, while the WSTF (2012) estimates 66% groundwater recharge from irrigation return flows and 18% from direct rainfall. WSTF (2012) also provided province-wise estimates of groundwater balance which suggest that groundwater is in balance for all provinces other than KPK. However, these WSTF estimates are different from others, including those based on GRACE satellite data, and assessments based on in situ observations with robust regional-scale assessments, which contrarily revealed significant groundwater depletion across Pakistan (MacDonald et al. 2016).

11.3 Gravity Recovery and Climate Experiment (GRACE) Data

Generally, a dense network of groundwater monitoring wells is required for proper investigation of groundwater variations (Sediqi et al. 2019). However, the installation of such a network is costly and often not established in developing countries (Krishanan et al. 2009). The resulting lack of data is a major hurdle for the proper investigation of groundwater variations in most groundwater use regions (Han et al. 2016). The advancement of computing technologies has opened an era for institutes to use data based on satellite information (Rodell et al. 2009). Several institutes have developed groundwater storage variations data under the GRACE project umbrella (Sediqi et al. 2019). GRACE data products are effective tools for investigating groundwater resources (Bhanja et al. 2016, 2018; Frappart and Ramillien 2018). The twin GRACE satellites were first launched in 2002 to measure the mean and time-varying component of Earth's gravity field. They measure minute fluctuations of the earth's gravity field by monitoring variations between the two satellites as they orbit Earth (Tapley et al. 2004b). The change in distance between the two satellites and the time variations of the gravity field on the earth is due to water mass variations (Tapley et al. 2004a). In a very short time, different institutes produced numerous data products at different spatial scales. In the present study, three data products obtained from the Center for Space Research (CSR), Jet Propulsion Laboratory (JPL), and GeoforschungsZentrum Potsdam (GFZ) were used.

11.4 Methodology

The present study was conducted in four major steps. In the first step, the data of GRACE CSR, GFZ, and JPL with a spatial resolution of 1° × 1° were downloaded and extracted over 127 grid points covering the data period of 2003 to 2016. Any missing data were filled using mean values from corresponding months. In the second step, the mean minimum values of groundwater storage for each year from all data products were individually extracted to develop a series. Next, the magnitude of change and the significance of trends were investigated by applying Sen's slope and modified Mann-Kendall (MMK) trend respectively at a 95% level of confidence at individual grid cells. Finally, spatial patterns of groundwater storage change and magnitude of change with significant trends were mapped using ArcMap 10.3. The details of the methods used in the present study are given below.

11.4.1 Sen's Slope Estimator

The Sen's slope estimator (Sen 1968) is a robust method frequently applied for the investigation of the magnitude of change (Ahmed et al. 2018). The method is non-parametric, relies on the median of the slopes of data, and can be calculated as follows:

$$Q_s = \begin{cases} Q_{(n+1)/2} & \text{when n is odd} \\ \dfrac{Q_{n/2} + Q_{(n+2)/2}}{2} & \text{when n is even} \end{cases} \qquad (11.1)$$

where Q_s is Sen's slope, Q is the slopes between two equally spaced data, and n is the total data.

11.4.2 Modified Mann-Kendall (MK) Test

In the modified Mann-Kendall (MMK) test (Hamed 2008), the existence of a trend is initially evaluated using the classical Mann-Kendall test (Mann 1945). If the trend exists in series, the MMK test eliminates the trend and estimates the normal variants of ranks on the de-trended series. Later, the autocorrelation (ρ_l) of lag (l) for an assumed Hurst coefficient (*H*) is estimated using the following equation as suggested by Koutsoyiannis (2003):

$$\rho_l = \frac{1}{2}\left(|l+1|^{2H} - 2|l|^{2H} + |l-1|^{2H}\right) \tag{11.2}$$

The mean and standard deviation of *H* at 0.5 is used to assess the significance. If the estimated *H* is found significant then the biased variance of MMK test is calculated using Eq. 11.3 and then the bias is corrected with bias correction (*B*) using Eq. 11.4:

$$V(S)^{H'} = \sum_{i<j}\sum_{k<l}\frac{2}{\pi}\sin^{-1}\left(\frac{\rho|j-i| - \rho|i-l| - \rho|j-k| + \rho|i-k|}{\sqrt{(2-2\rho|i-j|)(2-2\rho|k-l|)}}\right) \tag{11.3}$$

$$V(S)^{H} = V(S)^{H'} \times B \tag{11.4}$$

The significance of trend is confirmed using Z statistics as below:

$$Z = \{ \begin{array}{ll} \frac{S-1}{\sqrt{V(S)^H}} & \text{when S} > 0 \\ 0 & \text{when S} = 0 \\ \frac{S-1}{\sqrt{V(S)^H}} & \text{when S} < 0 \end{array} \tag{11.5}$$

More details on the MMK test can be found in the study by Ahmed et al. (2017).

11.4.3 Assessment of Spatial Patterns of Changes

Spatial patterns of hydro-climatic variables are often mapped to understand the dynamics of variables over time and space. Nevertheless, understanding the spatial patterns of any variable requires a variety of information over a large area. A geographic information system (GIS) is widely used to demonstrate the spatial patterns of a variable on the map. One such GIS is ArcMap 10.3, which provides numerous options to map spatial data. However, in the present study, spatial patterns of groundwater storage and the magnitude of changes with significance are displayed at each grid point with the use of symbols. Colour patterns are applied to better understand the spatial patterns of groundwater storage and the magnitude of changes, and the significance of these trends are displayed using plus and minus signs.

11.5 Results and Discussion

11.5.1 Monthly and Yearly Variations in Groundwater Mean Storage

In order to understand the fluctuations of groundwater mean storage in different months and years, the data of groundwater variations obtained from different GRACE products for the period 2003 to 2016 in different months are summarised. Groundwater storage data for a random grid cell are shown in Fig. 11.3 as an example, and show how groundwater storage varies across months and years. In this case, groundwater storage varied from −15.7 to +16.1, −13.9 to +15.4 and −13.2 to +15.9 cm/month against the mean for CSR, GFZ and JPL data respectively.

Monsoon and winter precipitation are very important for agriculture in Pakistan (Chaudhry and Rasul 2012; Afzal et al. 2013). The results of the present study confirm that groundwater storage variations are a result of precipitation events and agricultural activities throughout the country. For instance, groundwater storage in the month of July showed an increase after the commencement of the monsoon season in June.

The data also demonstrated drastic declines in groundwater storage at all locations in May and June, being the start of the main *Kharif* cropping season, which suggests extractions exceed recharge during these months. The interannual variations in groundwater showed a large change in groundwater storage in the years 2007 and 2016 in all products. It is worth mentioning that severe droughts in 2014–2016 affected almost the whole country (Ahmed et al. 2018). Lack of precipitation during these years, coupled with ongoing extraction of groundwater for irrigation, triggered a major reduction in groundwater storage.

11.5.2 Spatial Patterns of Groundwater Storage Variations

The spatial characteristics of groundwater storage variations using CSR, GFZ and JPL data are displayed in Fig. 11.4. The figures were prepared by presenting groundwater storage values at each grid point using ArcMap 10.3. The groundwater

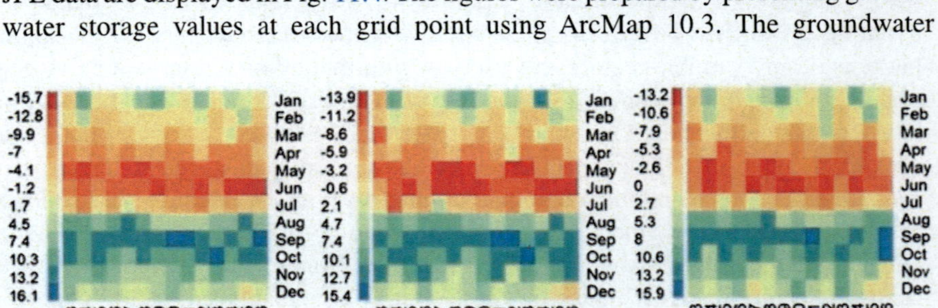

Fig. 11.3 Seasonal variations in groundwater storage in different months during 2003–2016 estimated using three GRACE products (**a**) CSR, (**b**) GFZ and (**c**) JPL at a random grid point

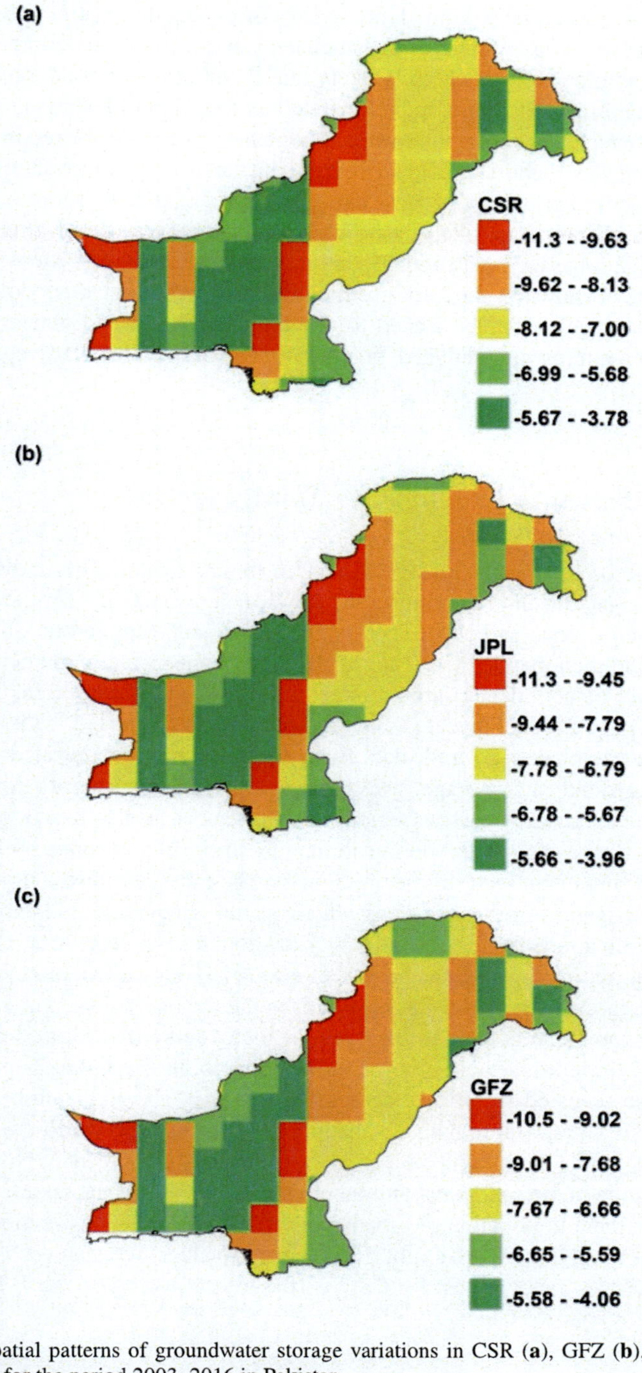

Fig. 11.4 Spatial patterns of groundwater storage variations in CSR (**a**), GFZ (**b**), and JPL (**c**) GRACE data for the period 2003–2016 in Pakistan

storage values are equally divided into five classes using the natural breaks available in ArcMap 10.3. The figures reveal that changes in groundwater storage are uneven across the whole country, which is likely due to variations in topography, the suitability of land for agriculture, and climate conditions across Pakistan.

The lowest changes in groundwater storage have mainly occurred in areas in the south, while the highest changes occurred in the west. The northern regions of Pakistan, where agriculture is most intense, show a change of about −6 to −8 cm/year across all products, likely to be caused by extensive use of groundwater for irrigation (Chinnasamy et al. 2015). A storage change of about −7 to −9 cm/year in all products is shown in the very north of Pakistan, where glaciers cover the landmass. These results may be linked to climate change-induced glacier melts from rising temperatures and reduced precipitation in the Himalayan region (Moiwo et al. 2011).

11.5.3 Trends in Groundwater Storage Decline

The magnitude of change and significance of trends in annual groundwater storage were investigated using Sen's slope and MMK test respectively for the period 2003 to 2016. Sen's slope and MMK tests were individually applied to CSR, GFZ and JPL data and then mapped (see Fig. 11.5). Significant increase and decrease trends (>95% level of confidence) are indicated with plus and minus signs respectively. Data from each product revealed a similar magnitude of change, with minor differences in a few places. Overall, data from all three products revealed a significant decline in groundwater storage, with greatest concentrations across the agricultural lands of the central and eastern regions of the country, as also found by Khan et al. (2008), as well as across the border in neighbouring Indian states by Bhanja et al. (2018). The major cause of this declination may be excessive abstractions of groundwater for irrigation in these regions which, if not adequately addressed, can have serious implications for the sustainability of groundwater resources, and hence for irrigated agriculture. Van Steenbergen et al. (2015) showed stern concerns on the over-exploitation of groundwater resources in the region. A few significant decreasing trends were also noticed in the south of the country. The highest decrease was noticed on the northwestern side of the country in all the GRACE products. The figures also revealed significant decreasing trends at a higher number of grids in CSR and GFZ data (more than 27 grid points) compared with JPL, which showed significant trends at 35 grid points. It is important to note that positive change mainly occurs in the southwest and the northeast corners of the country.

It is pertinent to note that the spatial patterns of all products are more or less the same even though there are slight differences in values between each product (i.e. from −11.3 to −3.78 cm/year for CSR, −10.5 to −4.06 cm/year for JPL, and −11.3 to −3.96 cm/year for GFZ).

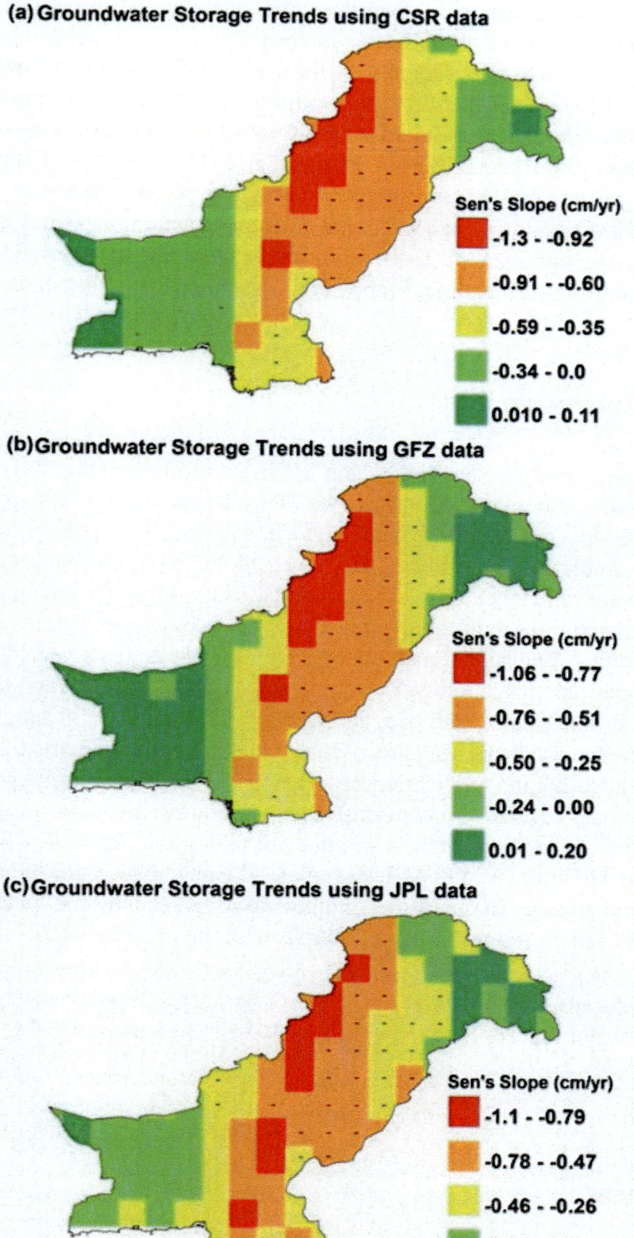

Fig. 11.5 Spatial patterns of groundwater storage trends in CSR (**a**) GFZ (**b**) and JPL (**c**) GRACE data for the period 2003–2016 in Pakistan. Negative (−) mark indicates reducing trends at 95% confidence level estimated using the MMK test

The results of the study confirm the rapid reduction in groundwater storage in Pakistan. The decline in groundwater forces farmers to abstract groundwater from greater depths which, consequently, will raise irrigation costs and overall production costs. The rise in irrigation costs would have a direct impact on farmer profitability and livelihoods, and should pressure farmers and others to change practices. In particular, governments across Pakistan need to identify measures that will encourage sustainable management of available water resources and increase groundwater recharge through structural and non-structural measures. Farmers need support and encouragement to grow crops that require less water and to use water more efficiently. Cultivation of high water-consuming crops should be restricted.

11.6 Conclusions

This study has applied Sen's slope and MMK test to investigate the changing patterns of groundwater storage in Pakistan. This investigation was carried out using the GRACE data of CSR, JPL, and GFZ for the period 2003 to 2016, with relatively consistent trends and results. The study revealed that monthly changes in groundwater storage are influenced by seasonal fluctuations in precipitation and the extent of agricultural activities in the study area. Groundwater storage declines the most from January to June. The spatial patterns of groundwater storage revealed that there are uneven variations in groundwater storage across the country, which may be due to variations in landscape, land use, and climate. The trends in groundwater storage revealed the lowest change in parts of the south and highest in parts of the west. The study revealed that the application of GRACE data over rough topography and the diverse climate of Pakistan is useful where groundwater-monitoring stations are scarce. The present study was conducted using the data of CSR, JPL, and GFZ. However, recently, GRACE has released a new dataset for the assessment of groundwater storage. In the future, another study can also be conducted to compare the results of the latest versions of GRACE with the previous ones.

Acknowledgments The work is supported by the International Foundation for Science (IFS) and the Organisation of Islamic Cooperation's Standing Committee on Scientific and Technological Cooperation (COMSTECH) via grant no I-2-W-6266-1. This work is also supported by the Higher Education Commission of Pakistan via grant no. No: 9568/Balochistan/NRPU/R&D/HEC/2017.

References

Afzal M, Haroon M, Rana A, Imran A (2013) Influence of North Atlantic oscillations and Southern oscillations on winter precipitation of Northern Pakistan. Pak J Meteorol 9(18):1–8

Ahmad I, Shah SAH, Zahid MS (2004) Why the Green Revolution was short-run phenomena in the development process of Pakistan: a lesson for future. J Rur Dev Admin 4(35):89–104

Ahmed K, Shahid S, bin Harun S et al (2015) Assessment of groundwater potential zones in an arid region based on catastrophe theory. Earth Sci Inf 8:539–549. https://doi.org/10.1007/s12145-014-0173-3

Ahmed K, Shahid S, Chung E-S et al (2017) Spatial distribution of secular trends in annual and seasonal precipitation over Pakistan. Clim Res 74:95–107. https://doi.org/10.3354/cr01489

Ahmed K, Shahid S, Nawaz N (2018) Impacts of climate variability and change on seasonal drought characteristics of Pakistan. Atmos Res 214:364–374. https://doi.org/10.1016/j.atmosres.2018.08.020

Ahmed K, Shahid S, Wang X et al (2019) Spatiotemporal changes in aridity of Pakistan during 1901–2016. Hydrol Earth Syst Sci 23:3081–3096. https://doi.org/10.5194/hess-23-3081-2019

Bhanja SN, Mukherjee A, Saha D et al (2016) Validation of GRACE based groundwater storage anomaly using in-situ groundwater level measurements in India. J Hydrol 543(B):729–738. https://doi.org/10.1016/j.jhydrol.2016.10.042

Bhanja SN, Mukherjee A, Rodell M (2018) Groundwater storage variations in India. In: Mukherjee A (ed) Groundwater of South Asia. Springer, Singapore, pp 49–59. https://doi.org/10.1007/978-981-10-3889-1_4

Briscoe J, Qamar U, Contijoch M et al (2005) Pakistan's water economy: running dry. World Bank, Washington, DC

Byerlee D, Siddiq A (1994) Has the green revolution been sustained? The quantitative impact of the seed-fertilizer revolution in Pakistan revisited. World Dev 22(9):1345–1361. https://doi.org/10.1016/0305-750X(94)90008-6

Chaudhry QZ, Rasul G (2012) Pakistan summer monsoon. In: Tiyagi A, Asnani GC, De US et al (eds) Monsoon monograph. India Meteorological Department, Ministry of Earth Sciences, Government of India, pp 120–131

Chinnasamy P, Maheshwari B, Prathapar S (2015) Understanding groundwater storage changes and recharge in Rajasthan, India through remote sensing. Water 7(10):5547–5565. https://doi.org/10.3390/w7105547

Döll P, Müller Schmied H, Schuh C et al (2014) Global-scale assessment of groundwater depletion and related groundwater abstractions: combining hydrological modeling with information from well observations and GRACE satellites. Water Resour Res 50(7):5698–5720. https://doi.org/10.1002/2014WR015595

Falcon WP, Gotsch CH (1968) Lessons in agricultural development: Pakistan. In: Papanek GF (ed) Development policy: theory and practice. Harvard University Press, Cambridge, MA, pp 269–315

FAO (2016) AQUASTAT main database. Food and Agriculture Organization of United Nations. http://www.fao.org/nr/water/aquastat/data/query/index.html. Accessed 9 Mar 2020

FAO (2020) AQUASTAT. Food and Agriculture Organization of United Nations. http://www.fao.org/nr/water/aquastat/data/query/index.html. Accessed 9 Mar 2020

Frappart F, Ramillien G (2018) Monitoring groundwater storage changes using the gravity recovery and climate experiment (GRACE) satellite mission: a review. Remote Sens 10(6):829. https://doi.org/10.3390/rs10060829

FWM (2019) Federal Water Management Cell. Ministry of National Food Security and Research. Islamabad

Hamed KH (2008) Trend detection in hydrologic data: the Mann–Kendall trend test under the scaling hypothesis. J Hydrol 349:350–363. https://doi.org/10.1016/j.jhydrol.2007.11.009

Han D, Currell MJ, Cao G (2016) Deep challenges for China's war on water pollution. Environ Pollut 218:1222–1233. https://doi.org/10.1016/j.envpol.2016.08.078

Iqbal Z, Shahid S, Ahmed K et al (2019) Spatial distribution of the trends in precipitation and precipitation extremes in the sub-Himalayan region of Pakistan. Theor Appl Climatol 137:2755–2769. https://doi.org/10.1007/s00704-019-02773-4

Kahlown MA, Azam M (2002) Individual and combined effect of waterlogging and salinity on crop yields in the Indus basin. Irrig Drain 51:329–338. https://doi.org/10.1002/ird.62

Kazmi SI, Ertsen MW, Asi MR (2012) The impact of conjunctive use of canal and tube well water in Lagar irrigated area. Pak Phys Chem Earth Parts A/B/C 47:86–98. https://doi.org/10.1016/j.pce.2012.01.001

Kelleners T, Chaudhry M (1998) Drainage water salinity of tubewells and pipe drains: a case study from Pakistan. Agric Water Manag 37(1):41–53. https://doi.org/10.1016/S0378-3774(98)00037-7

Khan S, Rana T, Gabriel H, Ullah MK (2008) Hydrogeologic assessment of escalating groundwater exploitation in the Indus Basin, Pakistan. Hydrogeol J 16:1635–1654. https://doi.org/10.1007/s10040-008-0336-8

Khan N, Shahid S, Tb I, Wang X-J (2018) Spatial distribution of unidirectional trends in temperature and temperature extremes in Pakistan. Theor Appl Climatol 136:899–913. https://doi.org/10.1007/s00704-018-2520-7

Kijne JW (1999) Improving the productivity of Pakistan's irrigation: the importance of management choices. International Water Management Institute, Colombo

Koutsoyiannis D (2003) Climate change, the Hurst phenomenon, and hydrological statistics. Hydrol Sci J 48(1):3–24. https://doi.org/10.1623/hysj.48.1.3.43481

Krishanan S, Islam A, Machiwal D et al (2009) Using the living wisdom of well drillers to construct digital groundwater data bases across Indo-Gangetic basin. In: Mukherji A, Villholth KG, Sharma BR, Wang J (eds) Groundwater governance in the Indo-Gangetic and Yellow River basins: realities and challenges. CRC Press, Boca Raton, pp 247–264

MacDonald A, Bonsor H, Ahmed K et al (2016) Groundwater quality and depletion in the Indo-Gangetic Basin mapped from in situ observations. Nat Geosci 9:762–766

Mann HB (1945) Nonparametric tests against trend. Econometrica 13(3):245–259. https://doi.org/10.2307/1907187

Margat J, Van der Gun J (2013) Groundwater around the world: a geographic synopsis. CRC Press, Boca Raton

Meinzen-Dick RS (1996) Groundwater markets in Pakistan: participation and productivity. Research report 105, International Food Policy Research Institute, Washington DC

Mohammad G (1964) Some strategic problems in agricultural development in Pakistan. Pak Dev Rev 4:223–260

Mohammad G (1965) Private tubewell development and cropping patterns in West Pakistan. Pak Dev Rev 5:1–53

Moiwo JP, Yang Y, Tao F et al (2011) Water storage change in the Himalayas from the gravity recovery and climate experiment (GRACE) and an empirical climate model. Water Resour Res 47(7):W07521. https://doi.org/10.1029/2010WR010157

Nulty L (1972) The green revolution in West Pakistan. Implications of technological change. Praeger, New York

Qureshi R, Ashraf M (2019) Water security issues of agriculture in Pakistan. Pakistan Academy of Sciences (PAS), Islamabad. https://www.paspk.org/wp-content/uploads/2019/06/PAS-Water-Security-Issues.pdf

Qureshi AS, McCornick PG, Qadir M, Aslam Z (2008) Managing salinity and waterlogging in the Indus Basin of Pakistan. Agric Water Manag 95(1):1–10. https://doi.org/10.1016/j.agwat.2007.09.014

Qureshi AS, McCornick PG, Sarwar A, Sharma BR (2010) Challenges and prospects of sustainable groundwater management in the Indus Basin, Pakistan. Water Resour Manag 24(8):1551–1569. https://doi.org/10.1007/s11269-009-9513-3

Rasul G (2016) Managing the food, water, and energy nexus for achieving the sustainable development goals in South Asia. Environ Dev 18:14–25. https://doi.org/10.1016/j.envdev.2015.12.001

Rodell M, Velicogna I, Famiglietti JS (2009) Satellite-based estimates of groundwater depletion in India. Nature 460:999–1002. https://doi.org/10.1038/nature08238

Sediqi MN, Shiru MS, Nashwan MS et al (2019) Spatio-temporal pattern in the changes in availability and sustainability of water resources in Afghanistan. Sustainability 11(20):5836. https://doi.org/10.3390/su11205836

Sen PK (1968) Estimates of the regression coefficient based on Kendall's tau. J Am Stat Assoc 63(324):1379–1389

Siebert S, Burke J, Faures JM et al (2010) Groundwater use for irrigation–a global inventory. Hydrol Earth Syst Sci 14(10):1863–1880. https://doi.org/10.5194/hess-14-1863-2010

Tapley BD, Bettadpur S, Ries JC et al (2004a) GRACE measurements of mass variability in the earth system. Science 305(5863):503–505. https://doi.org/10.1126/science.1099192

Tapley BD, Bettadpur S, Watkins M, Reigber C (2004b) The gravity recovery and climate experiment: mission overview and early results. Geophys Res Lett 31:L09607. https://doi.org/10.1029/2004GL019920

van Steenbergen F (2002) Local groundwater regulation water praxis documents. Water praxis document no. 14, Land & Water Product Management Group, ARCADIS Euroconsult. http://www.groundwatermanagement.org/documents/14locgrounwregulationpraxis.pdf

van Steenbergen F, Gohar S (2005) Groundwater development and management. In: Briscoe J, Qamar U (eds) Pakistan water economy running dry – background papers. Oxford University Press, Oxford

van Steenbergen F, Kaisarani AB, Khan NU, Gohar MS (2015) A case of groundwater depletion in Balochistan, Pakistan: enter into the void. J Hydrol Reg Stud 4:36–47. https://doi.org/10.1016/j.ejrh.2014.11.003

Wada Y, Van Beek LP, Van Kempen CM et al (2010) Global depletion of groundwater resources. Geophys Res Lett 37(20):L20402. https://doi.org/10.1029/2010GL044571

Watto MA, Mugera AW (2016) Groundwater depletion in the Indus Plains of Pakistan: imperatives, repercussions and management issues. International Journal of River Basin Management 14:447–458. https://doi.org/10.1080/15715124.2016.1204154

WSTF (2012) A productive and water-secure Pakistan: infrastructure, institutions, strategy. The report of the water sector task force (WSTF) of the friends of democratic Pakistan, Islamabad, Pakistan

Young W, Anwar A, Bhatti T et al. (2019) Pakistan: getting more from water. World Bank, Washington, DC. https://openknowledge.worldbank.org/handle/10986/31160

Chapter 12
Impacts of Water Quality on Human Health in Pakistan

Safdar Bashir, Zubair Aslam, Nabeel Khan Niazi, Muhammad Imran Khan, and Zhongbing Chen

Abstract Pakistan is ranked third among countries facing severe water shortage. Yet the country's water resources are being depleted with such intensity that it has become a country with physical water deficiency, with water contamination another contributing factor. Most of the population is exposed to hazards of drinking unsafe and polluted surface and groundwater due to unavailability of clean water. Over 60% of the population get their drinking water from hand or motor pumps, with the figure in rural areas being over 70%. Inadequate quantity and quality of potable water and poor sanitation facilities and practices are associated with a host of illnesses such as diarrhoea, typhoid, intestinal worms and hepatitis resulting in national income losses of Rs. 25–28 billion annually, or approximately 0.6–1.44% of the country's GDP. Studies conducted by national and international agencies suggest that all key quality parameters for Pakistan's drinking water are exceeding acceptable limits. This chapter provides an updated review of literature on water quality issues in Pakistan, sources of contamination, and health impacts. The review leads to a set of strategies that Pakistan could adopt to reduce the level of waterborne health impacts.

Keywords Water contamination · Carcinogenic · Waterborne diseases · Heavy metals · Microbial pathogens

---

S. Bashir (✉)
Department of Soil and Environmental Sciences, Ghazi University, D.G. Khan, Punjab, Pakistan
e-mail: safdar.bashir@gudgk.edu.pk

Z. Aslam
Department of Agronomy, University of Agriculture, Faisalabad, Faisalabad, Pakistan

N. K. Niazi · M. I. Khan
Institute of Soil and Environmental Sciences, University of Agriculture, Faisalabad, Faisalabad, Pakistan

Z. Chen
Faculty of Environmental Sciences, Czech University of Life Sciences Prague, Prague, Czech Republic
e-mail: chenz@fzp.czu.cz

M. A. Watto et al. (eds.), *Water Resources of Pakistan*, World Water Resources 9,
https://doi.org/10.1007/978-3-030-65679-9_12

12.1 Introduction

12.1.1 Water Quality Definition and Global Context

"Water quality" is a term used to express the suitability of water for various purposes or to sustain various water-dependent processes. Any specific use of water requires measurement against acceptable levels of physical, chemical, and biological properties of water. In particular, drinking water must clear global standards of limits on the concentrations of these elements. The quality of water bodies is thus assessed in terms of their hydrology, physico-chemistry, and biology. A range of variables are used to reduce threats to human health, food production, ecosystem functions and economic benefits (Bartram and Balance 1996). Because of the complexity of factors determining water quality and large choices of the variable used to describe the status of water bodies in quantitative terms, it is difficult to provide a simple definition but in broader terms, water quality can be defined as follows:

> "Set of concentrations, speciations, and physical partitions of inorganic or organic substances, or composition and state of aquatic biota in the water body, or description of temporal and spatial variations due to factors internal and external to the water body" (Chapman 1996, p. 6).

Water quality can be assessed in relative terms as it not only depends on the function of its conditions or components but also its usefulness and usability (WWAP 2015). It is for these reasons that the World Health Organisation (WHO) has established a world standard set of water quality guidelines (e.g. WHO 2008) that is regulated by different countries according to the specific conditions.

In the global context, 785 million people lack basic drinking water services, out of which 144 million people use surface water for drinking purposes. Globally at least 2 billion people use drinking water contaminated with faeces resulting in transmission of diseases such as diarrhoea, cholera, dysentery, typhoid, and polio. Out of these diseases, diarrhoea counts for 485,000 deaths every year. By 2025, half of the world's population will be living in water-stressed areas. In the least developed countries, 22% of health care facilities have no water service, 21% no sanitation service, and 22% no waste management service (WHO and UNICEF 2014).

12.1.2 Water Quality and Anthropogenic Impacts

Water is essential for human survival. Once it is contaminated, it is difficult and expensive to remediate. Natural processes and anthropogenic activities influence the quality of both surface and groundwater resources. Natural essential components of good quality water include dissolved substances, insoluble particulate matter, and microbes which help to maintain biogeochemical cycles as well. Exceptionally, there are possibilities that naturally occurring substances trigger any water quality changes detrimental to human health. But anthropogenic activities such as domestic use, agricultural production, mining, industrial production, power generation, and other factors alter the natural composition of water which threaten its safe use. For example, industry is responsible for dumping an estimated 300–400 million tonnes of heavy metals, solvents, and toxic sludge into water each year (UN Water 2011). Therefore, every contaminant entering the water body has an impact on water quality.

Increasing population and industrialisation has increased the demands of water, especially good quality water for drinking, personal hygiene, agriculture, and other industrial activities. Each water use leads to a considerable impact on water quality and several human activities have undesirable impacts on water quality. For example, around 80% of the global wastewater produced due to human activities goes untreated, containing everything from human faeces to highly toxic industrial discharges (van Vliet et al. 2017). Managing water quality will become the main challenge societies will face during the twenty-first century, as most human activities that use water ultimately produce wastewater (WWAP 2017). Untreated wastewater poses significant risks of diarrhoea, microbial infections, and malnutrition, accounting for 1.7 million deaths annually, of which over 90% are in developing countries and almost half are children (Fig. 12.1) (WHO 2002).

12.1.3 Water Quality Status of Pakistan

Although Pakistan arguably had adequate surface and groundwater resources at the time of independence, population growth, unsustainable urbanisation, and excessive water use has placed stress on quality as well as quantity of water resources in the

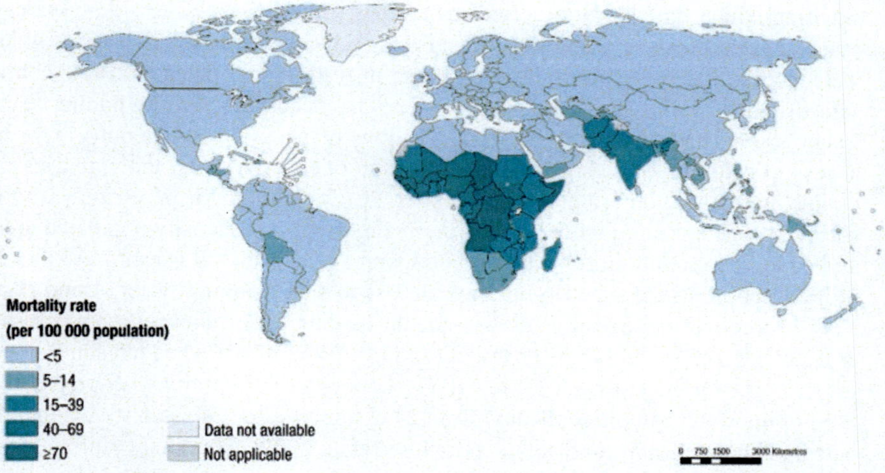

Fig. 12.1 Deaths from unsafe water, sanitation and hygiene (WHO 2016, p. 73). (Reproduced with WHO permission from http://gamapserver.who.int/mapLibrary/Files/Maps/Global_WASH_Mortality_2012.png)

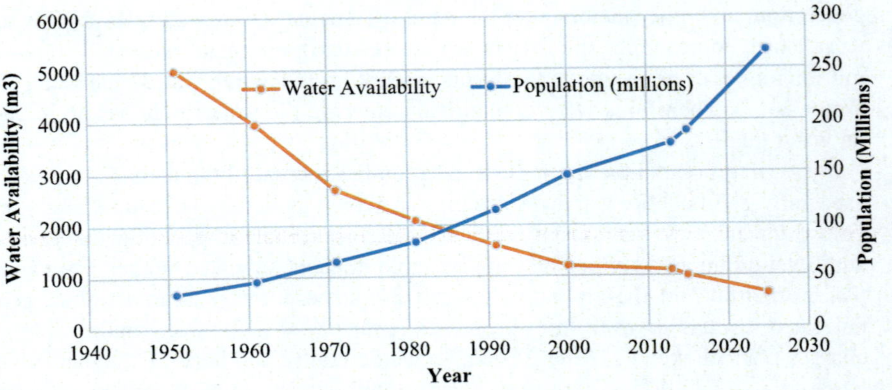

Fig. 12.2 Water availability per capita in Pakistan. (Authors, using data from Jabeen et al. 2015, Table 12.1, p. 1517)

country (Kamran and Omran 2020). Total and per capita water availability is continuously declining in Pakistan (Fig. 12.2). The quality of water is directly related to the quality of human health due to its routine consumption. Most of the population in the country is exposed to the hazards of drinking unsafe and polluted water from both surface and groundwater sources (Daud et al. 2017).

According to UK-based charity WaterAid, providing basic water services in a rapidly urbanising and populous country like Pakistan is extremely challenging – a situation which the charity describes as a crisis (https://www.wateraid.org/pk/the-crisis). Like most developing countries, Pakistan has failed to supply safe drinking water to its citizens resulting in an increase in water-borne diseases. Provision of safe drinking water, sanitation, and hygiene are becoming serious public health issues in Pakistan (Amin et al. 2019). Millions of people in the country have no access to safe drinking water, making Pakistan one of the worst ten countries for people living without safe water (Cooper 2018). Results from various studies showed that the water quality in Pakistan in almost all high-density populated areas and/or sources are not fit for drinking (Podgorski et al. 2017). It is estimated that 21 million people in Pakistan do not have access to safe drinking water (Daud et al. 2017). Moreover, water-borne diseases are the leading cause of mortality in Pakistan especially in infants and children (WHO and UN Water 2017). The Pakistan Council for Research on Water Resources (PCRWR) concluded in their five-year project that 84% of the water samples collected from 23 major cities of Pakistan were contaminated and hence not fit for drinking (Kahlown et al. 2002). This chapter provides an updated overview of water major contamination issues in Pakistan and their impacts on human health.

12.1.4 *Major Pollutants in Water Resources of Pakistan and Their Sources*

Water pollution can be categorised as point source and non-point source. Point sources are those which discharge effluents into water bodies directly e.g. from industrial units, factories, power plants, etc. Non-point source pollutants mainly come from agriculture run-off, drainage, seepage, etc. In the case of Pakistan, industrial effluents, solid wastes, and agriculture runoff are the major sources of water contamination. Almost 90% of domestic and industrial wastes in Pakistan are dumped into water bodies which can ultimately leach to groundwater (Mustafa et al. 2013). Around 92% of sewage is disposed of untreated and about 50% of human excreta produced in urban areas goes directly into water bodies (Pakistan Economic Survey 2013). These untreated waste effluents not only add toxic contaminants but also result in widespread disease incidents.

Poor living conditions and mismanagement are the main reasons for water contamination. Industrial waste in Pakistan often receives minimal treatment resulting in increased contamination of surface as well as groundwater. For example, Sial et al. (2006) found that of over 6000 industries registered in Pakistan's two largest industrial estates in Karachi, 1228 were highly polluting because they discharged heavy metals, oils, and other contaminants. The Ministry of Environment (2005) reported that 80,000 m^3 of industrial effluents were being discharged directly into the Kabul River posing a serious threat to human health (as cited by Mulk et al. 2016).

The agriculture sector is also another source of pollutants in water bodies in Pakistan, including from agriculture runoff. Agricultural pollutants mainly comprise sediments, pesticides, nutrients, nitrates, phosphorous and heavy metals. Runoff from livestock farms directly into water bodies is also an important contributor to water contamination, a matter of particular concern given that livestock farms can be closely located to populated areas. These point and non-point contaminants make drinking water unsafe for human consumption. As a result, the drinking water quality is deteriorating in both urban and rural areas (Ashraf 2016). Major pollutants prevailing in Pakistan's water resources are discussed in the following sections.

12.2 Status of Microbial Pathogens in Water Resources of Pakistan and Their Health Impacts

The consumption of contaminated water with human and animal faeces in drinking water is a major threat. Pathogenic bacteria, viruses, and parasites are responsible for the most common infectious diseases, as detailed in Table 12.1. One of the main reasons for this type of contamination is the breakdown of water supply safety which leads to large-scale disease outbreaks, especially diarrhoea. While significant efforts have been made to reduce diarrheal diseases, diarrhoea remains a leading cause of global mortality and morbidity, causing an estimated 1,655,944 deaths in

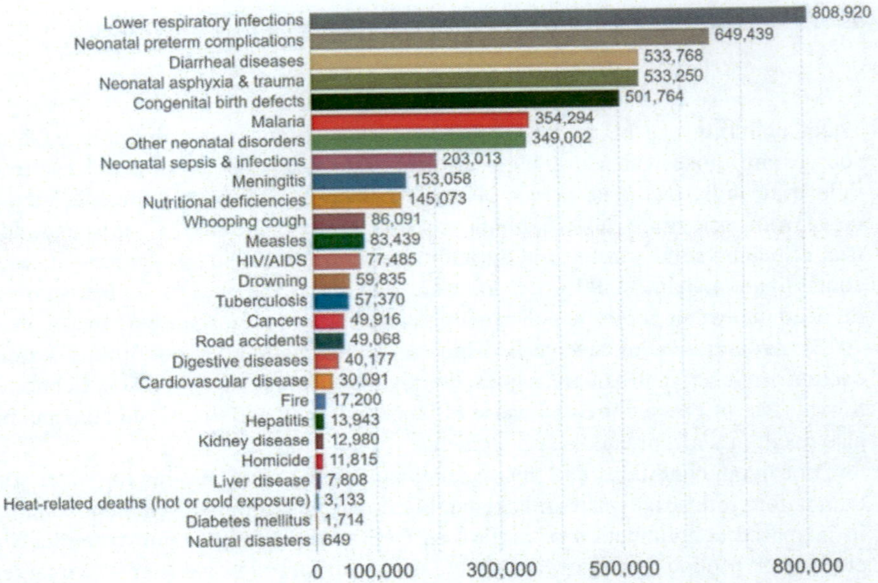

Fig. 12.3 Cause of death globally in children under the age of 5 in year 2017. (Reproduced from Ritchie and Roser (2018) https://ourworldindata.org/causes-of-death)

2016 globally, out of which 533,259 were children under the age of 5 (Troeger et al. 2018, and see Fig. 12.3). These deaths were mainly due to drinking unsafe water. Epidemics caused by waterborne pathogens and health hazards in developing countries are mainly associated with improper management of water resources. Waterborne diseases cost an estimated 12 billion dollars globally (Ramírez-Castillo et al. 2015).

The poor quality of Pakistan's municipal water supply, especially in small towns and rural areas, is due to main sewers being laid at depths of 30–50 feet and are only 10–15 feet away from drinking water supply lines (Arshad and Imran 2017, citing a 2011 conference paper by Benjamin). Their poor management results in contaminated water mixing with drinking water supplies as well as groundwater contamination. Citing a World Bank report, Cooper (2018) also noted that 42% of households in rural Punjab and 60% in rural Khyber Pakhtunkhwa have no drains, and that where drains exist, they are open and managed without treatment. This ultimately increases the risks of water contamination.

Out of different water pollutants, microbial contamination has been regarded as a serious threat to public health in Pakistan. While there are few investigations to determine overall status of microbial pathogens in Pakistan, many individual studies have identified microbial contamination in Pakistan's drinking water (see Table 12.1). Studies conducted in major cities of Pakistan reveal coliform and faecal coliform are frequently found in drinking water (Kahlown et al. 2002). Ahmad et al. (2018) found high levels of enteroviruses in river water samples that had been

Table 12.1 List of pathogens distributed through drinking water with reports confirming their presence in Pakistan

Pathogen	Health Effects	Persistence in water supplies	Disease	Reported in Pakistan	Identified in
Bacteria					
Burkholderia pseudomallei	High	May multiply	Melioidosis	Naureen et al. (2010)	Affected equine
				Shabbir et al. (2015)	Soil samples
				Ali et al. (2017)	Soil samples
Campylobacter jejuni, C. coli	High	Moderate	Diarrhoea	Noreen et al. (2020)	Children with diarrhoea
				Nisar et al. (2018)	Retail meat
				Mahmood et al. (2009)	Retail milk
				Kanwal et al. (2019)	Wastewater & bird droppings
Escherichia coli – Pathogenic	High	Moderate	Serious food poisoning, septic shock, meningitis, urinary tract infections	Shah et al. (2016)	Tap water
				Huma et al. (2019)	Children with diarrhoea
E. coli – Enterohaemorrhagic	High	Moderate	Diarrhoea or haemorrhagic colitis	Bano and Ali (2019)	Sewage waste
				Iqbal et al. (2017)	Drinking water
Legionella spp.	High	May multiply	Legionnaires' disease	Zahir et al. (2016)	Dental water flush
Non-tuberculous mycobacteria	Low	May multiply	Non-tuberculous mycobacterial (NTM), lung disease	Iqbal et al. (2016)	Lung disease patients
Pseudomonas aeruginosa	Moderate	May multiply	Pneumonia, urinary tract infections (UTIs), bacteremia	Saleem and Bokhari (2019)	Hospital patients
Salmonella typhi	High	Moderate	Typhoid	Tagg et al. (2020)	Hospital patients
Other salmonellae	High	May multiply	Diarrhoea, stomach cramps, abdominal pain	Yasmin et al. (2019)	Poultry
Shigella spp.	High	Short	Shigellosis	Nisa et al. (2020)	Human faeces

(continued)

Table 12.1 (continued)

Pathogen	Health Effects	Persistence in water supplies	Disease	Reported in Pakistan	Identified in
Vibrio cholerae	High	Short to long	Cholera	Sarwar et al. (2016)	A child
				Hussain et al. (2020)	Tap water
Yersinia enterocolitica	High	Long	Versiniosis	Ullah et al. (2019)	Sheep & goats
				Mengal et al. (2019)	Salad & water
Viruses					
Adenoviruses	High	Long	Common cold, conjunctivitis, bronchitis, pneumonia	Ahmad et al. (2016)	Drinking water
Enteroviruses	High	Long	Poliomyelitis	Ahmad et al. (2016)	Drinking water
				Ahmad et al. (2018)	Drinking water
Hepatitis A	High	Long	Liver disease	Ahmad et al. (2018)	Drinking water
Hepatitis E	High	Long	Liver disease	Butt and Sharif (2016)	Jaundice patients
Noroviruses and Sapoviruses	High	Long	Gastroenteritis	Alam et al. (2016)	Children with diarrhoea
Rotavirus	High	Long	Gastroenteritis	Yousuf et al. (2017)	Drinking water
Protozoa					
Acanthamoeba spp.	High	May multiply	Granulomatous amoebic encephalitis (GAE)	Tanveer et al. (2013)	Drinking water
				Yousuf et al. (2017)	Tap water
Cryptosporidium parvum	High	Long	Cryptosporidiosis	Iqbal et al. (1999) and Khan et al. (2017)	Children with diarrhoea
				Haseeb et al. (2017)	Tap water
				Aslam et al. (2014)	Tap water
Cyclospora cayetanensis	High	Long	Cyclosporiasis	Haseeb et al. (2017)	Tap water

(continued)

Table 12.1 (continued)

Pathogen	Health Effects	Persistence in water supplies	Disease	Reported in Pakistan	Identified in
Entamoeba histolytica	High	Moderate	Amoebic dysentery (bloody diarrhoea)	Tasawar et al. (2010) and Zeb et al. (2018)	Human faeces
				ul Akbar et al. (2014)	Tap water
Giardia intestinalis	High	Moderate	Giardiasis	Tayyab et al. (2017)	Tap water
Naegleria fowleri	High	May multiply	Primary amoebic meningo-encephalitis (PAM)	Naqvi et al. (2016)	Affected patients
				Tanveer et al. (2017)	Drinking water
				Yousuf et al. (2017)	Tap water
Toxoplasma gondii	High	Long	Toxoplasmosis	Khan et al. (2013a) and Ayaz et al. (2011)	Tap water
				Ajmal et al. (2013)	Tap water, food & soil
Helminths					
Dracunculus medinensis	High	Moderate	Dracunculiasis	Eradicated in Pakistan (Shah et al. 2017)	
Schistosoma spp.	High	Short	Snail fever, bilharzia	Asghar (2017)	Livestock

The first three columns are reproduced with permission from Table 7.1 in WHO 2017, p. 119

polluted from sewerage in Lahore (40%), Islamabad (29%) and Rawalpindi (33%). These pathogens are responsible for digestive, diarrheal, liver and kidney diseases, which were among the top ten causes of death during the year 2017 (Fig. 12.4). A study in 2006 found that 20–40% of bed occupancy in Pakistan hospitals was a result of waterborne diseases, and accounted for one-third of deaths (Pak-SECA 2006).

Diarrheal disease is a threat associated with contaminated water, especially in children under the age of five. During the year 2017, diarrheal and digestive diseases associated with contaminated water use accounted for 74,647 and 59,787 deaths respectively in Pakistan in total (Ritchie and Roser 2018, and see Fig. 12.4). Almost 50% of deaths associated with diarrhoea were of children under the age of five from 2007–2017 (Ritchie and Roser 2018, and see Fig. 12.5). Several studies reported occurrence in food products of the kinds of microbial pathogens in Pakistan that cause diarrhoea (e.g. Mahmood et al. 2009; Nisar et al. 2018), while Nisa et al. (2020) identified a link between such pathogens and patients with

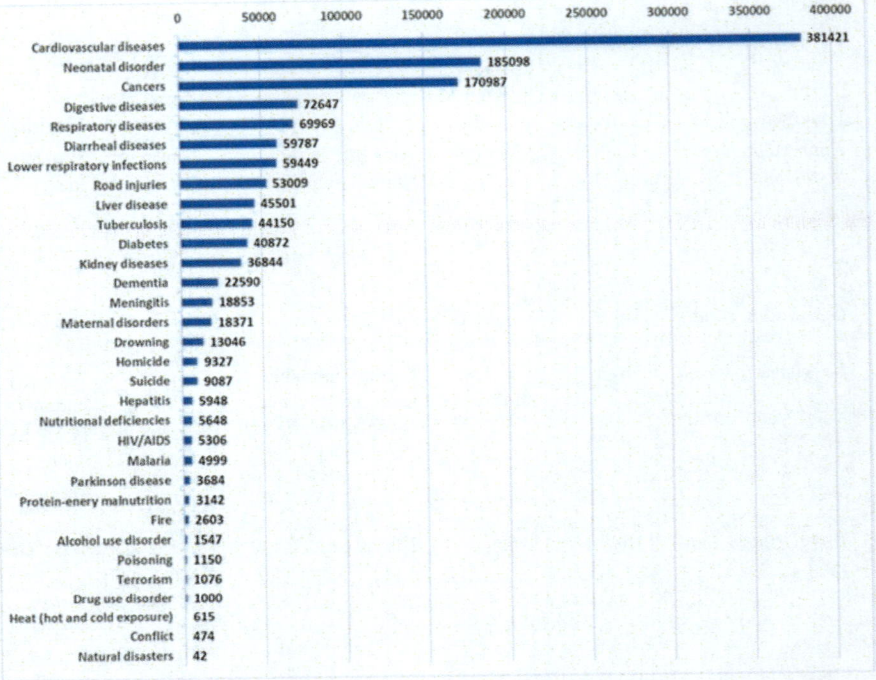

Fig. 12.4 Causes of death in Pakistan during the year 2017. (Reproduced from Ritchie and Roser (2018) https://ourworldindata.org/causes-of-death)

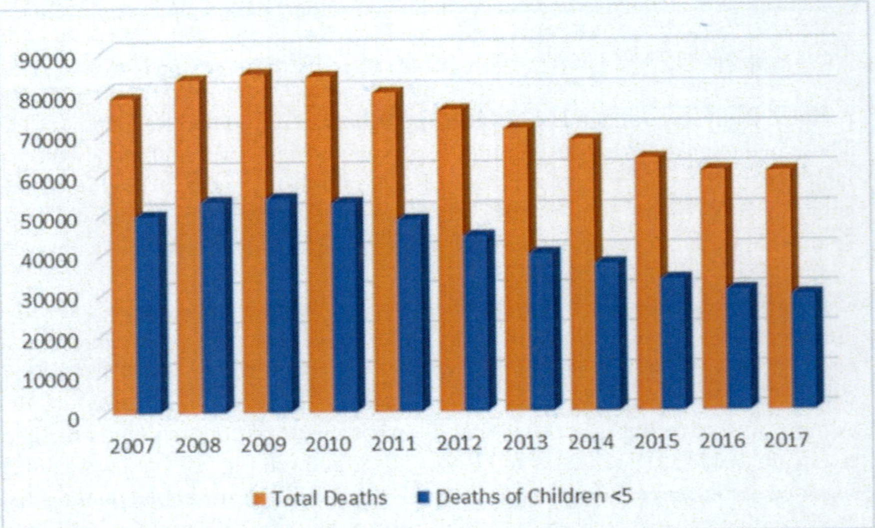

Fig. 12.5 Deaths in Pakistan due to diarrhoea during 2007–17 (Ritchie and Roser 2018)

diarrhoea. The survey conducted by Pakistan Council for Research in Water from 24 major cities found that more than 80% of samples of drinking water were unsafe for human consumption, with most samples positive for coliform bacteria (Ashraf 2016).

Various individual studies conducted in different parts of the country support the findings of PCRWR. Water quality is serious even in Pakistan's main cities. A study conducted in Islamabad found that 14.5% of the water resources in Islamabad did not meet safe limits for microbial pathogens in its drinking water as 21 samples out of 55 were found contaminated with total coliform (Ahmed et al. 2015a). Even microbial contamination was detected in water from filtration plants in Islamabad (Hisam et al. 2014). Similarly, water samples from pumping stations at the twin cities of Islamabad and Rawalpindi were positive for microbial pathogens (Farooq et al. 2008) as were samples taken from food snacks and drinking water in Islamabad schools (Saddozai et al. 2009).

An investigation in a particular riverside community in Lahore found 92% of 50 drinking water samples contained pathogens and were unfit for drinking, with the most common illness in the area being diarrhoea (Qureshi et al. 2011), while another larger sampling of Lahore city as a whole identified 37.2% of 530 samples were unfit due to the presence of microbial pathogens, especially coliform (Anwar et al. 2010). Other Lahore-based investigations revealed similar findings (Hannan et al. 2010; Haydar et al. 2016), with Yousaf and Chaudhry (2013) and Syed et al. (2014) even finding pathogens in bottled water. Other parts of Punjab were also found to have unfit levels of pathogens in drinking water, including Gujrat (Ahmed et al. 2017), Jampur (Rafique et al. 2014), Alipur (Shehzadi et al. 2015), and the Salt Range Wetlands (Ullah et al. 2012). An earlier study involving residents of Gujrat found that 90% of people interviewed suffered from water borne diseases frequently (Tanwir et al. 2003).

Like other big cities of Pakistan, the quality of drinking water is very poor in Karachi. An investigation in five administrative districts in Karachi found that 96% of the samples collected were positive for microbial pathogens, especially total coliform, with a related high prevalence of waterborne diseases: diarrhoea and vomiting, skin problems, malaria, prolonged fever, eye problems and jaundice in that order (Amin et al. 2019). Similarly, an investigation into six of Karachi's filtration plants reported *Acanthamoeba* spp. in 35% of drinking water samples collected (Yousuf et al. 2017). A recent study of drinking water in Sindh schools found around half of samples to be contaminated with *Escherichia coli*, *Shigella* spp., *Salmonella* spp., and *Vibrio cholerae*, and an associated varying risk of infection and illness across the province, with schoolchildren in Karachi experiencing the highest probability of waterborne illnesses (Ahmed et al. 2020).

Water is also not safe in the provincial capital of Khyber Pakhtunkhwa province due to the presence of bacterial pathogens (Nabeela et al. 2014), including in samples taken from drinking water in schools (Ali et al. 2011), and in its surrounding rural areas where 50% of handpump water samples were contaminated with *E. coli* (Ali et al. 2013a). Similar results were found in the province's Hazara division and in the city of Kohat, where high loads of pathogens were found in wells used for drinking, including *Shigella spp*. fecal coliform and *staphylococci* (Muhammad et al. 2017). A prevalence of waterborne diseases was found in Abottabad, north of Islamabad, including diarrhoea, skin infections, typhoid and hepatitis in the urban population (Jabeen et al. 2011), and among students of various academic institutes (Ahmed et al. 2015b). In both cases these illnesses were found to be linked to drinking contaminated water.

The above-mentioned accounts of widespread microbial contamination in drinking water pose a great threat to public health in Pakistan, with digestive diseases and diarrhoea being the fourth and sixth major causes of deaths in Pakistan (see Fig. 12.4). Pakistan has also faced epidemics associated with water-borne pathogens, including an outbreak of drug-resistant typhoid fever in Sindh from 2016 (WHO 2018), and of Hepatitis C, where the estimated 7 million people found to be chronically affected in Pakistan in 2013 represented one-tenth of the global Hepatitis C burden (Lim et al. 2018).

12.3 Status of Toxic Metals in Water Resources of Pakistan and Their Health Impacts

Heavy metals and potentially toxic elements are among the chemicals posing a great threat to humans when present in water (Martin and Griswold 2009). Like microbial contaminations, heavy metals and other toxic elements pose a great threat to the water resources of Pakistan. Heavy metal contamination in surface water and groundwater has increased due to increased population, industrialisation and urbanisation. Because of the grave threats being faced in Pakistan, numerous studies have investigated heavy metal presence in drinking water sourced from both groundwater and surface water. A selection of these studies is summarised in Table 12.2.

Arsenic is a major threat to public health in Pakistan, where many regions have arsenic concentration in drinking water above safe limits (10μg/L) recommended by the World Health Organization (WHO). A meta-analysis of 43 published studies in Pakistan revealed 73% had mean arsenic levels above the WHO limit, and 41% were higher than Pakistan's much less strict limit of 50 mg/L (Shahid et al. 2018b). Podgorski et al. (2017) estimated that 50–60 million people in the Indus Bain use groundwater with a high concentration of arsenic (see Fig. 12.6). Several studies have reported serious health impacts on the human population exposed to high concentrations of arsenic in water in Pakistan (Nickson et al. 2005; Shakoor et al. 2015), including cancer (Wadhwa et al. 2013; Waqas et al. 2017), melanosis, leucomelanosis, keratosis, hyperkeratosis, dorsum, non-pitting oedema, gangrene (Ali et al. 2013b), decrement in lung function (Nafees et al. 2011) and skin problems (Hussain et al. 2016).

Chromium is another threat to the human population in Pakistan and has been reported extensively in Pakistan's water resources (Azizullah et al. 2011), with several studies identifying chromium concentrations above safe limits set by WHO

Table 12.2 Concentration of heavy metals in groundwater used for drinking in selected cities across Pakistan

Metal	Concentrations in groundwater sourced drinking water		
	City	Concentration (μg/L)	References
Arsenic (As)	Jamshoro	13–106	Baig et al. (2009)
	Lahore	5.2–80	Bibi et al. (2015)
	Mailsi	11–828	Rasool et al. (2016a)
	Sheikhupura	40–65	Abbas and Cheema (2015)
	Tharparkar	523–2350	Brahman et al. (2016)
	Vehari	32.5–61.5	Shahid et al. (2018a)
Chromium (Cr)	Karachi	18–145	Ul Haq et al. (2009)
	Kasur	50–9800	Tariq et al. (2008)
	Sialkot	1048-3182	Rafique et al. (2010)
	Swat	244–606	Khan et al. (2013b)
Lead (Pb)	Chitral	0.3–32	ur Rehman et al. (2020)
	Karachi	14–320	Ul Haq et al. (2009)
	Mailsi	10–230	Rasool et al. (2016b)
	Muzaffarabad	ND-665	Ali et al. (2019)
	Zhob	1–63	Chandio et al. (2020)
Nickel (Ni)	Lahore	4–7190	Hussain et al. (2019)
	Sialkot	10–220	Ullah et al. (2009)

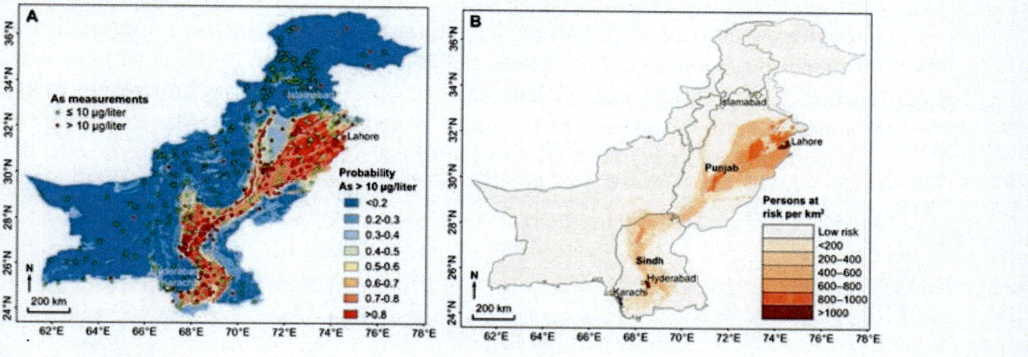

Fig. 12.6 Arsenic prediction and risk models. (**a**) Probability (hazard) map of the occurrence of arsenic concentrations in groundwater exceeding the WHO as guideline of 10μg/litre along with the aggregated arsenic data points used in modelling (n = 743). (**b**) Density of population at risk of high levels of arsenic in groundwater using the WHO As guideline of 10μg/litre. (Reproduced from Podgorski et al. 2017 with permission of American Association for the Advancement of Science (AAAS). © Podgorski et al. 2017, some rights reserved; exclusive licensee AAAS. Distributed under a Creative Commons Attribution NonCommercial License 4.0 (CC BY-NC) http://creativecommons.org/licenses/by-nc/4.0/)

(Table 12.2). Hexavalent compounds from chromium cause various diseases including cancer, skin irritations and those associated with the digestive, excretory, and reproductive system (Azizullah et al. 2011). Contact with nickel creates allergic reactions, with WHO recommending safe limits in drinking water being below 20μg/L whereas Pakistan has cases well above these levels (see Table 12.2).

Lead occurs naturally and it is normal to have traces of lead in drinking water. However, anthropogenic activities have led to toxic concentrations of lead in water. In Pakistan lead is often found to be above safe limits recommended by WHO (10μg/L) (see examples in Table 12.2).

The status of heavy metals in water resources of Pakistan is highly variable in concentration and frequency. Poor sanitation, improper management and lack of treatment facilities are major issues which expose the human population to various disease-causing contaminants. In addition to the above-mentioned contaminants, nitrate in groundwater has also been found to be higher than permissible limits, and this can lead to blue baby syndrome in infants (Soomro et al. 2017).

12.4 Impact of Water-Borne Diseases on the Pakistan Economy

Pakistan not only has issues of drinking water contamination with industrial wastes and municipal sewage, but also a lack of water disinfection practices and quality monitoring at treatment plants, which ultimately result in waterborne diseases. Lack of record maintenance in hospitals makes it hard to quantify the burden of waterborne diseases in Pakistan. One estimate of the cost to the economy relates to inadequate sanitation being Rs. 343.7 billion equivalent or 3.94% of the country's GDP (Khalid and Khaver 2019), while another suggests water-linked diseases result in income losses of Rs 25–28 billion annually, or about 0.6–1.44% of the country's GDP (Khwaja and Aslam 2018).

12.5 Environmental Regulations in Pakistan

Pakistan's 1983 Environmental Protection Ordinance established an institution to develop the country's environmental program. This institution later became the Pakistan Environmental Protection Council with environmental protection agencies at federal and provincial levels. Negligence in the implementation of environmental legislation is evident in that it took 10 years to organise the first meeting of the Pakistan Environmental Protection Council after its establishment (i.e. in 1993). After this meeting, the National Environmental Quality Standards (NEQS) were established to develop permissible limits for municipal effluents and industry discharges. It took another seven more years to bring them to the implementation phase. But still, governments were hesitant to implement such laws either due to lack of monitoring capacity or in response to political pressure. After devolution in 2010, the environment became a provincial responsibility. All provinces established a water and sanitation policy but implementation and legislation regarding water safety are still lingering.

Pakistan's National Water Policy was approved in 2018, yet its implementation and shortcomings are still a huge debate between policymakers and stakeholders. There are several challenges related to the financial stability of service delivery organisations in the water sector (Khalid and Khaver 2019). Water supply schemes became dysfunctional as the provincial government did not allocate sufficient resources to their operation and maintenance (World Bank 2016). While laws and regulations on wastewater treatment and disposal have been formulated, the actual problem is the implementation of these laws and regulations due to lack of resources and workforce skills. As a result, their effectiveness in practice, despite the existence of adequate and necessary administrative capacity on paper, is significantly weakened due to these shortcomings.

12.6 Recommended Strategies to Reduce Water-Borne Health Issues

Pakistan requires strategic planning to reduce the health impacts associated with water contamination. Water-borne diseases are a huge burden on Pakistan's economy and the health sector in Pakistan. Pakistan has made significant progress in improving access to safe drinking water and sanitation, with 90% of Pakistan's population having access to safe drinking water (UNICEF Pakistan 2018). However, reality is on the ground, with UNICEF finding that around 70% of households drink contaminated water (https://www.unicef.org/pakistan/wash-water-sanitation-and-hygiene-0).

Another major issue is the sustainability of existing water supply systems in Pakistan, where lack of proper maintenance of pumps and supply systems results in deterioration of supplied water quality. Acute water shortages and unreliable water supply has left communities with little option other than to buy water at high cost or to drink contaminated water, causing an increased disease burden in the country. Another factor is that management and governance of water and sanitation has been allocated through the 18th amendment to provincial councils and corporations, who in turn have devolved responsibility to lower levels of district administration, who have failed to implement policies due to severely limited budgets and capacity (Khalid and Khaver 2019). This makes Pakistan off-track to meet the UN Sustainable Development Goal of universal access to safe water and sanitation services.

Given the gravity of the problem and the level of investment required, a targeted and multi-tiered approach is required to cope with this serious issue. There is an urgent need for national-level actions including allocation of specific public sector budget allocations for safe water supply and sanitation programs. The national government must establish a separate ministry to monitor the progress of provincial governments to improve people's access to safe water, sanitation, and hygiene services. Pakistan's long-term aim must be adequate and safe piped water supply with metering, realistic tariffs and the safe removal of faecal waste away from human settlements.

Pakistan can learn from case studies in other countries where water and sanitation services have been improved (e.g. Salian and Anton 2010). A community-based natural resource management program can help to keep water resources safe from contamination. This can be implemented with the concepts of sanitation improvement, income generation, poverty alleviation, and sustainable resource management. Such an approach can also incorporate ecological sanitation systems, where treated urine and dry faecal matter is used as fertiliser for agriculture and kitchen gardening. This approach can also be combined with composting of organic wastes, rainwater harvesting, use of grey water, thus ultimately reducing the burden of contaminants on the surface as well as groundwater resources.

Agricultural nutrients and other essential elements recovered from wastewater has gained increased attention in recent years (Ye et al. 2020). Countries like Pakistan, which is already importing huge amount of fertilisers, can gain substantial economic return by adopting this technique to treat wastewater. Wastewater contains nutrients essential for human food production (nitrogen, phosphorous and potassium, or NPK) allowing agriculture to reduce its dependence on the use of mineral NPK fertilisers. It has been documented that the rate of phosphorous recovery from wastewater influent plants is 10–60%, from wastewater slug 35–70% and from sludge 70–98% (Cornel and Schaum 2009). Thus, this approach must be implemented in water management policies and strategies at local as well as national levels for economic gain and to keep water resources safe from contamination.

12.7 Conclusion

Water resources in Pakistan are highly contaminated and not fit for human consumption as most of the disease-causing agents are present. Biological as well as chemical contamination is a great threat to human health in Pakistan. One of the main reasons for water quality deterioration in Pakistan is the mixing of sewerage water with drinking water, which results in the occurrence of disease-causing pathogens and chemicals. The second source of water contamination is industrial effluents which introduce toxic chemicals and elements in drinking water resources. Thus, to reduce the impact of water-borne diseases there is a need to upgrade and maintain regular monitoring campaigns from source to end-user levels and strict compliance of the environmental legislation is required to reduce contaminant burden in water resources. Severe economic impacts can be quantified if regular records are maintained in hospitals and diagnostic facilities are improved. The Pakistan Government must conduct comprehensive surveys and continuous monitoring campaigns to assess the severity of water-borne diseases and their possible sources.

Acknowledgements Authors acknowledge Michael Mitchell for his valuable suggestions to improve this chapter.

References

Abbas M, Cheema K (2015) Arsenic levels in drinking water and associated health risk in District Sheikhupura. J Anim Plant Sci 25(Supp2):719–724

Ahmad T, Arshad N, Adnan F et al (2016) Prevalence of rotavirus, adenovirus, hepatitis A virus and enterovirus in water samples collected from different region of Peshawar, Pakistan. Ann Agric Environ Med 23(4):576–580

Ahmad T, Adnan F, Nadeem M et al (2018) Assessment of the risk for human health of Enterovirus and hepatitis a virus in clinical and water sources from three metropolitan cities of Pakistan. Ann Agri Environ Med 25(4):708–713. https://doi.org/10.26444/aaem/99590

Ahmed T, Imdad S, Butt NM (2015a) Bacteriological assessment of drinking water of Islamabad capital territory, Pakistan. Desalin Water Treat 56:2316–2322. https://doi.org/10.1080/19443994.2014.963154

Ahmed T, Pervez A, Mehtab M, Sherwani SK (2015b) Assessment of drinking water quality and its potential health impacts in academic institutions of Abbottabad (Pakistan). Desalin Water Treat 54:1819–1828. https://doi.org/10.1080/19443994.2014.890133

Ahmed AB, Barq MG, Javed S (2017) Bacteriological and physiochemical analysis of drinking water from different areas of district Gujrat, Pakistan. Int J Appl Biol Foren 1(1):25–33

Ahmed J, Wong LP, Chua YP et al (2020) Quantitative microbial risk assessment of drinking water quality to predict the risk of waterborne diseases in primary-school children. Int J Environ Res Public Health 17(8):2774. https://doi.org/10.3390/ijerph17082774

Ajmal A, Maqbool A, Qamar MF et al (2013) Detection of toxoplasma gondii in environmental matrices (water, soil, fruits and vegetables). Afr J Microbiol Res 7(16):1505–1511. https://doi.org/10.5897/AJMR12.925

Ali J, Hussain A, Abid H (2011) The occurrence of coliforms bacteria in water reservoir at different schools of district Peshawar. Pak J Biochem Mol Biol 44(2):64–67

Ali J, Ullah N, Khan FA et al (2013a) Bacteriological quality analysis of drinking water of rural areas of Peshawar. Am Eur J Agri Environ Sci 13(9):1202–1206

Ali SS, Karim N, Munshi AB, Siddqui I, Khan FA (2013b) Health hazards among coastal villagers of Pakistan due to arsenic contaminated drinking water. J Water Res Prot 5:1235–1241. https://doi.org/10.4236/jwarp.2013.512132

Ali MA, Muhammad K, Rabbani M et al. (2017) Spatial distribution of Burkholderia mallei genome in Punjab, Pakistan. In Proceedings of the 2017 14th international Bhurban conference on applied sciences and technology, Islamabad, Pakistan, 10–14 January 2017, pp 197–213. https://doi.org/10.1109/IBCAST.2017.7868056

Ali U, Batool A, Ghufran M, Asad-Ghufran M, Sabahat-Kazmi S, Hina-Fatimah S (2019) Assessment of heavy metal contamination in the drinking water of muzaffarabad, Azad Jammu and Kashmir, Pakistan. Int J Hydro 3(5):331–337

Alam A, Qureshi SA, Vinjé J, Zaidi A (2016) Genetic characterization of norovirus strains in hospitalized children from Pakistan. Journal of medical virologay 88(2):216–223

Amin R, Zaidi MB, Bashir S et al (2019) Microbial contamination levels in the drinking water and associated health risks in Karachi, Pakistan. J Water Sanit Hyg Dev 9(2):319–328. https://doi.org/10.2166/washdev.2019.147

Anwar MS, Lateef S, Siddiqi GM (2010) Bacteriological quality of drinking water in Lahore. Biomedica 26:66–69

Arshad N, Imran S (2017) Assessment of arsenic, fluoride, bacteria, and other contaminants in drinking water sources for rural communities of Kasur and other districts in Punjab, Pakistan. Environ Sci Pollut Res 24(3):2449–2463. https://doi.org/10.1007/s11356-016-7948-7

Asghar M (2017) Epidemiological Studies of Schistosomiasis in buffaloes and cows, in district Bahawalpur, Pakistan. J Anim Plant Sci 21(2):841–843

Ashraf M (2016) Managing water scarcity in Pakistan: moving beyond rhetoric. In: Proceedings of AASSA-PAS regional workshop on challenges in water security to meet the growing food requirement. Pakistan Academy of Sciences, Islamabad, pp 3–14

Aslam MS, Khan SU, Ayaz S et al (2014) Molecular detection of Giardia lamblia and Cryptosporidium parvum in different water sources of District Bannu, Khyber Pakhtunkhwa, Province of Pakistan. Microbiol Res J Int 4(1):80–88. https://doi.org/10.9734/BMRJ/2014/3947

Ayaz S, Khan S, Khan SN et al (2011) Prevalence of zoonotic parasites in drinking water of three districts of Khyber Pakhtunkhwa Province Pakistan. Pak J Life Soc Sci 9(1):67–69

Azizullah A, Khattak MNK, Richter P, Häder D (2011) Water pollution in Pakistan and its impact on public health: a review. Environ Int 37:479–497. https://doi.org/10.1016/j.envint.2010.10.007

Baig JA, Kasi TG, Arain MB et al (2009) Evaluation of arsenic and other physico-chemical parameters of surface and ground water of Jamshoro, Pakistan. J Hazard Mater 166:662–669. https://doi.org/10.1016/j.jhazmat.2008.11.069

Bano A, Ali B (2019) Isolation and culture dependent characterization of Escherichia coli from the sewage waste water of Lahore, Pakistan. South Asian J Res Microbiol 5(1):1–13. https://doi.org/10.9734/sajrm/2019/v5i130122

Bartram J, Balance R (1996) Water quality monitoring: a practical guide to the design and implementation of freshwater quality studies and monitoring programmes. Chapman and Hall, London

Bibi S, Farooqi A, Ramzan M, Javed A (2015) Health risk of arsenic in the alluvial aquifers of Lahore and Raiwind, Punjab Province, Pakistan: an investigation for safer well water. Toxicol Environ Chem 97:888–907. https://doi.org/10.1080/02772248.2015.1066175

Brahman KD, Kazi TG, Afridi HI et al (2016) Exposure of children to arsenic in drinking water in the Tharparkar region of Sindh, Pakistan. Sci Total Environ 544:653–660. https://doi.org/10.1016/j.scitotenv.2015.11.152

Butt AS, Sharif F (2016) Viral hepatitis in Pakistan: past, present, and future. Eur J Hepato Gastroenterol 6(1):70–81. https://doi.org/10.5005/jp-journals-10018-1172

Chandio TA, Khan MN, Muhammad MT et al (2020) Co-exposure of neurotoxic contaminants (Pb and Mn) in drinking water of Zhob District, Baluchistan Pakistan. Environ Nanotechnol Monit Manage 14:100328. https://doi.org/10.1016/j.enmm.2020.100328

Chapman DV (1996) Water quality assessments: a guide to the use of biota, sediments, and water in environmental monitoring. Routledge, New York

Cooper R (2018) Water, sanitation and hygiene services in Pakistan. K4D helpdesk report. Institute of Development Studies, Brighton. https://opendocs.ids.ac.uk/opendocs/handle/20.500.12413/14245

Cornel P, Schaum C (2009) Phosphorus recovery from wastewater: needs, technologies and costs. Water Sci Technol 59(6):1069–1076. https://doi.org/10.2166/wst.2009.045

Daud M, Nafees M, Ali S et al (2017) Drinking water quality status and contamination in Pakistan. Biomed Res Int 2017:7908183. https://doi.org/10.1155/2017/7908183

Farooq S, Hashmi I, Qazi IA et al (2008) Monitoring of coliforms and chlorine residual in water distribution network of Rawalpindi, Pakistan. Environ Monit Assess 140:339–347. https://doi.org/10.1007/s10661-007-9872-2

Hannan A, Shan S, Arshad U (2010) Bacteriological analysis of drinking water from 100 families of Lahore by membrane filtration technique and chromagar. Biomedica 26:152–156

Haseeb A, ur Rehman H, Zakir M et al (2017) Comparative study of different types of parasites present in water sources of district DI Khan. J Entomol Zool Stud 5(2):1013–1016

Haydar S, Arshad M, Aziz J (2016) Evaluation of drinking water quality in urban areas of Pakistan: a case study of Southern Lahore. Pak J Eng Appl Sci 5:16–23

Hisam A, Rahman MU, Kadir E et al (2014) Microbiological contamination in water filtration plants in Islamabad. J Coll Phys Surg Pak 24:345–350

Huma Z, Tareen AM, Ullah K et al (2019) Incidence of diarrheagenic Escherichia coli pathotypes in children suffering from diarrhea in tertiary care hospitals, Quetta, Pakistan. Pak J Zool 51(6):2015–2021. https://doi.org/10.17582/journal.pjz/2019.51.6.2015.2021

Hussain Y, Dilawar A, Ullah SF, Akhter G et al (2016) Modelling the spatial distribution of arsenic in water and its correlation with public health, central Indus Basin, Pakistan. J Geosci Environ Prot 4(2):63349. https://doi.org/10.4236/gep.2016.42003

Hussain S, Habib-Ur-Rehman M et al (2019) Health risk assessment of different heavy metals dissolved in drinking water. Int J Environ Res Public Health 16(10):1737. https://doi.org/10.3390/ijerph16101737

Hussain Z, Abbasi AZ, Ahmad R et al. (2020) Vibrio cholerae dynamics in drinking water: mathematical and statistical analysis. Appl Nanosci (online first). https://doi.org/10.1007/s13204-020-01292-3

Iqbal J, Munir MA, Khan MA (1999) Cryptosporidium infection in young children with diarrhea in Rawalpindi, Pakistan. Am J Trop Med Hyg 60(5):868–870. https://doi.org/10.4269/ajtmh.1999.60.868

Iqbal N, Irfan M, Farooqi J, Jabeen K, Khan J (2016) Clinical features, spectrum and susceptibility pattern of non tuberculous mycobacterium isolated from pulmonary samples at tertiary care center, Pakistan. Eur Respir J 48(Suppl 60):PA2570. https://doi.org/10.1183/13993003.congress-2016.PA2570

Iqbal MS, Ahmad MN, Hofstra N (2017) The relationship between hydro-climatic variables and E. coli concentrations in surface and drinking water of the Kabul River Basin in Pakistan. AIMS Environ Sci 4(5):690–708. https://doi.org/10.3934/environsci.2017.5.690

Jabeen S, Mahmood Q, Tariq S et al (2011) Health impact caused by poor water and sanitation in district Abbottabad. J Ayub Med Coll Abbottabad 23(1):47–50

Jabeen A, Huang X, Aamir M (2015) The challenges of water pollution, threat to public health, flaws of water Laws and Policies in Pakistan. J Water Res Prot 7:1516–1526. https://doi.org/10.4236/jwarp.2015.717125

Kahlown M, Majeed A, Tahir M (2002) Water quality status in Pakistan. Pakistan Council of Research in Water Resources (PCRWR), Islamabad, Pakistan

Kamran HW, Omran A (2020) Water contamination and health hazards in Pakistan: an overview of the current scenario and contemporary challenges. In: Omran A, Schwarz-Herion O (eds) Sustaining our environment for better future. Springer, Singapore, pp 75–84. https://doi.org/10.1007/978-981-13-7158-5_6

Kanwal S, Noreen Z, Aalam V et al (2019) Variation in antibiotic susceptibility and presence of type VI secretion system (T6SS) in Campylobacter jejuni isolates from various sources (2019) comparative immunology. Microbiol Infect Diseas 66:101345. https://doi.org/10.1016/j.cimid.2019.101345

Khalid IS, Khaver AA (2019) Political economy of water pollution in Pakistan: an overview, SDIP working paper no. 170. Sustainable Development Policy Institute (SDPI), Islamabad

Khan I, Khan AM, Ayaz S et al (2013a) Molecular detection of Toxoplasma gondii in water sources of district Nowshehra, Khyber Pakhtunkhwa, Pakistan. J Toxic Environ Health A 76(14):837–841. https://doi.org/10.1080/15287394.2013.821962

Khan K, Lu Y, Khan H et al (2013b) Health risks associated with heavy metals in the drinking water of Swat, northern Pakistan. J Environ Sci 25(10):2003–2013. https://doi.org/10.1016/S1001-0742(12)60275-7

Khan NU, Saleem MH, Durrani AZ et al (2017) Cryptosporidium: an emerging zoonosis in southern Khyber Pakhtunkhwa (KPK), Pakistan. Pak J Zool 49(4):1455–1461. https://doi.org/10.17582/journal.pjz/2017.49.4.1455.1461

Khwaja MA, Aslam A (2018) Comparative assessment of Pakistan national drinking water quality standards with selected Asian countries and World Health Organization. Sustainable Development Policy Institute (SDPI), Islamabad, Pakistan. http://hdl.handle.net/11540/8388

Lim AG, Qureshi H, Mahmood H et al (2018) Curbing the hepatitis C virus epidemic in Pakistan: the impact of scaling up treatment and prevention for achieving elimination. Int J Epidemiol 47(2):550–560. https://doi.org/10.1093/ije/dyx270

Mahmood MS, Hussain I, Arshad MI et al (2009) Seasonal prevalence of Campylobacter species in milk and milk products in Pakistan. Pak J Zool Suppl Ser 9:227–231

Martin S, Griswold W (2009) Human health effects of heavy metals. Environmental science and technology briefs for citizens issue 15. Centre for Hazardous Substance Research, Kansas State University, Manhattan, Kansas

Mengal H, Samad A, Asmat MT et al (2019) The study to evaluate the prevalence of Yersinia species in salad sold and water supplied in Quetta City, Pakistan. Pure Appl Biol 8(3):1902–1908. https://doi.org/10.19045/bspab.2019.80133

Ministry of Environment (2005) State of the environment report 2005 (draft). Ministry of Environment, Government of Pakistan, Islamabad

Muhammad N, Banoori N, Akbar A et al (2017) Microbial and toxic metal contamination in well drinking water: potential health risk in selected areas of Kohat, Pakistan. Urban Water J 14(4):394–400. https://doi.org/10.1080/1573062X.2016.1173218

Mustafa D, Akhter M, Nasrallah N (2013) Understanding Pakistan's water-security nexus. United States Institute of Peace Washington, DC

Mulk S, Korai AL, Azizullah A, Khattak MNK (2016) Decreased fish diversity found near marble industry effluents in River Barandu, Pakistan. Ecotoxicology 25(1):132–140. https://doi.org/10.1007/s10646-015-1575-9

Nabeela F, Azizullah A, Bibi R et al (2014) Microbial contamination of drinking water in Pakistan: a review. Environ Sci Pollut Res 21:13929–13942. https://doi.org/10.1007/s11356-014-3348-z

Nafees AA, Kazi A, Fatmi Z et al (2011) Lung function decrement with arsenic exposure to drinking groundwater along River Indus: a comparative cross-sectional study. Environ Geochem Health 33:203–216. https://doi.org/10.1007/s10653-010-9333-7

Naqvi AA, Yazdani N, Ahmad R et al (2016) Epidemiology of primary amoebic meningoencephalitis-related deaths due to Naegleria fowleri infections from freshwater in Pakistan: An analysis of 8-year dataset. Arch Pharm Pract 7(4):119–129. https://doi.org/10.4103/2045-080X.191924

Naureen A, Saqib M, Muhammad F et al (2010) Antimicrobial susceptibility of 41 Burkholderia mallei isolates from spontaneous outbreaks of equine glanders in Punjab, Pakistan. J Equine Vet 30(3):134–140. https://doi.org/10.1016/j.jevs.2010.01.056

Nickson R, McArthur J, Shrestha B et al (2005) Arsenic and other drinking water quality issues, Muzaffargarh District, Pakistan. Appl Geochem 20(1):55–68. https://doi.org/10.1016/j.apgeochem.2004.06.004

Nisa I, Qasim M, Driessen et al (2020) Molecular epidemiology of Shigella flexneri isolated from pediatrics in a diarrhea-endemic area of Khyber Pakhtunkhwa, Pakistan. Eur J Clin Microbiol Infect Dis 39:971–985. https://doi.org/10.1007/s10096-020-03811-0

Nisar M, Mushtaq MH, Shehzad W et al (2018) Occurrence of Campylobacter in retail meat in Lahore, Pakistan. Acta Trop 185:42–45. https://doi.org/10.1016/j.actatropica.2018.04.030

Noreen Z, Siddiqui F, Javed S et al (2020) Transmission of multidrug-resistant Campylobacter jejuni to children from different sources in Pakistan. J Glob Antimicrob Resist 20:219–224. https://doi.org/10.1016/j.jgar.2019.07.018

Pak-SECA (2006) Pakistan strategic country environmental assessment report: rising to the challenges. South Asia Environment and Social Development Unit, Islamabad, Pakistan

PES (2013) Pakistan Economic Survey 2012–13. Islamabad, Pakistan. http://www.finance.gov.pk/survey/chapters_14

Podgorski JE, Eqani SA, Khanam T et al (2017) Extensive arsenic contamination in high-pH unconfined aquifers in the Indus Valley. Sci Adv 3(8):e1700935. https://doi.org/10.1126/sciadv.1700935

Qureshi EMA, Khan AU, Vehra S (2011) An investigation into the prevalence of water borne diseases in relation to microbial estimation of potable water in the community residing near river Ravi, Lahore, Pakistan. Afr J Environ Sci Technol 5:595–607
Rafique U, Ashraf A, Khan AK et al (2010) Toxic chromium from tanneries pollute water resources and soils of Sialkot (Pakistan). J Chem Soc Pak 32:644–649
Rafique H, Abbas I, Sohl MA et al (2014) Appraisal of drinking water quality of tehsil Jampur, Pakistan. Desalin Water Treat 52:4641–4648. https://doi.org/10.1080/19443994.2013.803936
Ramírez-Castillo FY, Loera-Muro A, Jacques M et al (2015) Waterborne pathogens: detection methods and challenges. Pathogens 4(2):307–334. https://doi.org/10.3390/pathogens4020307
Rasool A, Farooqi A, Masood S, Hussain K (2016a) Arsenic in groundwater and its health risk assessment in drinking water of Mailsi, Punjab, Pakistan. Hum Ecol Risk Assess Int J 22(1):187–202. https://doi.org/10.1080/10807039.2015.1056295
Rasool A, Farooqi A, Xiao T et al (2016b) Elevated levels of arsenic and trace metals in drinking water of Tehsil Mailsi, Punjab, Pakistan. J Geochem Explor 169:89–99. https://doi.org/10.1016/j.gexplo.2016.07.013
Ritchie H, Roser M (2018) Causes of death. Publisehed online by Our World in Data. https://ourworldindata.org/causes-of-death
Saddozai AA, Samina K, Tabassum H (2009) Microbial quality of food snacks and drinking water in Islamabad schools and colleges. Pak J Agric Res 22(3/4):144–149
Saleem S, Bokhari H (2019) Resistance profile of genetically distinct clinical Pseudomonas aeruginosa isolates from public hospitals in Central Pakistan. J Infect Public Health 3(4):598–605. https://doi.org/10.1016/j.jiph.2019.08.019
Salian P, Anton B (2010) Making urban water management more sustainable: achievements in Berlin. ICLEI European Secretariat, Freiburg. Available from http://www.switchurbanwater.eu/
Sarwar S, Hannan A, Sultana Q et al (2016) Non-O1Vibrio cholerae bacteremia in an infant, first case report from Pakistan. J Infect Dev Count 10(2):188–189. https://doi.org/10.3855/jidc.6554
Shabbir MZ, Jamil T, Ali AA et al (2015) Prevalence and distribution of soil-borne zoonotic pathogens in Lahore district of Pakistan. Front Microbiol 6:917. https://doi.org/10.3389/fmicb.2015.00917
Shah M, Eppinger M, Ahmed S et al (2016) Flooding adds pathogenic Escherichia coli strains to the water sources in southern Khyber Pakhtunkhwa, Pakistan. Indian J Med Microbiol 34(4):483–488
Shah A, Wahid S, Khan NH, Jamal Q (2017) Follow-on to eradication of Dracunculiasis in district Dera Ismail Khan, Khyber Pakhtunkhwa, Pakistan. Pak J Zool 49(3):1135–1137. https://doi.org/10.17582/journal.pjz/2017.49.3.sc8
Shahid M, Khalid M, Dumat C et al (2018a) Arsenic level and risk assessment of groundwater in Vehari, Punjab Province, Pakistan. Expos Heal 10:229–239. https://doi.org/10.1007/s12403-017-0257-7
Shahid M, Niazi NK, Dumat C et al (2018b) A meta-analysis of the distribution, sources and health risks of arsenic-contaminated groundwater in Pakistan. Environ Pollut 242(Part A):307–319. https://doi.org/10.1016/j.envpol.2018.06.083
Shakoor MB, Niazi NK, Bibi I et al (2015) Unraveling health risk and speciation of arsenic from groundwater in rural areas of Punjab, Pakistan. Int J Environ Res Public Health 12(10):12371–12390. https://doi.org/10.3390/ijerph121012371
Shehzadi R, Rafique H, Abbas I et al (2015) Assessment of drinking water quality of Tehsil Alipur, Pakistan. Desalin Water Treat 55(8):2253–2264. https://doi.org/10.1080/19443994.2014.934735
Sial RA, Chaudhary MF, Abbas ST et al (2006) Quality of effluents from Hattar Industrial Estate. J Zhejiang Univ Sci B 7:974–980. https://doi.org/10.1631/jzus.2006.B0974
Soomro F, Rafique T, Michalski G et al (2017) Occurrence and delineation of high nitrate contamination in the groundwater of Mithi sub-district, Thar Desert, Pakistan. Environ Earth Sci 76:355. https://doi.org/10.1007/s12665-017-6663-0

Syed WS, Shahzad KA, Khan FA, Abid K (2014) Bacteriological analysis of tap, processed and filtered water samples. Sci Lett 2(1):24–27

Tagg KA, Amir A, Ikram A et al (2020) Sequencing and characterization of five extensively drug-resistant *Salmonella enterica* serotype Typhi isolates implicated in human infections from Punjab, Pakistan. Microbiol Res Announce 9(13):e01466–e01419. https://doi.org/10.1128/MRA.01466-19

Tanveer T, Hameed A, Muazzam AG et al (2013) Isolation and molecular characterization of potentially pathogenic Acanthamoeba genotypes from diverse water resources including household drinking water from Khyber Pakhtunkhwa, Pakistan. Parasitol Res 112:2925–2932. https://doi.org/10.1007/s00436-013-3465-5

Tanveer T, Hameed A, Bin-Dukhyil A-AA et al (2017) Isolation of thermotolerant pathogenic *Naegleria australiensis* from diverse water resources including household drinking water in Pakistan. Acta Protozool 56:317–322. https://doi.org/10.4467/16890027AP.17.027.7829

Tanwir F, Saboor A, Shan M (2003) Water contamination, health hazards and public awareness: a case of the urban Punjab, Pakistan. Int J Agric Biol 5(4):560–562

Tariq SR, Shah MH, Shaheen N et al (2008) Statistical source identification of metals in groundwater exposed to industrial contamination. Environ Monit Assess 138:159–165. https://doi.org/10.1007/s10661-007-9753-8

Tasawar Z, Kausar S, Lashari M (2010) Prevalence of Entamoeba histolytica in humans. Pak J Pharm Sci 23(3):344–348

Tayyab M, Haseeb A, ur Rehman H, Saeed K (2017) Detection of *Giardia lamblia* by microscopy in different water sources of district DI Khan, KP, Pakistan. J Entomol Zool Stud 5(3):1–5

Troeger C, Blacker BF, Khalil IA et al (2018) Estimates of the global, regional, and national morbidity, mortality, and aetiologies of diarrhoea in 195 countries: a systematic analysis for the global burden of disease study 2016. Lancet Infect Dis 18(11):1211–1228. https://doi.org/10.1016/S1473-3099(18)30362-1

ul Akbar N, Ayaz S, Rahman S et al. (2014) Molecular detection of Entamoeba histolytica in different water sources of district Peshawar, Pakistan. Ann Res Rev Biol, 4(9):1461–1470. doi: https://doi.org/10.9734/ARRB/2014/7935

Ul Haq N, Arain MA, Haque Z et al (2009) Drinking water contamination by chromium and lead in industrial lands of Karachi. J Pak Med Assoc 59:270–274

Ullah I, Ahmad K, Ali Z (2012) Coliforms and halophiles pollution in surface and sub-surface water of salt range wetlands, Punjab, Pakistan. Rec Zool Surv Pak 21:42–46

Ullah R, Malik RN, Qadir A (2009) Assessment of groundwater contamination in an industrial city, Sialkot, Pakistan. African Journal of Environmental Science and Technology 3(12)

Ullah Q, Jamil T, Hussain MH et al (2019) High seroprevalence of pathogenic Yersinia spp. in sheep and goats across nine government farms in the Pakistani Punjab. J Infect Dev Count 13(9):843–846. https://doi.org/10.3855/jidc.11289

UN Water (2011) UN Water policy brief: water quality. https://www.unwater.org/publications/un-water-policy-brief-water-quality/

ur Rehman I, Ishaq M, Ali L et al (2020) Potentially toxic elements' occurrence and risk assessment through water and soil of Chitral urban environment, Pakistan: a case study, Environ Geochem Heal (online first). https://doi.org/10.1007/s10653-020-00531-4

Unicef (2017) Progress on drinking water and sanitation: 2014 update. World Health Organization

van Vliet MT, Flörke M, Wada Y (2017) Quality matters for water scarcity. Nat Geosci 10:800–802. https://doi.org/10.1038/ngeo3047

Wadhwa SK, Kazi TG, Afridi HI et al (2013) Arsenic in water, food and cigarettes: a cancer risk to Pakistani population. J Environ Sci Health A 48(14):1776–1782. https://doi.org/10.1080/10934529.2013.823332

Waqas H, Shan A, Khan YG et al (2017) Human health risk assessment of arsenic in groundwater aquifers of Lahore, Pakistan. Hum Ecol Risk Assess Int J 23(4):836–850. https://doi.org/10.1080/10807039.2017.1288561

WHO (2002) The world health report 2002: reducing risks, promoting healthy life. World Health Organization, Geneva

WHO (2008) Guidelines for drinking water quality. World Health Organization, Geneva

WHO (2016) World health statistics: monitoring health for the SDGs. World Health Organization, Geneva. https://www.who.int/gho/publications/world_health_statistics/2016/en/

WHO (2017) Guidelines for drinking-water quality: fourth edition incorporating the first addendum. World Health Organization, Geneva. https://www.who.int/water_sanitation_health/publications/drinking-water-quality-guidelines-4-including-1st-addendum/en/

WHO (2018, 27 December) Disease outbreak news: typhoid fever: Islamic Republic of Pakistan. World Health Organization, Geneva. https://www.who.int/csr/don/27-december-2018-typhoid-pakistan/en/

WHO, UN Water (2017) UN-water global analysis and assessment of sanitation and drinking-water (GLAAS) 2017 report: financing universal water, sanitation and hygiene under the sustainable development goals. World Health Organization, Geneva

WHO, UNICEF (2014) Joint water supply, sanitation monitoring programme. Progress on drinking water and sanitation: 2014 update. World Health Organization, Geneva

World Bank (2016) Strengthening local providers for improved water supply in Pakistan. World Bank, Wasington, DC. https://doi.org/10.1596/24714

WWAP (2015) The United Nations world water development report 2015: water for a sustainable world. United Nations Water Water Assessment Programme (WWAP), UNESCO, Paris

WWAP (2017) The United Nations world water development report 2017: wastewater: the untapped resource; facts and figures. United Nations Water Water Assessment Programme (WWAP), UNESCO, Paris

Yasmin S, Nawaz M, Anjum AA et al (2019) Antibiotic susceptibility pattern of salmonellae isolated from poultry from different districts of Punjab, Pakistan. Pak Veter J 40:98–102. https://doi.org/10.29261/pakvetj/2019.080

Ye Y, Ngo HH, Guo W et al (2020) Nutrient recovery from wastewater: from technology to economy. Biores Technol Rep 11:100425. https://doi.org/10.1016/j.biteb.2020.100425

Yousaf S, Chaudhry MA (2013) Microbiological quality of bottled water available in Lahore City. J Pak Med Stud 3(2):110–112

Yousuf FA, Siddiqui R, Khan NA (2017) Presence of rotavirus and free-living amoebae in the water supplies of Karachi, Pakistan. Rev Inst Med Trop Sao Paulo 59:e32. https://doi.org/10.1590/s1678-9946201759032

Zahir S, Zahir H, Khurshid Z et al (2016) Is flushing of dental water lines effective in removing microbial contaminants? Pak Oral Dental J 36(2):305–307

Zeb A, Qureshi AW, Khan L et al (2018) Prevalence of Entamoeba histolytica in district Buner Khyber Pakhtunkhwa, Pakistan. J Entomol Zool Stud 6(1):185–188

Chapter 13
Improving Water Management in Pakistan Using Social-Ecological Systems Research

Michael Mitchell, Catherine Allan, Jehangir F. Punthakey, C. Max Finlayson, and Mobushir R. Khan

Abstract High quality research informed by systems thinking can contribute to positive outcomes in complex, dynamic situations related to managing natural resources such as water. This chapter refers to social-ecological systems thinking to identify characteristics of high quality transdisciplinary research that makes a lasting impact. We primarily draw on lessons from a four-year research for development project that focused on learning how to improve groundwater management in Pakistan. Uncontrolled and unmonitored use of groundwater for irrigation has resulted in declining water levels in parts of Punjab, Sindh and Balochistan. The project sought to address this by developing and supporting professional relationships among groundwater managers from government agencies, university researchers and farming communities. Six in-depth case studies, two from each province, enabled groundwater monitoring capacity, and understanding of the social and economic aspects of water use, to be developed together. Stakeholder forums ultimately developed as platforms for co-learning and collaborative planning around on-farm interventions and mobile applications. In this chapter we present the background literature that informed us, and what we did. We also reflect on what could be improved in similar future projects. We note the constraints of short term project funding on this type of collaborative learning based project, and highlight where structured and consistent investment in water resources planning is required. We also suggest that projects such as this would be improved by incorporating ecological perspectives alongside technical, social and economic aspects.

Keywords Transdisciplinary research · Systemic co-inquiry · Groundwater management · Capacity building · Co-learning

---

M. Mitchell (✉) · C. Allan · J. F. Punthakey · C. M. Finlayson · M. R. Khan
Institute for Land, Water and Society, Charles Sturt University, Albury, NSW, Australia
e-mail: mimitchell@csu.edu.au; callan@csu.edu.au; eco@ecoseal.com; mfinlayson@csu.edu.au; mobkhan@csu.edu.au

M. A. Watto et al. (eds.), *Water Resources of Pakistan*, World Water Resources 9, https://doi.org/10.1007/978-3-030-65679-9_13

13.1 Introduction

High quality research informed by systems thinking can contribute to positive outcomes in complex, dynamic situations related to managing natural resources such as water (Christen et al. 2019). To create positive change in such contexts requires building capacity for learning at societal scales (Pahl-Wostl et al. 2008) and a managerial capacity to be adaptable to that learning and to constantly changing circumstances (Allen et al. 2011). This chapter identifies characteristics that constitute high quality research that engages and enhances capacity of social actors to learn and adapt. We primarily draw on our recent experience leading a four-year research for development project, funded by the Australian Centre for International Agricultural Research (ACIAR) as a collaboration between Australia and Pakistan, and focused on how groundwater management in Pakistan can be improved (https://www.aciar.gov.au/project/LWR-2015-036). Henceforth we refer to this as the Improving Groundwater Management Project.

The chapter draws on ideas associated with social-ecological systems research as an aid to understanding and articulating the underlying causes affecting groundwater management. Consideration of social-ecological systems explains why our research project was needed, and what kind of practice changes can make a positive difference. Here we explain that a social-ecological systems approach to research is interdisciplinary, transdisciplinary, and involves co-inquiry and interventions. We conclude by echoing the concluding point of the second chapter of this volume. Specifically, we argue that a key aspect to advance research to improve water resources management in Pakistan is to improve understanding of how investment in water resources information systems, groundwater management, basin planning, water resources institutions, and water-related ecological outcomes can contribute to enhanced development outcomes for Pakistan.

13.2 Why Water Management in Pakistan Needs Improving: Groundwater as a Case Study

Agricultural, urban, and industrial development in Pakistan is concentrated in the Indus River plain. The Indus Basin is water-stressed and has long been recognised as facing a pending water crisis. Demand for water in all sectors is growing – the major user of surface and groundwater is irrigated agriculture. In Punjab's agricultural areas about 50% of irrigation depends on groundwater, and about 20% in Sindh. This development has been driven by a combination of poor access to surface water, increased cropping intensities, and policies to encourage groundwater use to reduce waterlogging. The uncontrolled and unmonitored use of groundwater for irrigation has resulted in declining water levels in the eastern doabs of Punjab, southern Punjab and parts of Sindh, where groundwater is exploited from shallow freshwater lenses. It has also led to declining water levels in the irrigated lands of Balochistan, where using groundwater is often the only option for agriculture.

When designing the Improving Groundwater Management Project, we took the view that research aiming to improve groundwater management is best undertaken in collaboration with groundwater users and managers. Our strategy to cement this collaboration involved using six in-depth case studies across three provinces of Pakistan: Balochistan, Punjab and Sindh (Fig. 13.1). In the case study areas within Punjab and Sindh, groundwater is primarily used as a supplement to the surface water supplied through the Indus Basin Irrigation System (IBIS), while irrigators in the case study areas of Balochistan are almost totally dependent on groundwater due to the low and highly variable rainfall patterns they experience.

The project team consisted of researchers from Australia partnering with staff from universities, provincial water management agencies and national and international research organisations in Pakistan. A foundational objective of the research project was to develop a shared understanding among the project team and those with whom we collaborated of the groundwater situation, and the need for improved management. One aspect of co-developing this shared understanding involved team

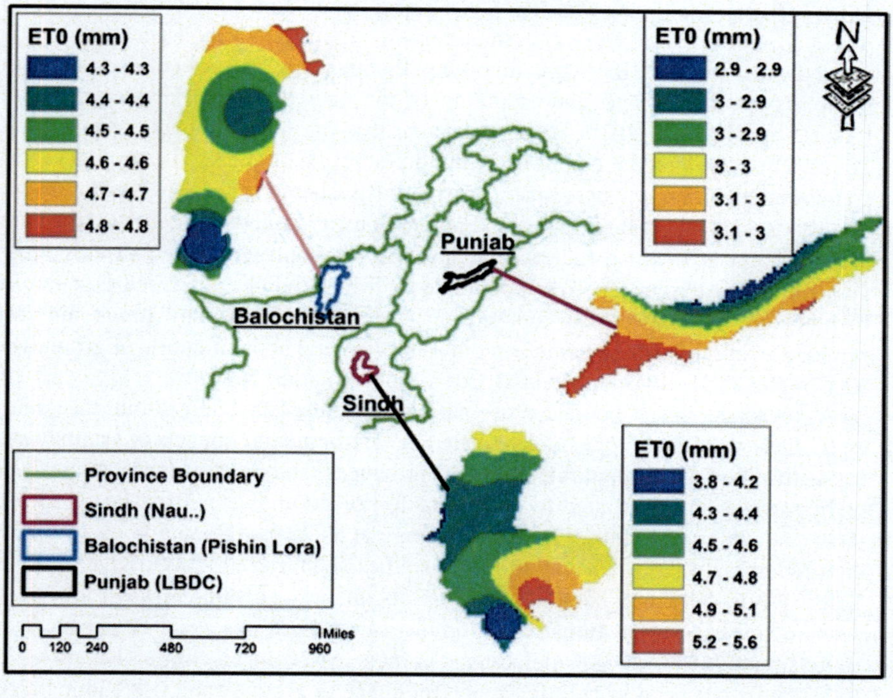

Fig. 13.1 Map displaying location of case study areas (Kuchlagh and Pishin Lora sub-basins in Balochistan; 1R and 11L distributaries in the Lower Bari Doab Canal (LBDC) command area in Punjab; and Malwa and Chiho distributaries in Shaheed Benazirabad and Naushahro Feroze districts respectively in Sindh). ET0 is a measure of evapotranspiration, the amount of water loss from plants and soil

members collaborating to review the literature related to groundwater use and management for agricultural production in Pakistan (Mitchell et al. forthcoming). This was followed by extensive engagement with groundwater users and managers in the case study areas through participatory rural appraisals and the formation of stakeholder forums. The following two sub-sections summarise key themes emerging from the literature review to help understand ecological and social impacts of groundwater use related to the project's case study areas.

13.2.1 Ecological Impacts

Literature exploring ecological impacts from groundwater use and management in Pakistan is very limited. Most examples we found would be better defined as being broadly 'environmental' impacts; i.e., they describe how impacts on the natural environment are affecting society and the economy. Our most focused search, in March 2017, used the Scopus database and Google Scholar search engine and the names of our potential case study locations: "Pishin Lora" (Balochistan), "Bari Doab" (Punjab), and "Nawabshah", "Khairpur" and "Naushshro Feroze" (Sindh). These searches revealed a predominant research focus on social aspects among the groundwater relevant literature involving Pishin Lora, in particular a focus on exploring awareness and understanding of the causes of groundwater depletion (Ashraf and Routray 2013; Jilani and Khair 2014), and its impacts (Kakar et al. 2014). This was also the emphasis of the literature involving the Lower Bari Doab Canal command area, where spatial variation (Basharat et al. 2014), groundwater quality (Basharat and Tariq 2013) and potential for groundwater recharge (Basharat and Basharat 2019) also received attention. It was only our potential case study areas for Sindh that yielded research related to what could be described as 'environmental', but this was mostly focused on concerns for local populations and crop production in relation to arsenic and other heavy metal contamination of groundwater (Baig et al. 2011; Brahman et al. 2016; Rabbani et al. 2017).

At a broader scale, beyond those studies specifically related to our case study areas, there were also more studies related to environmental impacts of groundwater use focused on Sindh than the other two provinces, and most of these studies also highlighted concerns related to contamination of groundwater from heavy metals (Haq et al. 2005; Siddique et al. 2012; Alamgir et al. 2016). These aspects were also the predominant focus of Punjab-based studies (Ullah et al. 2009; Khattak et al. 2012). While there have been some studies documenting decline in water levels and water quality of internationally important wetlands (Ramsar sites) (Khan et al. 1996; Mastoi et al. 2008), including dammed reservoirs in Sindh (Kazmi et al. 2006), there has been very little, if any, reference to groundwater use and management. However, there were some studies that identified community benefits arising from conservation of wetlands and mangroves (Khan et al. 1996, 2014; Rasool et al. 2002). There is also a growing recognition that saline intrusion in the Lower Indus Basin Aquifer results from overuse of groundwater (Chandio and Lee 2012) and reduced outflows from the Indus River (IUCN 2003; Kamal et al. 2012). Our conclusion is that there is currently insufficient research to improve our understanding of how groundwater use is impacting the ecology and ecological assets of the project's case study areas, as well as of Pakistan more broadly, and how an improvement of these assets can in turn offer improvements for society.

13.2.2 Societal Impacts

While it is evident that development of irrigated agriculture across Pakistan has raised productivity and profitability of the agricultural industry as a whole, it is apparent that is that the way groundwater is being used and managed exacerbates existing social inequalities across the three provinces (Mitchell et al. forthcoming). This theme came through most clearly in the literature related to groundwater use for irrigated agriculture in Balochistan, which has received significant academic attention (van Steenbergen 1995; Mustafa and Qazi 2007, 2008; Khair et al. 2015; van Steenbergen et al. 2015). In particular, the longitudinal study led by van Steenbergen articulates the classic common pool resource dilemma being experienced in Balochistan due to a 'socio-institutional void' – or lack of effective management. This 'socio-institutional void' was described by van Steeenbergen (1995) when the 'groundwater rush' was in full swing, and the lack of any government-led or informal community-led institutional responses represented what he described as a wild west 'frontier' problem with a predictable outcome: groundwater over-exploitation.

Access to groundwater in Pakistan is usually tied to land ownership, with no restrictions imposed, and no cost to the farmer apart from the costs of extraction. Extraction is generally through small tube-wells – the bores used to pump groundwater, consisting of a long tube drilled into the ground and sunk to a depth below the water table. Tube-well owners have exclusive rights to the groundwater, and have no restrictions on selling their groundwater to other farmers, which benefits wealthier farmers over smallholder farmers and tenants (Meinzen-Dick 1996). Such easy access to groundwater in Balochistan created a boom time 'apple economy' for farming families, especially during the 1980s and 1990s, when returns were four to five times better than previously achieved (Khair et al. 2015). Yet Balochistan remains the most impoverished and least developed province in Pakistan, with an annual economic growth rate of 2.5% compared with a national growth rate of 4.4% for the period 1999–2000 to 2014–2015, and per capita income almost half the national average (Pasha 2015). The apple growing bonanza led to a massive increase in the number of tube-wells, and this in turn led to depletion of aquifers (van Steenbergen et al. 2015), with groundwater levels dropping by 2–5 m annually (Khair and Culas 2013). This situation had been predicted by van Steenbergen (1995), who warned of consequences of the socio-institutional void. No initiatives were taken to reverse the trend, and depletion of aquifers continued, despite the costs of drilling and pumping from ever greater depths (van Steenbergen et al. 2015).

The impacts for farmers in the Kuchlagh sub-basin of Balochistan have been well detailed by van Steenbergen et al. (2015). Many farmers were forced to sell their land and migrate, or lease land to survive. For those who retained their land, the scarcity of water and costs of extraction meant most had to abandon their apple orchards or remove the trees to be replaced with low water consuming crops. Some farmers sold their groundwater to private truck owners to be on-sold for domestic household use, especially given the massive population increase due to influx of migrants escaping the conflict in Afghanistan. The unfolding situation being described is reminiscent of a 'desakota'-like rural to urban transformation, where livelihoods based on agriculture become subsumed by the need to find other ways of making a living (van Steenbergen et al. 2015, citing Desakota Study Team 2008). Not only does this situation create a profound shift away from a sense of community towards individual competitiveness, but it also exacerbates the extent of inequality between rich and poor.

In the Indus Basin irrigation command areas of Punjab and groundwater dependent areas of Sindh, the situation is different from that in Balochistan, but the resulting groundwater over-extraction also exacerbates existing social inequalities. A large canal network to carry surface water was established by the British initially to protect against crop failure, but is now extensively used for increasing agricultural production (Narain 2008). The canal water is low cost but insufficient, so groundwater is accessed by farmers via tube-wells to supplement it. Groundwater levels are usually highest closer to the main canals of the command areas, meaning farmers whose lands are located at the tail-end of canal distributaries have deeper groundwater and greater extraction costs than those whose lands are located at the head of canals, as was found to be the case for the Lower Bari Doab Canal command area of Punjab (Basharat and Tariq 2015). It is also the tendency that wealthier farmers own land at the head of canal distributaries, and the extent of inequality across the Pakistan rural landscape is compounded by existing socio-political dynamics of land ownership, tenancy and dependency (Tagar et al. 2016). While surface water allocations are regulated by provincial governments, local power dynamics often mean that wealthier farmers living near the head of distributaries have greater access to surface water supplies, with many distributaries running dry at the tail-end, especially in southern Punjab and across the Indus delta areas of Sindh (personal observations – see, for example Mitchell et al. [2020] – though Shah et al.'s [2016] efforts to validate such observations with analysed data paint a more complex picture; also see Anwar and Ul Haq [2013] for an analysis of inequity along the length of a canal). The resulting increased dependency on groundwater at tail-ends inflicts a greater cost burden on farmers, increases rates of depletion, and results in groundwater of decreasing quality being used, with associated salinisation and lower productivity of the land for these farmers (Latif 2007; Latif and Ahmad 2009; Basharat 2012; Punthakey et al. 2015). Variations in access to both surface and groundwater feed a vicious cycle that exacerbates existing inequalities.

13.2.3 Causes: A Social-Ecological Systems Analysis

The literature review provided some context, but developing a shared understanding of the need for improved groundwater management also required ongoing dialogue among the project team and our other collaborators on what is driving the situation causing the adverse social and environmental impacts described above. This dialogue sought to avoid focusing only on the visible events and symptoms by ongoing inquiries seeking to understand underlying causes for the observed phenomena. Over time, such a dialogue both reflects and influences the system of interest. Here, we take stock of where our understandings have reached as a result of the dialogue, and we apply a social-ecological systems framing to articulate underlying causes of systemic issues as co-evolving dynamics between the biophysical and social worlds (Folke et al. 2005, 2016).

As Fig. 13.2 demonstrates, an examination of underlying causes can include investigations into patterns and processes of human behaviour, and how the natural environment responds. In the groundwater for agricultural use context, such an examination will often focus on resource users – those humans directly farming the land and using water for productive benefit. This often results in farmers being blamed for the observable events and symptoms of groundwater depletion, waterlogging, and the effect of the use of poor water quality on crop production. The simplistic narrative of assuming farmers are either unaware, or unwilling to change, can result in calls for 'education', greater regulation and incentives to modify the behaviours of irrigators.

There are other, however, more systematic aspects that drive the patterns and processes resulting in the observable phenomena. Systems and structures have

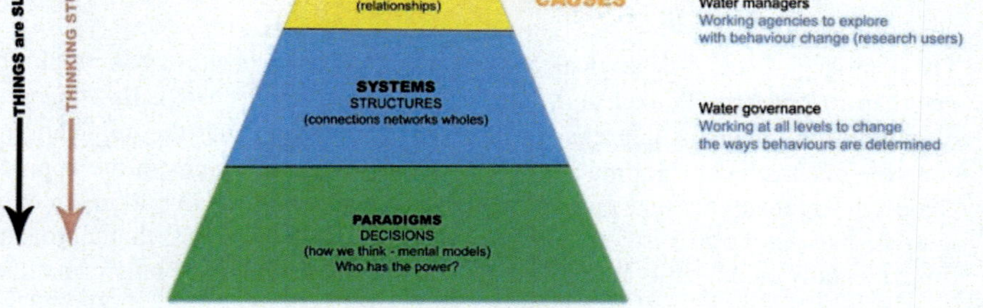

Fig. 13.2 Understanding underlying causes from a systems perspective using an iceberg metaphor. (Image adapted from that used by Paul Ryan, Australian Resilience Centre, with permission)

developed and reified over time to influence whether and how those directly managing the land and water can adopt practices that use resources more sustainably, and with less adverse impacts on society and the environment.

Systemic change is needed if Pakistan is to use groundwater in a way that approaches sustainable development, i.e., "development that meets the needs of the present without compromising the ability of future generations to meet their own needs" where the concept of needs focuses particularly on "the essential needs of the poor, to which overriding priority should be given" (World Commission on Environment and Development 1990, p. 87). This classic articulation of sustainable development emphasises that its pursuit involves both sustaining the environment and challenging social injustice. There is a social justice dimension to achieving sustainability, especially because of the poverty trap for those eking out a living through further degradation of the marginal lands where they are forced to live (Dobson 1999).

Reflecting on the above led our project development team to a critical realisation. Transformation cannot occur by focusing solely on behavioural change among water users. Rather, it is imperative to work with those responsible for managing water and its use to explore how behavioural changes at the system level can lead to more sustainable use of resources. Equally important was to put learning and adaption at the heart of the project, such that project members and participants were supported to be co-researchers as much as possible.

In the groundwater management context of Pakistan, our Improving Groundwater Management Project went beyond merely having academic institutions as project partners. The project team has benefited from having formal partnership relationships established with provincial irrigation departments as water managers (the project's intended next users of the research outputs), as well as a set of co-inquiry case studies with irrigator communities as water users (the project's ultimate intended research beneficiaries). How this collaborative, multi-player project worked is described in the next section.

13.3 Contributing to Improved Water Management Using the Practice of Social-Ecological Systems Research

The approach of social-ecological systems research has been explored more in theory than in practice (Walker and Salt 2012; Sellberg et al. 2018). Its conceptual origins are in understanding change in ecological systems and the application of adaptive management (Holling 1973, 1978). Significant advances in the approach came with realisations that ecological dynamics generally evolve as interactions with changes in social systems (Berkes and Folke 1998), and that the approach is fundamentally about embracing change – both adaptation and transformation – through processes of active adaptive management (Folke et al. 2005). These advances were in part driven by the work of Ostrom (1990) based on her investigations into groundwater management as "governing the commons", which in turn led to the need for adaptive governance as a change in paradigm (Dietz et al. 2003). It is also interesting to note that one of the earliest applications of the approach involved an investigation to understand the underlying causes driving change of an Australian catchment affected by waterlogging (Walker et al. 2009). There is also a large body of research in cybersystemics (e.g. Ison 2016) exploring ways to facilitate transformation within institutions for more just and adaptive governance, for example Blackmore et al. (2007).

While the Improving Groundwater Management Project was not framed as social-ecological systems research, the approach informed its origins so that it became a project with integrating leadership from both the biophysical and social sciences. The project's design thus offers guidance for how social-ecological systems research can be effectively put into practice. At the core of the social-ecological systems concept is that it involves interdisciplinary research, i.e., pursuing a shared research objective by actively traversing and enabling learning and growth across academic disciplinary boundaries (Tress et al. 2005). We would also assert that for social-ecological systems research to have practical influence, it needs to go beyond interdisciplinarity to become transdisciplinary, i.e., to develop and pursue a shared research objective with actors in society (such as land managers and other stakeholders), by "unsettling the distinction between research providers and research users" (Mitchell et al. 2017, p. 2). For transdisciplinary research to become an effective process of learning through active adaptive management requires intervening in the systems being studied (Midgley 2003) through co-inquiry with those who are part of these systems (Foster et al. 2019; Allan et al. 2020). Reflection on practice is also essential (Stringer et al. 2006) to enhance learning and identify what could be improved. The following sub-sections use this framing for how to deliver high quality social-ecological research to critically evaluate the research performance of the Improving Groundwater Management Project.

13.3.1 Interdisciplinary Research: Building Partnerships Across Disciplines

13.3.1.1 What We Have Been Doing

The core of the Improving Groundwater Management Project involved the development of groundwater management tools and options that would have the potential of improving livelihoods for farming families. Delivering on this objective would require, at the very least, expertise in hydrogeology (for the development of groundwater management tools and options) and agricultural economics (to explore how these tools could influence farming family livelihoods). The project relied on the development of both groundwater and socio-economic models, and regular efforts to integrate them as they were developed, so that their inputs and outputs were informed by each other's inputs and outputs.

The project's overall aim, however, involved building capacity of managers and practitioners, which required research strategies informed by the social sciences. As further described below, in the case of groundwater management tools and options, the priority was that these be developed with intended research users – the provincial irrigation departments – to ensure they had the capacity to use and further develop the tools and options after the four-year project had concluded. In addition, as these tools and options are meant to inform and influence groundwater use behaviour, their further use and development is best achieved by irrigation department staff working in collaboration with groundwater using community representatives and a broader array of service providers with whom these communities work. The provincial department staff were developing their own capacity with technical experts and research beneficiaries simultaneously, and were thus in a key position to benefit from as well as facilitate learning. The research strategies adopted were thus influenced by social learning theory, which involves challenging assumptions about how society functions, and where "ideas and attitudes learned … must diffuse outwards to wider social units or communities" (Reed et al. 2010, p. 4; and see Steyaert and Jiggins 2007; Faysse et al. 2014).

13.3.1.2 Suggested Improvements

On reflection, the Improving Groundwater Management Project's understanding of the issues and their underlying causes could have benefited from greater integration with environmental science expertise to offer additional appreciation for how the groundwater situation has emerged as a result of interlinking social and ecological dynamics. The ecosystem services framework, such as that used to define the provisioning, regulating, cultural and supporting service roles provided by wetlands (Millennium Ecosystem Assessment 2005; Finlayson et al. 2011), offers one approach to exploring these interactions for resource management contexts. Restoring and maintaining the health of the environment and water related ecosystems is also a priority set out in Pakistan's National Water Policy. Future research and capacity development would need to incorporate environmental objectives to support improved management of freshwater ecosystems and, in particular, the benefits that ecosystem services bring, often free of charge, to local people (Finlayson et al. 2019).

Similarly, developing a shared understanding of how the groundwater situation being faced can be articulated as a challenge for sustainable development can be informed by the growing body of literature that explores integration using the United Nation's Sustainable Development Goals (Stafford-Smith et al. 2017; Velis et al. 2017), including that which specifically promotes multiple use of aquatic systems and their resources (Lynch et al. 2019) and the impacts of climate change (Pittock et al. 2019).

13.3.2 Transdisciplinary Research: Building Partnerships Between Researchers and Research Users

13.3.2.1 What We Have Been Doing

The organisation that funded the research, ACIAR, encourages research that creates and maintains genuine partnerships. Its core business is to ensure research is developed collaboratively with partners in the countries with development needs that it seeks to support as part of the country's overseas aid program (ACIAR 2018). Key to the strategy is to secure co-investment with partner organisations, which it primarily achieves through in-kind contributions of these partners' staff time.

Because our project sought to have groundwater management tools and options developed as a collaborative activity in each province, securing staff time commitments from researcher and research user organisations to work in collaboration with each other has been essential. For example, the Punjab Irrigation Department (PID) committed both senior-level and middle-level professional staff to work with academics from the University of Agriculture, Faisalabad to develop the groundwater model for the Lower Bari Doab. Examples of collaboration include PID sharing its groundwater monitoring data, and working with researchers to enhance its collection of that data. Similarly, in Sindh, early career academics at Mehran University of Engineering and Technology took the lead in working with officials at the Sindh Irrigation Department (SID). This collaboration has resulted in the two organisations co-investing in further groundwater modelling training programs to disseminate that capacity more broadly across SID. A comparative dearth of groundwater monitoring data availability in Balochistan meant collaboration there has focused on co-investment to drill monitoring bores and install loggers in the project's case study areas. An added benefit in this case is that decisions involving these monitoring bores were undertaken in collaboration with the communities as intended beneficiaries, members of whom made the space available and built protective structures. Balochistan Irrigation Department staff have also benefited from having their GIS skills improved.

As a project team, our efforts to build a transdisciplinary approach have thus gone beyond establishing partnerships with research users. Core to our strategy has been to nurture ongoing partnerships between university researchers and government staff and their respective organisations and ensure mentoring and training was available where needed. These and other collaborations developed and were supported by occasional gatherings of the whole project team – traversing provinces, organisations, discipline areas and social hierarchy. These events, while complicated and costly, provided space to share and develop new framings of situations, and to build confidence and trust in the team network. The collaborative relationships between the community, irrigation departments and academic institutions has, for example, facilitated the uptake of EC meters to monitor groundwater quality and a desire to improve water management practices.

13.3.2.2 Suggested Improvements

Our transdisciplinary strategy was intended to have been informed by an institutional analysis. However, this research activity did not proceed due to difficulty accessing the necessary resources and expertise to deliver the analysis we were proposing. Institutions include the rules, norms and strategies shaping decision-making by individuals and organisations (Scott 2014). A critical institutional analysis undertaken with social actors as research partners can identify institutional arrangements that are performing well, as well as those that are not delivering well according to the purpose we would seek from them (Cleaver and de Koning 2015; Clement et al. 2017). The approach facilitates social learning towards institutional designs that can more effectively deliver on, for example, sustainable groundwater development and use.

A four-year project inevitably imposes limits on what can be achieved. We were aware from the outset that our project would focus on improvements to groundwater management (principally the realm of hydrologists and water managers – Mukherji and Shah 2005), rather than being able to influence governance arrangements (about how people in society share power with governments in decision-making and program delivery – Stoker 1998) related to groundwater. Research that would investigate and intervene in groundwater governance arrangements was seen as beyond the scope of the project.

13.3.3 Co-inquiry and Interventionist Research: Building Partnerships Between Researchers, Research Users and Research Beneficiaries

13.3.3.1 What We Have Been Doing

Co-inquiry is a deliberative research process to institutionalise collaborative action by research providers and users as a community in response to the complex situations they are navigating. It relies on iterative social learning as the means to improve outcomes (Foster et al. 2019). In our case, the actions undertaken as a collaboration among researchers, research users and research beneficiaries used the methodology of intervention: purposefully co-creating change in the system being studied (Midgley 2003). Such collaborative action with social actors in the case study areas was championed and facilitated by academic team members as a form of participatory action research (Fals Borda 2006; Woodward and Hetley 2007), drawing on support from collaborative evaluation, reflection and sense making (Enfors-Kautsky et al. 2018) as the case study co-inquiries evolved.

The process of co-inquiry began with a series of participatory rural appraisals (Chambers 1994; Allan and Curtis 2002) in our proposed study areas to identify communities to collaborate with, establish partnerships and networks, learn about the situation and identify problems to be addressed, secure local stakeholder ownership and benefit, and establish pathways for their inclusion in determining case study research activities, design and outcomes. These activities identified six case study communities to work with, based partly on biophysical and spatial criteria, and partly on community potential and willingness for involvement. The activities led to the local establishment of stakeholder forums for each case study with representation from the community, farmer organisations, non-government organisations, irrigation departments, agricultural extension and on farm water management units and local agribusinesses. Each stakeholder forum held a workshop to determine collaborative research actions.

The workshops were used to investigate four potential interventions with the project team:

1. Improving use of a range of monitoring tools, such as data loggers and EC meters, to enhance co-learning among irrigators, extension agents and researchers about issues and potential solutions associated with groundwater use and management.
2. Accessing the new groundwater modelling information that shows trends in groundwater levels and quality over time, and explores scenarios related to changed pumping regimes and future climatic conditions.
3. Using a mobile App and web-based tool developed by the project team to record information from tube-wells, including water levels and quality, and to access other land and weather information that can inform decision-making.
4. Accessing social and economic data created and analysed by the project team to offer information, guidance and generate discussion with case study communities about their future pathways for agricultural development, resulting in more detailed decision-making and associated justification for on-farm interventions to be investigated, such as to trial high efficiency irrigation methods in the Balochistan case studies, low water use crops in Punjab, and raised bed and integrated farming in Sindh.

Action plans were developed at each workshop, and are currently being implemented. The COVID-19 pandemic has extended the length of time for the project team to be engaged with communities on these action plans, but has also undermined opportunities for interaction and co-inquiry.

While formal action plans are an example of direct outcomes for irrigator communities as intended research beneficiaries, many of the outcomes of developing habits of learning together are less direct and tangible, but no less important. One outcome is the creation of new shared operating spaces among participants of the Improving Groundwater Management Project team, outside of this actual project. These spaces allow for faster development and implementation, such as when a new project from a national partner was established in record time in Balochistan because of networks developed among multiple agency staff and farmers.

These new operating spaces have provided opportunities for women to share in inquiry and design of future actions. The stakeholder forums are deliberately inclusive, driven and supported by strong female project members from universities and NGOs. The stakeholder forum activities have showcased women's involvement in learning and planning for water management. For example, at a stakeholder forum in Sindh, female councillors were included in deliberating on options, bringing local expertise that might otherwise have been missed.

A Representative Agricultural Pathways (RAPs) framework was introduced to the stakeholder forums in Punjab as a means to explore future scenarios building on information provided by the groundwater and socio-economic models. The RAPs framework is developed on the premise that both biophysical and socio-economic drivers are essential components of agricultural pathways. Its methodology is based on capturing plausible farm-level improvements as climate change and other impacts such as water scarcity affect the future of farming operations. By exploring shared visions, it enables a farming community to consider and adopt new adaptation strategies. Using the RAPs methodology with stakeholders also allowed the identification of climate change scenarios that would benefit both the farming community and the operational responses of the Punjab Irrigation Department.

In Sindh, the involvement of the Sindh Irrigation Department from the outset has resulted in improved knowledge across the Sindh-based team of groundwater issues, as well as enhanced team capacity for groundwater modelling. Appreciation of the importance of diffusing this knowledge and capacity has, for example, led to partner co-funding of training programs for staff in the Sindh Irrigation Department.

In Balochistan, the irrigation department has collaborated with researchers and the provincial agriculture extension department, who together have been active in co-inquiry and engagement with the irrigator community. This enabled decisions on where to site monitoring bores to be made jointly with the community, and ensured their assistance in safeguarding and learning from the bore and loggers.

13.3.3.2 Mobile Application "Apna Pani" for Groundwater Data Collection and Visualisation

Apni Pani (our water) is a mobile application which is used to collect and report groundwater depth and water salinity (EC) data (Fig. 13.3). Through this application, users are able to monitor their groundwater. They can also access daily agro-meteorological data from world weather online (temperature, wind speed and humidity) to help them in their daily/seasonal operations such as input applications (pesticide and irrigation). Groundwater and meteorological data are integrated into another mobile and web-based application – *Apna Farm*. *Apna Pani* and *Apna Farm* applications are easy to use and provide up-to-date data, offering improvements on other available applications (e.g. NASA 2020). Moreover, groundwater use can be better managed as users can get information on when there is a need for further irrigation and how much water is needed (Patel 2018). *Apna Farm* can be accessed at http://mriazkhan.com/aciar/dss/ and provides data on land characteristics, enabling users to compute multiple crops' water requirements. Such data can inform decision-making about which crops to grow in relation to the on-farm surface and groundwater situation (Fig. 13.4). The following data are provided:

Fig. 13.3 Data reported by users of *Apna Pani* mobile application

- Current weather conditions.
- Soil properties data including soil organic matter, potassium, pH, phosphorus, and EC.
- Crop profitability.

The information provided by *Apna Pani* and *Apna Farm* also provide the potential for the practice of co-inquiry to continue beyond the life of the project. Their use is being currently championed by younger members of farming families as well as extension agents. Together with shared understanding of future pathways for agricultural development, farming communities are better placed to consider ongoing adaptations in practice in discussion with extension agents and others with expertise and experience in irrigation and on-farm water management.

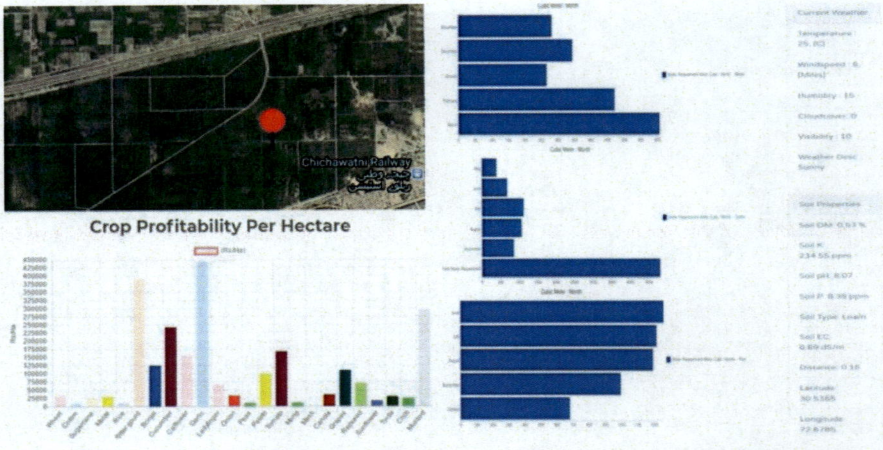

Fig. 13.4 Computing crop water requirements for multiple crops and accessing weather, soil and crop profitable data for optimal decision-making

13.3.3.3 Suggested Improvements

The project's mission of establishing the means for ongoing co-inquiry is to have farming families seen as having a particular set of expertise that is just as important as the expertise offered by extension agents, water resource managers, agricultural economists and biophysical scientists. Key to this strategy is to provide the space and means to enable discussion that combines the experiential knowledge of farming families with other types of knowledge. While the project's interventions offer steps forward through use of integrating tools such as *Apna Pani* and *Apna Farm*, a key aspect that has been missing from these discussions is a set of perspectives that ecology could bring. This was a product of decisions made at the project's design stage, when we were steered away from adopting a social-ecological systems approach which did not seem to fit well with the funding agency's focus on agricultural research. On reflection, it might have been beneficial if we had been able to explore how a social-ecological systems approach could extend and embellish the established agricultural perspectives and approach that underpinned the research. Extending this approach by more overtly linking with the concepts of social-ecological systems, such as the multiple benefits for local people that can be obtained from agricultural landscapes, could provide new perspectives and further opportunities to combine knowledge to improve how groundwater and other natural resources are used and managed.

13.3.4 Investment Planning

Individual projects can only go so far. To improve management of groundwater in the Indus Basin and to meet the policy requirements set out in the National Water Policy and the Provincial Water Policy in Punjab will require improved information, technical capacity and regulation to assist farmers to adapt to a water scarce environment. Based on what we have learned from the Improving Groundwater Management Project, we identify five areas for future investments, as detailed in Table 13.1.

13.4 Conclusion

In the earlier chapter in this volume by Westcoat et al., a vision is presented of the "Indus Basin as a garden", offering a transformation from its current state. Our chapter has provided a framework and examples of strategies that can assist such a transformation – that is, to enable systemic change.

The Improving Groundwater Management Project aims to encourage and support transdisciplinary research and implementation. Useful and hopefully ongoing partnerships have been formed, with shared operating spaces enabling new situation framing, questions and practices. Collaborating, learning, capacity building and reflecting enable this project to go beyond imposition of technical concerns and solutions onto groundwater users who needed to be 'changed'. The Improving Groundwater Management Project embraced change at institutional and organisational levels, and some innovations have emerged that are likely to have some traction in the communities within which they were developed.

It is important to note that this research, like almost all research in this field, is funded on a short-term project basis, meaning that researchers and funds come into an area and then leave, often, as is the case with this project, just when momentum of collaborating and innovation has built to levels where real change may occur. New funding may be sought, and maybe even gained, but in the interim momentum is lost and trust is reduced. We are trying to transcend this pattern of project by engaging and building the capacity of those who have ongoing responsibility for water resources management in Pakistan. By simultaneously building capacities in groundwater and economic modelling, as well as collaborative inquiry and reflection, project team members will be able to continue to use and benefit from the research investment long after the research project has ended. Mobile applications facilitate continued logging, mapping and analysing of groundwater data. All of these aspects combined with enhanced capacity and engagement of practitioners is the project's legacy for societal change. For this to have deep and lasting impact, agency and other organisational staff must have the support of their organisations, in the form of time to participate, if the new operating spaces are to have longevity.

Table 13.1 Suggested areas and actions for investment planning

Priority areas for investment	Strategic actions
Water resources information management	Conduct needs assessment for monitoring and management at community, canal command, sub-basin and basin scales. Develop robust information management systems, data archiving and access. Map spatial and temporal trends to evaluate the state of groundwater. Make strategic and robust monitoring systems operational for use in groundwater assessment, planning and management (quantity, quality, uses and users). Improve access to data for all researchers and institutions.
Sustainable management of groundwater resources	Delineate groundwater management areas and develop strategies to manage depletion and water quality degradation to improve sustainability of the groundwater resource. Develop groundwater models to improve understanding of sustainable extraction and scenario analysis for adaptation to climate change. Delineate groundwater depletion, waterlogging and salinity and groundwater contamination areas (hotspots).
Basin scale planning	Identify and manage risks to basin water resources (water use for irrigation, environmental flows, water quality and salinity management plans). Introduce sustainable diversion limits at sub-basin and canal command scale. Sustain freshwater lakes and environmental assets. Establish a basin management authority similar to the Murray-Darling Basin Authority and other successful models.
Agricultural water productivity	Establish agro-ecological zones for improved agricultural water productivity. Identify best options for conjunctive water use in agriculture. Reassess canal duties within and between canal command areas. Support adoption of high value – low water use crops and associated value chains to improve agricultural productivity and farming livelihoods. Build capacity of farming communities to improve water management practices.
Water resource institutions	Build water resources management institutions (not just organisations) with capacity to manage surface and groundwater in an integrated manner. Build capacity of staff to use technical tools, undertake monitoring, mapping and modelling, and provide policy support. Seek and support opportunities for collaboration across organisational and physical boundaries, modelled on the project's stakeholder forums as spaces for collaborative learning and design of interventionist investigation.

The same reasoning also reinforces why it is important for research projects to have a legacy for environmental change. Water resource management involves social and ecological dynamics, and people working with both social and ecological factors need enhanced capacity for adaptation and transformation. The social-ecological system may need to be able to bounce back from disruptive changes, or it may need to be able to bounce forward to new ways of operating if the old ways are no longer tenable given the changed conditions.

A broad systems approach to situations such as declining groundwater resources highlights the potential benefit of investing in collaborative research with institutional, social and environmental aspects to improve water resource use and management outcomes. It also helps develop a platform for extending the reach of the research and the development and implementation of additional approaches that can benefit the communities within which they were developed.

Acknowledgements This chapter is based on research funded by the Australian Centre for International Agricultural Research (project LWR-2015-036), with additional funds provided by Charles Sturt University.

References

ACIAR (2018) 10-year strategy 2018–2027: research that works for developing countries and Australia. ACIAR, Canberra. https://www.aciar.gov.au/publication/Ten-Year-Strategy

Alamgir A, Khan MA, Schilling J et al (2016) Assessment of groundwater quality in the coastal area of Sindh province, Pakistan. Environ Monit Assess 188(2):1–13. https://doi.org/10.1007/s10661-015-5061-x

Allan C, Curtis A (2002) Participatory rural appraisal: using it to understand rural communities. Nat Resour Manag 5(1):28–34

Allan C, Ison RL, Colliver R et al (2020) Jumping off the treadmill: transforming NRM to systemic governing with systemic co-inquiry. Policy Stud 41(4):350–371. https://doi.org/10.1080/01442872.2020.1726312

Allen CR, Fontaine JJ, Pope KL, Garmestani AS (2011) Adaptive management for a turbulent future. J Environ Manag 92(5):1339–1345. https://doi.org/10.1016/j.jenvman.2010.11.019

Anwar AA, Ul Haq Z (2013) An old–new measure of canal water inequity. Water Int 38(5):536–551. https://doi.org/10.1080/02508060.2013.832124

Ashraf M, Routray JK (2013) Perception and understanding of drought and coping strategies of farming households in north-west Balochistan. Int J Disaster Risk Reduct 5:49–60. https://doi.org/10.1016/j.ijdrr.2013.05.002

Baig JA, Kazi TG, Shah AQ et al (2011) Evaluation of arsenic levels in grain crops samples, irrigated by tube well and canal water. Food Chem Toxicol 49(1):265–270. https://doi.org/10.1016/j.fct.2010.11.002

Basharat M (2012) Spatial and temporal appraisal of groundwater depth and quality in LBDC command: issues and options. Pak J Eng Appl Sci 11:14–29

Basharat M, Basharat M (2019) Developing Sukh-Beas as a potential recharge site during wet years for Bari Doab. Appl Water Sci 9(7):163. https://doi.org/10.1007/s13201-019-1041-6

Basharat M, Tariq AUR (2013) Long-term groundwater quality and saline intrusion assessment in an irrigated environment: a case study of the aquifer under the LBDC irrigation system. Irrig Drain 62(4):510–523. https://doi.org/10.1002/ird.1738

Basharat M, Tariq AUR (2015) Groundwater modelling for need assessment of command scale conjunctive water use for addressing the exacerbating irrigation cost inequities in LBDC irrigation system, Punjab, Pakistan. Sustain Water Resour Manag 1(1):41–55. https://doi.org/10.1007/s40899-015-0002-y

Basharat M, Alia SU, Azhar AH (2014) Spatial variation in irrigation demand and supply across canal commands in Punjab: a real integrated water resources management challenge. Water Policy 16(2):397–421. https://doi.org/10.2166/wp.2013.060

Berkes F, Folke CF (eds) (1998) Linking social and ecological systems: management practices and social mechanisms for building resilience. Cambridge University Press, Cambridge

Blackmore C, Ison RL, Jiggins J (2007) Social learning: an alternative policy instrument for managing in the context of Europe's water. Environ Sci Policy 10(6):493–498. https://doi.org/10.1016/j.envsci.2007.04.003

Brahman KD, Kazi TG, Afridi HI et al (2016) Exposure of children to arsenic in drinking water in the Tharparkar region of Sindh, Pakistan. Sci Total Environ 544:653–660. https://doi.org/10.1016/j.scitotenv.2015.11.152

Chambers R (1994) The origins and practice of participatory rural appraisal. World Dev 22(7):953–969. https://doi.org/10.1016/0305-750X(94)90141-4

Chandio AS, Lee TS (2012) Managing saline water intrusion in the lower Indus Basin aquifer. Water Resour Manag 26(6):1555–1576. https://doi.org/10.1007/s11269-011-9972-1

Christen EW, Mitchell M, Roth C, Rowley E (2019) Addressing research complexity: analysing pathways to impact and using transdisciplinary approaches. Agric Sci 30/31(2/1):32–43

Cleaver F, de Koning J (2015) Furthering critical institutionalism. Int J Commons 9(1):1–18. https://doi.org/10.18352/ijc.605

Clement S, Moore SA, Lockwood M, Mitchell M (2017) Fit-for-purpose institutions? An evaluation of biodiversity conservation in the agricultural landscape of the Tasmanian Midlands, Australia. J Environ Policy Plan 19(2):135–155. https://doi.org/10.1080/1523908X.2016.1162708

Desakota Study Team (2008) Re-imagining the rural–urban continuum: understanding the role ecosystem services play in the livelihoods of the poor in desakota regions undergoing rapid change. Institute for Social and Environmental Transition – Nepal, Kathmandu. https://www.i-s-e-t.org/resource-rural-urban-continuum. Accessed 17 July 2020

Dietz T, Ostrom E, Stern PC (2003) The struggle to govern the commons. Science 302(5552):1907–1912. https://doi.org/10.1126/science.1091015

Dobson A (1999) Introduction. In: Dobson A (ed) Fairness and futurity: essays on environmental sustainability and social justice. Oxford University Press, Oxford, pp 1–18

Enfors-Kautsky E, Järnberg L, Quinlan A, Ryan P (2018) Evaluation, reflection and sense making module. In: Enfors-Kautsky E, Järnberg L, Quinlan A, Ryan P Wayfinder: a resilience guide for navigating towards sustainable futures. GRAID programme, Stockholm Resilience Center, Stockholm. https://wayfinder.earth/the-wayfinder-guide/learning-your-way-forward/phase-closure-evaluation-reflection-and-sense-making/. Accessed 17 July 2020

Fals Borda O (2006) Participatory (action) research in social theory: origins and challenges. In: Reason P, Bradbury H (eds) Handbook of action research. Sage, London, pp 27–37

Faysse N, Errahj M, Imache A et al (2014) Paving the way for social learning when governance is weak: supporting dialogue between stakeholders to face a groundwater crisis in Morocco. Soc Nat Resour 27(3):249–264. https://doi.org/10.1080/08941920.2013.847998

Finlayson CM, Davidson N, Pritchard D et al (2011) The Ramsar Convention and ecosystem-based approaches to the wise use and sustainable development of wetlands. J Int Wildl Law Policy 14(3–4):176–198. https://doi.org/10.1080/13880292.2011.626704

Finlayson CM, de Groot RS, Hughes FMR, Sullivan CA (2019) Freshwater ecosystem services and functions. In: Hughes FMR (ed) Freshwater ecology and conservation: approaches and techniques. Oxford University Press, Oxford, pp 321–337

Folke C, Hahn T, Olsson P, Norberg J (2005) Adaptive governance of social-ecological systems. Annu Rev Environ Resour 30(1):441–473. https://doi.org/10.1146/annurev.energy.30.050504.144511

Folke C, Biggs R, Norström AV et al (2016) Social-ecological resilience and biosphere-based sustainability science. Ecol Soc 21(3):41. https://doi.org/10.5751/es-08748-210341

Foster N, Ison R, Blackmore C, Collins K (2019) Revisiting deliberative policy analysis through systemic co-inquiry: some experiences from the implementation of the Water Framework

Directive in England. Policy Stud 40(5):510–533. https://doi.org/10.1080/01442872.2019.1618816
Haq MU, Khattak RA, Puno HK et al (2005) Surface and ground water contamination in NWFP and Sindh provinces with respect to trace elements. Int J Agric Biol 7(2):214–217
Holling CS (1973) Resilience and stability of ecological systems. Annu Rev Ecol Syst 4(1):1–23. https://doi.org/10.1146/annurev.es.04.110173.000245
Holling CS (1978) Adaptive environmental management and assessment. Wiley, Chichester
Ison RL (2016) Governing in the Anthropocene: what future systems thinking in practice? Syst Res Behav Sci 33(5):595–613. https://doi.org/10.1002/sres.2421
IUCN (2003) Indus Delta, Pakistan: economic costs of reduction in freshwater flows (Case Studies in Wetland Evaluation # 5). https://cmsdata.iucn.org/downloads/casestudy05indus.pdf. Accessed Feb 2018
Jilani G, Khair SM (2014) Evaluating farmers' perceptions on causes of water scarcity and coping strategies: a case study example from Tehsil Karazat District Pishin. J Appl Emerg Sci 5(2):144–154
Kakar Z, Khair SM, Khan MZ, Khan MA (2014) Socio-economic impact of water scarcity on the economy of Pishin Lora Basin in Balochistan. J Appl Emerg Sci 5(2):90–96
Kamal S, Amir P, Mohtadullah K (2012) Development of Integrated River Basin Management (IRBM) for Indus Basin: challenges and opportunities. WWF-Pakistan, Lahore
Kazmi SJH, Qureshi S, Siddiqui MU, Arsalan MH (2006) Depleting wetlands of lower Sindh, Pakistan: a spatio-temporal study through satellite remote sensing. Paper presented at the 2006 International Conference on Advances in Space Technologies, ICAST. https://doi.org/10.1109/ICAST.2006.313786
Khair SM, Culas RJ (2013) Rationalising water management policies: tube well development and resource use sustainability in Balochistan region of Pakistan. Int J Water 7(4):294–316. https://doi.org/10.1504/IJW.2013.056673
Khair SM, Mushtaq S, Reardon-Smith K (2015) Groundwater governance in a water-starved country: public policy, farmers' perceptions, and drivers of tubewell adoption in Balochistan, Pakistan. Groundwater 53(4):626–637. https://doi.org/10.1111/gwat.12250
Khan AA, Sheik KM, Wright G (1996) A perspective on community-based management at Lake Zangi Nawar, Baluchistan, Pakistan. Lakes Reserv Res Manag 2(3–4):153–155. https://doi.org/10.1111/j.1440-1770.1996.tb00058.x
Khan MZ, Jabeen T, Ghalib SA et al (2014) Effect of Right Bank Outfall Drain (RBOD) on biodiversity of the wetlands of Haleji Wetland Complex, Sindh. Can J Pure Allied Sci 8(2):2871–2900
Khattak MA, Ahmed N, Qazi MA et al (2012) Evaluation of ground water quality for irrigation and drinking purposes of the areas adjacent to Hudiara industrial drain, Lahore, Pakistan. Pak J Agric Sci 49(4):549–556
Latif M (2007) Spatial productivity along a canal irrigation system in Pakistan. Irrig Drain 56(5):509–521. https://doi.org/10.1002/ird.320
Latif M, Ahmad MZ (2009) Groundwater and soil salinity variations in a canal command area in Pakistan. Irrig Drain 58(4):456–468. https://doi.org/10.1002/ird.417
Lynch A, Baumgartner LJ, Boys C et al (2019) Speaking the same language: can the sustainable development goals translate the needs of inland fisheries into irrigation decisions? Mar Freshw Res 70(9):1211–1228. https://doi.org/10.1071/MF19176
Mastoi GM, Shah SGS, Khuhawar MY (2008) Assessment of water quality of Manchar Lake in Sindh (Pakistan). Environ Monit Assess 141(1):287–296. https://doi.org/10.1007/s10661-007-9895-8
Meinzen-Dick RS (1996) Groundwater markets in Pakistan: participation and productivity (IFPRI Research Report No. 105). International Food Policy Research Institute, Washington, DC
Midgley G (2003) Science as systemic intervention: some implications of systems thinking and complexity for the philosophy of science. Syst Pract Action Res 16(2):77–97. https://doi.org/10.1023/a:1022833409353

Millennium Ecosystem Assessment (2005) Ecosystems and human well-being: wetlands and water synthesis. World Resources Institute, Washington DC

Mitchell M, Moore SA, Clement S et al (2017) Biodiversity on the brink: evaluating a transdisciplinary research collaboration. J Nat Conserv 40:1–7. https://doi.org/10.1016/j.jnc.2017.08.002

Mitchell M, Allan C, Punthakey J et al (2020) Living with salinity in the Indus Basin: SRA 2: Final report. ACIAR, Canberra. https://www.aciar.gov.au/project/WAC-2019-102

Mitchell M, Awan UK, Iqbal N, Punthakey J (eds) (forthcoming) Improving groundwater management to enhance agriculture and farming livelihoods: literature review

Mukherji A, Shah T (2005) Groundwater socio-ecology and governance: a review of institutions and policies in selected countries. Hydrogeol J 13(1):328–345. https://doi.org/10.1007/s10040-005-0434-9

Mustafa D, Qazi MU (2007) Transition from karez to tubewell irrigation: development, modernization, and social capital in Balochistan, Pakistan. World Dev 35(10):1796–1813. https://doi.org/10.1016/j.worlddev.2007.06.002

Mustafa D, Qazi MU (2008) Karez versus tubewell irrigation: the comparative social acceptability and practicality of sustainable groundwater development in Balochistan, Pakistan. Contemp South Asia 16(2):171–195. https://doi.org/10.1080/09584930701733514

Narain V (2008) Warabandi as a sociotechnical system for canal water allocation: opportunities and challenges for reform. Water Policy 10(4):409–422. https://doi.org/10.2166/wp.2008.057

NASA (2020) NASA data used to track groundwater in Pakistan. https://www.nasa.gov/feature/jpl/nasa-data-used-to-track-groundwater-in-pakistan. Accessed 4 July 2020

Ostrom E (1990) Governing the commons: the evolution of institutions for collective action. Cambridge University Press, Cambridge

Pahl-Wostl C, Tabara D, Bouwen R et al (2008) The importance of social learning and culture for sustainable water management. Ecol Econ 64(3):484–495. https://doi.org/10.1016/j.ecolecon.2007.08.007

Pasha HA (2015) Growth of the provincial economies (IPR Brief). Institute for Policy Reforms, Lahore, Pakistan. http://ipr.org.pk

Patel K (2018) Smart phones bring smart irrigation. https://earthobservatory.nasa.gov/images/92903/smart-phones-bring-smart-irrigation. Accessed 4 July 2020

Pittock J, Finlayson CM, Linke S (2019) Freshwater ecosystem security and climate change. In: Hughes JMR (ed) Freshwater ecology and conservation: approaches and techniques. Oxford University Press, Oxford, pp 359–377

Punthakey JF, Khan MR, Riaz M et al (2015) Optimising canal and groundwater management to assist water user associations in maximizing crop production and managing salinisation in Pakistan: assessment of groundwater resources for Rechna Doab, Pakistan (Report to ACIAR). Ecoseal, Roseville, Australia

Rabbani U, Mahar G, Siddique A, Fatmi Z (2017) Risk assessment for arsenic-contaminated groundwater along River Indus in Pakistan. Environ Geochem Health 39(1):179–190. https://doi.org/10.1007/s10653-016-9818-0

Rasool F, Tunio S, Hasnain S, Ahmad E (2002) Mangrove conservation along the coast of Sonmiani, Balochistan, Pakistan. Trees 16(2):213–217. https://doi.org/10.1007/s00468-001-0151-5

Reed MS, Evely AC, Cundill G et al (2010) What is social learning? Ecol Soc 5(4):r1. http://www.ecologyandsociety.org/vol15/iss4/resp1/

Scott WR (2014) Institutions and organizations: ideas, interests, and identities, 4th edn. Sage, Thousand Oaks

Sellberg MM, Ryan P, Borgström ST et al (2018) From resilience thinking to resilience planning: lessons from practice. J Environ Manag 217:906–918. https://doi.org/10.1016/j.jenvman.2018.04.012

Shah MAA, Anwar AA, Bell AR, ul Haq Z (2016) Equity in a tertiary canal of the Indus Basin Irrigation System (IBIS). Agric Water Manag 178:201–214. https://doi.org/10.1016/j.agwat.2016.09.018

Siddique A, Zaigham NA, Mohiuddin S et al (2012) Risk zone mapping of lead pollution in urban groundwater. J Basic Appl Sci 8(1):91–96

Stafford-Smith M, Griggs D, Gaffney O et al (2017) Integration: the key to implementing the Sustainable Development Goals. Sustain Sci 12(6):911–919. https://doi.org/10.1007/s11625-016-0383-3

Steyaert P, Jiggins J (2007) Governance of complex environmental situations through social learning: a synthesis of SLIM's lessons for research, policy and practice. Environ Sci Policy 10(6):575–586. https://doi.org/10.1016/j.envsci.2007.01.011

Stoker G (1998) Governance as theory: five propositions. Int Soc Sci J 50(155):17–28. https://doi.org/10.1111/1468-2451.00106

Stringer LC, Dougill AJ, Fraser E et al (2006) Unpacking "participation" in the adaptive management of social-ecological systems: a critical review. Ecol Soc 11(2):39. http://www.ecologyandsociety.org/vol11/iss2/art39/

Tagar HK, Bukhari BS, Qabtia M et al (2016) Sustainable development goals: performance and problems of agricultural labour in share cropping system and their impact on natural resource management and growth (a case of Sindh–Pakistan). Int J Innov Res Dev 5(6):443–449

Tress G, Tress B, Fry G (2005) Clarifying integrative research concepts in landscape ecology. Landsc Ecol 20(4):479–493. https://doi.org/10.1007/s10980-004-3290-4

Ullah R, Malik RN, Qadir A (2009) Assessment of groundwater contamination in an industrial city, Sialkot, Pakistan. Afr J Environ Sci Technol 3(12):429–446

van Steenbergen F (1995) The frontier problem in incipient groundwater management regimes in Balochistan (Pakistan). Hum Ecol 23(1):53–74. https://doi.org/10.1007/BF01190098

van Steenbergen F, Kaisarani AB, Khan NU, Gohar MS (2015) A case of groundwater depletion in Balochistan, Pakistan: enter into the void. J Hydrol Reg Stud 4(Part A):36–47. https://doi.org/10.1016/j.ejrh.2014.11.003

Velis M, Conti KI, Biermann F (2017) Groundwater and human development: synergies and trade-offs within the context of the sustainable development goals. Sustain Sci 12(6):1007–1017. https://doi.org/10.1007/s11625-017-0490-9

Walker B, Salt D (2012) Resilience practice: building capacity to absorb disturbance and maintain function. Island Press, Washington, DC

Walker B, Abel N, Anderies JM, Ryan P (2009) Resilience, adaptability, and transformability in the Goulburn-Broken Catchment. Aust Ecol Soc 14(1):12. http://www.ecologyandsociety.org/vol14/iss1/art12/

Woodward WR, Hetley RS (2007) Diffusion, decolonializing, and participatory action research. Integr Psychol Behav Sci 41(1):97–105. https://doi.org/10.1007/s12124-007-9000-4

World Commission on Environment and Development (1990) Our common future (the Brundtland report), Australian edn. Oxford University Press, Melbourne

Chapter 14
A Roadmap for a Comprehensive Water Resources Forecast System for Pakistan

Thomas E. Adams III

Abstract The transboundary Indus Basin covers approximately 1.12 million km^2 and spans four countries: Pakistan (47%), India (39%), China (8%), and Afghanistan (6%). Pakistan lacks a comprehensive flood prediction and water resource assessment modelling system. This is especially problematic given the need for transboundary hydrologic information to adequately model tributary flows into the Indus River originating from regions outside Pakistan. Information deficiencies complicate flood prediction and warning on the Indus River and its major tributaries within Pakistan's national borders. Inadequate hydrologic modelling also presents a major impediment to effective water resources management to support, in particular, irrigation of croplands and hydropower generation in Pakistan at multiple time scales: monthly, seasonal, annual, and climatic. Recent surveys of hydro-meteorological monitoring, data transmission, and data archiving in Pakistan suggest infrastructure modernisation and institutional capacity building are needed, including improved data analysis, modelling, and training of professional and technical staff at all levels. This chapter lays out a few general options that could be pursued by Pakistan's government to develop a comprehensive hydro-meteorological forecasting system. However, it is recommended, regardless of specific details, the modelling approach focus on the use of an ensemble hydro-meteorological modelling framework to capture the uncertainty inherent in the prediction of future meteorological variables used as hydrologic model inputs.

Keywords Flood forecasting · Forecasting system · Water resources · Pakistan · Indus River integrated

T. E. Adams III (✉)
TerraPredictions, LLC, Blacksburg, VA, USA

M. A. Watto et al. (eds.), *Water Resources of Pakistan*, World Water Resources 9,
https://doi.org/10.1007/978-3-030-65679-9_14

14.1 Introduction

The hydro-meteorological setting of the Indus Basin (Fig. 14.1) is highly variable, both spatially and seasonally, as discussed by Adams (2019). These complexities are derived from both natural influences and human activities. From a water resource management perspective these complexities clearly demonstrate the necessity for the development of a comprehensive hydro-meteorological modelling and forecast system for the Indus Basin within Pakistan. A forecast system is needed that spans forecast horizons from short lead-times (hours) to climate time scales (months, seasons, inter-annual). The system must also be comprehensive in that it meets flood forecasting, water resource management (including water supply, hydropower generation, water quality, and irrigation needs), and flash flood alert needs. Giupponi and Sgobbi (2013) emphasise the need for Integrated Water Resource Management (IWRM) tools and methods to improve natural resource management in a sustainable way. Such systems should include the capabilities of an early warning system for floods as well as meeting the needs for longer lead-times for water resource planning and management. A major challenge to Pakistan is that significant flow contributions from headwater regions of the Indus River originate in Afghanistan, China, and India, which currently leaves Pakistan vulnerable and dependent on the acquisition of real-time, observed and forecast streamflow information from

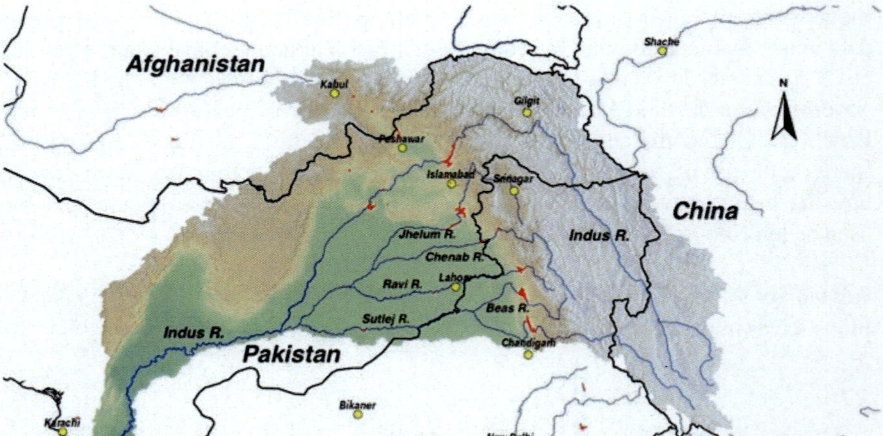

Fig. 14.1 Map of the Indus Basin (modified from Adams 2019) with major tributaries in blue, country borders in black, prominent reservoirs (from Lehner et al. 2011) in red and some major cities for reference in yellow. The Arabian Sea lies to the left of the map

neighbouring countries. As discussed by several authors in Khan and Adams (2019), contentious political conditions in the region make the likelihood of the development and implementation of a comprehensive, fully cooperative transboundary hydro-meteorological modelling system unlikely for the foreseeable future.

The hydro-climatology of the Indus River Basin is complex. Much of the basin falls within an arid zone that extends from West Africa to eastern India between about 10°N to 40°N latitude, as illustrated by the Köppen-Geiger climate classification (Köppen 1936; Kottek et al. 2006) globally, which is discussed in Adams (2019). The climate of South Asia is dominated by the summer monsoonal circulation system. The Himalayas commonly receive excessive rainfall, particularly over south-facing slopes, which leads to flooding episodes during the summer monsoons (June–July–August–September), often producing devastating landslides (Bhatt and Nakamura 2005). It is widely observed that orographic effects of major orogenic belts, such as the Himalayas, control the occurrence, enhanced intensity, and distribution of precipitation due to their prominent relief (Andermann et al. 2011).

Adams (2019) shows that nearly all flood-related deaths and the largest population displacements occur during the summer monsoon season. Interestingly, the peak of flood occurrences (August) and flood deaths (September) do not coincide. The greatest number of flooding deaths in the Indus Basin occurred from a single event in June 12–27, 2013, in India, in a glacially dominated region of the Himalayas. It is also noteworthy that population displacements can be massive, such as the 10,000,000 displaced in Pakistan in 2010 or the 742,000 displaced in Pakistan in 2012 from monsoonal rains and subsequent flooding. Many of the significant flooding events with substantial deaths occurred upstream of dams, where the dams offer no protection and in glacial regions where the threat of Glacial Lake Outburst Floods is significant (Adams 2019). Adams (2019) discusses how many flash flood related deaths occur with torrential rain or brief torrential rain indicated as the cause of the flooding. Some aid in forecasting the threat of flash floods will be provided by the World Meteorological Organisation's (WMO) Flash Flood Guidance System (FFGS), which is based on that developed by the Hydrologic Research Center (HRC) (see: http://www.wmo.int/pages/prog/hwrp/flood/ffgs/index_en.php). The WMO's FFGS is currently being implemented operationally for the South Asia region, including Pakistan. Clark et al. (2014) discusses important findings related to various FFGS approaches. With the aid of the WMO, the Pakistan Meteorological Department (PMD) was expecting to implement the WMO FFGS operationally by the end of 2017, but this has not yet executed. The WMO FFGS is designed to operate remotely at HRC, using both satellite-based and ground-based precipitation observations. FFGS products will be transmitted directly to PMD for issuing flash flood warnings. Extensive training is included in the FFGS implementation by the WMO and HRC.

The current chapter discusses the overall paradigm (Sect. 14.1.1) of hydrologic forecasting within the context of a comprehensive system that addresses real-time flood forecasting, and a range of water resources needs, such as reservoir management, flood control, hydro-electric power generation, domestic and industrial water supply, and irrigation planning and management. This leads to a discussion of model scale (Sect. 14.1.2), both temporally and spatially, that is critical to meeting various water resources needs, ranging from flash floods to monthly to seasonal water resources forecasting. These considerations, in turn, necessitate that we consider the role and importance of hydro-meteorological uncertainty in the forecast process (Sect. 14.1.3).

14.1.1 Forecast Paradigm

Adams (2019) discusses how hydrologic forecasting should be viewed in a broad context in terms of overlapping modelling needs and objectives. Hydrology is the scientific study of the movement, distribution and quality of water.

A hydrological service is an institution whose primary mission is to provide information about the water (or hydrological) cycle and the status and trends of a country's water resources, but, most typically, focuses on flood forecasting and warnings and drought monitoring and outlooks. As suggested by the World Meteorological Organisation (2008a, b), such services are commonly found as:

- A National Hydrological Service (NHS).
- Part of a National Hydro-meteorological Service or a National Hydrological and Meteorological Service.
- Sectoral hydrological services, of which there could be more than one in a single country.
- A federal hydrological service with many state/regional hydrological services.

National Hydrological Services or related agencies have been established in countries for systematic water resources data collection, archiving and dissemination of water resources related data and information. Their primary role is to provide information to decision makers on the status of and trends in water resources. Such information may be required, among other things, for the following purposes:

- Assessing a country's water resources (quantity, quality, distribution in time and space), the potential for water-related development, and the ability of the supply to meet actual or foreseeable demand.
- Planning, designing and operating water projects.
- Assessing the environmental, economic and social impacts of existing and proposed water resources management practices, and planning sound management strategies.
- Providing protection for people and property against water-related hazards, particularly floods and droughts.
- Allocating water among competing uses, both within the country and cross-border.
- Meeting regulatory requirements.

Most frequently, water resources information is collected as part of an ongoing monitoring program or for a specific purpose, such as the design of a hydroelectric power generation facility. However, increasing competition among users for scarce water requires that resources be managed in an integrated fashion, and that interactions among several projects and users are well understood. This places a much greater burden on the suppliers of water resources information, because a variety of types of information is simultaneously needed, and has to be presented in different forms for different users. This also makes it essential that NHSs and assessment agencies understand the needs of all of their users, not just those with whom they have traditionally dealt. Even more demanding is the need to look ahead to the possible needs of future users of data and services and to commence collecting the required information before an actual demand can be demonstrated with certainty (World Meteorological Organisation 2008a, b). One very good example of this is the increasing benefits of making use of seasonal climate outlooks presented in hydrological terms.

The underlying importance of water to human health and survival is indisputable. Civilisations have risen and fallen due to the availability or unavailability of water, and major population movements are often driven directly or indirectly by water scarcity. Lack of potable water is a problem in many regions. Recognising the irreplaceability of water for survival and development, the Joint Statement by the United Nations and the World Bank (2016) from the High Level Panel on Water called for a comprehensive and coordinated approach to water, as well as increased attention and investment in water-related services. The Panel promotes strengthened collective governance of water, suggesting a water data revolution, and recognises the unique social, economic, and environmental value of the water. Water resources cannot be properly managed without quantification of their physical extent, quality and associated variability, with projections into the foreseeable future. Data from hydrological networks are used by public and private sectors for planning, designing, operating and maintaining multipurpose water management systems, including: the preparation and distribution of flood forecasts and warnings aimed at protecting lives and property; the design of spillways, highways, bridges and culverts; flood plain mapping; determining and monitoring environmental or ecological flows; managing water rights and transboundary water issues; education and research; and protecting water quality and regulating pollutant discharges (USGS 2006).

The World Meteorological Organisation (2006) reports that, in the majority of countries, the functions of Hydrologic Services are typically served by several related water agencies, often with over-lapping responsibilities. Activities include systematic water resource monitoring, including data collection, processing, storage, and archiving, the production and dissemination of related water resource data and information, and hydrological forecasting spanning a range of time scales, with flood alerts and warnings issued as needed (Sene 2010; WMO 2011; Adams and Pagano 2016). Water resources cannot be properly managed without quantification of their current and potential future physical distribution, quality and variability. Data from hydrological networks are thus used by public and private sectors for a variety of other applications, including, reservoir management, hydropower generation, water supply, water allocations for irrigation, and Disaster Risk Management (DRM). The WMO Guide to Hydrological Practices (WMO 2008a, b, 2009) details how hydrological data are collected, the limitations and the reliability of the data, and how they are managed.

14.1.2 Model Scale

The need for hydro-meteorological prediction exists globally to provide flood warnings, for water resources planning and management, to meet regional and national development objectives, and to address hydrologic issues related to climate change impacts and adaptation. Examples representing the need for and commitment to modernisation of national flood forecasting systems are seen in the UK, European Union (EU) countries, US, Australia, Israel, Russia, China, and others (see Adams and Pagano 2016). Unfortunately, insufficient hydro-meteorological monitoring, modelling, and forecasting capabilities exist in most countries, including in developed countries like the US. The establishment of the National Water Center by the National Oceanic and Atmospheric Administration's (NOAA) National Weather Service (NWS) was made, in part, to address water resources and flood forecasting deficiencies in the US. For example, a clear need exists in the US to more broadly monitor and predict water quality issues in a real-time setting, which is not mandated as a core mission of the NWS. The disparity between forecasting needs and capabilities is especially evident in low and medium income countries within National Hydrologic Services programs (Herold and Rudari 2013). Significant gaps in hydro-meteorological forecasting capabilities have been identified in many low to medium income countries in Africa, Asia, and Central and South America. Recent literature (Ward et al. 2013) describes the development of Global Flood Risk Models to, in part, fill the hydrologic prediction gap in developing countries. The Sendai Framework for Disaster Risk Reduction (UN 2015) and the Warsaw International Mechanism for Loss and Damage Associated with Climate Change Impacts (UNFCCC 2013) were adopted internationally in response to growing global flood risk and the need to address water resources prediction issues.

Efforts to address the large disparity in hydro-meteorological forecasting capabilities, mostly for disaster risk management purposes, between rich and poor countries are emerging. Namely, with the rise of high resolution global geophysical data sets (Sampson et al. 2016), largely from satellite remote sensing, several international groups (Table 14.1) have been developed and implemented to address the need for hydrologic modelling systems for flood prediction and monitoring at continental and global scales.

While modelling efforts at large spatial scales are impressive and show utility, Emerton et al. (2016, p. 391) note that "Operational systems currently have the capability to produce coarse-scale discharge forecasts in the medium-range and disseminate forecasts and, in some cases, early warning products in real time across the globe, in support of national forecasting capabilities." This finding is also expressed by Ward et al. (2015). Sampson et al. (2015, p. 7378) illustrate the limitations of global and continental scale hydrologic modelling systems, stating: "These metrics

Table 14.1 Some global and continental scale modelling systems

European Flood Awareness System (EFAS)	The European Centre for Medium-range Weather Forecasting (ECMWF)	Continental	http://www.ecmwf.int/en/research/projects/efas
Global Flood Awareness System (GloFAS)	The European Centre for Medium-range Weather Forecasting (ECMWF)	Global	http://globalfloods.jrc.ec.europa.eu
Global Flood Forecasting Information System (GLOFFIS)	Deltares (The Netherlands)	Global	see Weerts et al. (2017)
Global Flood Monitoring System (GFMS)	University of Maryland	Global	http://flood.umd.edu; formerly at the NASA Goddard Space Flight Centre
Near Realtime Global Hydrological Simulation and Flood Monitoring Demonstration System	University of Oklahoma	Global	http://eos.ou.edu
Flooded Locations and Simulated Hydrographs (FLASH) Project NMQ-FLASH	University of Oklahoma	Continental	http://flash.ou.edu, http://eos.ou.edu/USA_Flood.html
Flooded Locations and Simulated Hydrographs (FLASH)	NOAA/National Severe Storms Laboratory (NSSL)	Continental	https://www.nssl.noaa.gov/projects/flash/
Dartmouth Flood Observatory	University of Colorado	Global	http://floodobservatory.colorado.edu
Snow Data Assimilation System (SNODAS)	US NOAA/National Weather Service, National Operational Hydrologic Remote Sensing Centre (NOHRSC) Continental	Global	https://www.nohrsc.noaa.gov
National Water Model (NWM)	US NOAA/National Weather Service, National Water Centre	Continental	http://water.noaa.gov/about/nwm
Integrated Flood Analysis System (IFAS)	International Centre for Water Hazard Analysis and Risk Management (ICHARM) with the United Nations Educational, Scientific and Cultural Organisation (UNESCO)	Global	http://www.icharm.pwri.go.jp/research/ifas/
Flash Flood Guidance System (FFGS)	Hydrologic Research Centre (HRC) with the World Meteorological Organisation (WMO)	Regional (with global coverage)	http://www.hrc-lab.org/giving/FFGS_index.php

indicate that the global model is typically able to capture between two thirds and three quarters of flooded area in the local benchmark data, and that along non-tidal reaches of large rivers, the skill is likely close to that of local models." It is also critical to note that the detailed street-level predictions that are found in national and regional scale flood forecast systems described in Adams and Pagano (2016) and Kauffeldt et al. (2016) are not feasible currently with global and continental scale modelling systems. Despite these limitations, there is clearly considerable utility with the use of global and continental scale models, especially where little to no flood information existed previously.

The Global Flood Forecasting Information System effort by Deltares is interesting because the forecast system uses its Flood Early Warning System (FEWS) software as its core modelling system. What is more, the NOAA/– National Weather Service 13 River Forecast Centers, Australian Bureau of Meteorology (BoM) Regional Forecast Centers, UK Environment Agency, Natural Resources Wales, and Scottish Environment Protection Agency (SEPA), as well as others, use FEWS as the foundation of their national flood forecasting systems. Deltares is an independent institute for applied research in the field of surface and subsurface hydrology throughout the world. Its main focus is on deltas, coastal regions and river basins, explaining that managing these densely populated and vulnerable areas is complex, and necessitates working closely with governments, businesses, other research institutes and universities in the Netherlands and abroad.

14.1.3 Forecast Uncertainty

All forecasts are uncertain. Hydrologic forecast uncertainty is driven by: (1) uncertainties in both observed and predicted meteorological model inputs (or forcings), as shown in Figs. 14.2 and 14.3, with hydrologic trace ensembles and ensemble precipitation fields, respectively; (2) model structural errors, since models do not perfectly represent physical systems; and (3) other uncertainties associated with model parameter estimates. Considerable research has focused on the use of ensemble methods to capture meteorological forcing prediction uncertainty. Recent work by Adams and Dymond (2018) has shown that the hydrologic ensemble median (or mean) forecast is comparable to deterministic approaches and superior for forecast lead-times beyond about 48 h.

There is considerable forecast uncertainty arising simply from the hydrometeorological inputs, both observed and forecast (Ward et al. 2013, 2015; Sampson et al. 2015; Emerton et al. 2016). Significant efforts have been made in the hydrologic prediction community to capture hydrologic forecast uncertainty, such as with the Hydrologic Ensemble Prediction Experiment (HEPEX). HEPEX was founded to advance the science and practice of hydrologic ensemble prediction and its use for risk-based decision-making. Examples of real-time hydrologic ensemble

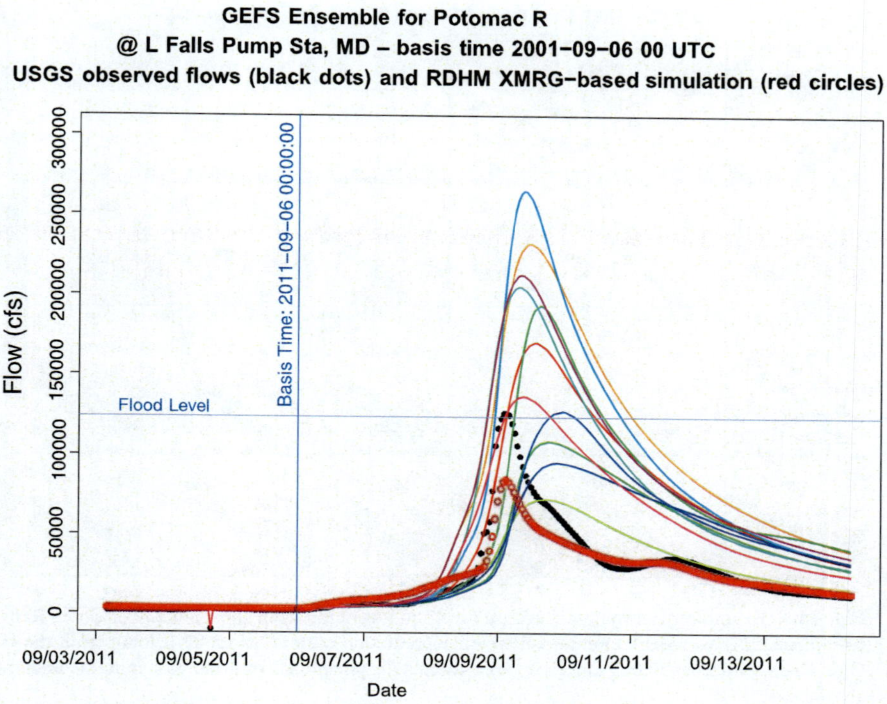

Fig. 14.2 Example of deterministic versus ensemble forecasts to capture meteorological forcings uncertainty; simulations are compared against observations

forecasting in the US by the NOAA/NWS are found with the Advanced Hydrologic Prediction Service (AHPS) Hydrologic Ensemble Forecast Service (HEFS) (Demargne et al. 2014) and Meteorological Model-based Ensemble Forecast System (MMEFS) (Adams and Dymond 2018).

Consequently, a strictly ensemble-based approach is recommended to capture forecast uncertainty that arises from model forcing errors, but ensemble median forecasts should be available as an alternative to a single-valued deterministic forecast of streamflow and river stage.

An additional issue with ensemble forecasts is that hydrologic forecast biases are present and should be accounted for, including underspread of ensembles (meaning that the spread of the ensemble traces is too narrow, implying greater confidence than really exists when the ensemble forecasts are statistically verified – see Demargne et al. 2014, for example). This forecast underspread is largely due to the need to account for model error that is independent of the meteorological forcings. Methods should be implemented to adjust model ensemble biases and anticipated underspread issues.

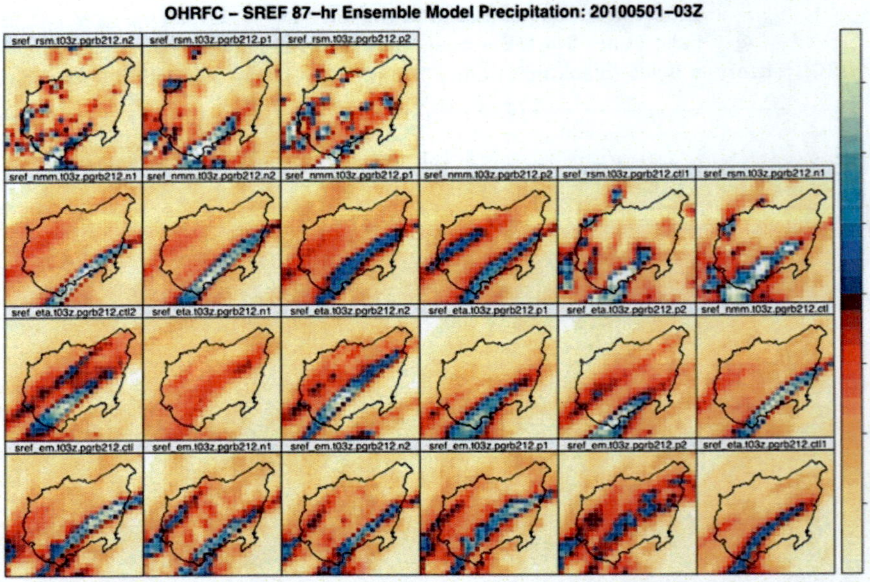

Fig. 14.3 Example of ensemble forecast precipitation fields from the 21-member Short Range Ensemble Forecast (SREF) numerical weather prediction system, as an 87 h total, from the US Ohio River Valley (~450,000 km^2 basin area, indicated by the black outline). Precipitation amounts are inches

14.2 Forecasting System Framework

The evolution of modern hydrologic forecast systems has progressed to the point that only a few basic forecast system paradigms are viable. These include:

1. Traditional approaches, involving considerable human forecaster interaction in model execution and adjustments (Sect. 14.2.1).
2. Fully automated forecast approach, with hands-off forecaster involvement, including data assimilation (Sect. 14.2.2).
3. Hybrid approaches, where human interaction is involved for manual data assimilation, followed by hands-off forecasting (Sect. 14.2.3).

A major drawback to the traditional forecast paradigm, discussed in Sect. 14.2.1, is the inability to account for hydrologic forecast uncertainty quantitatively, which is covered in Sect. 14.1.3.

14.2.1 Traditional Approach

Traditionally, operational streamflow forecasts, such as those produced by the NOAA/NWS River Forecast Centers, involve intensive manual interaction by forecasters, as described in Adams (2016). Delft-FEWS (Sect. 14.4.2), the US NOAA/NWS River Forecast System Interactive Forecast Program and FEWS-based Community Hydrologic Prediction System (Adams 2016), and other such systems, use graphical user interface (GUI) front-ends to permit real-time, interactive adjustments by hydrologic forecasters in a desktop workstation environment. These GUI-based systems allow forecasters to change model states, correct observational errors, and make other forecast adjustments interactively in a real-time setting (see Adams 1991, 2016; Adams and Smith 1993). This level of forecaster interaction has been termed an in-the-loop forecast process. Typically, this approach uses deterministic Quantitative Precipitation Forecast for forecasted (future) precipitation (Adams and Pagano 2016).

14.2.2 Automated Approach

In-the-loop forecast processes have strengths and limitations in their operational workflow. Forecasters use their expert knowledge on local basin conditions and hydrologic model performance to make manual adjustments to forecast data and conceptual hydrologic models at multiple stages during the operational forecast cycle. This practice usually leads to skilful predictions. Unfortunately, the process and forecast is often not reproducible, which limits opportunities to assess controlled variations, and the effort required to make forecasts in this way is an obstacle to expanding forecast service, that is, by adding new forecast locations, more frequent forecast updates, running more complex models, or producing short range ensembles and hindcasts that can support verification. This hands-off human involvement in the forecast workflow has been referred to as an over-the-loop process (see https://ral.ucar.edu/projects/assessing-the-viability-of-over-the-loop-streamflow-forecasting).

14.2.3 Hybrid Forecast Approach

A variation on the automated, over-the-loop forecast workflow is direct forecaster intervention using manual data assimilation of observed model forcings to establish initial model states prior to operational forecast run, combined with an over-the-loop forecast workflow (Sect. 14.2.2). An advantage of using this approach over a more traditional forecast workflow (Sect. 14.2.1) is the ability to use ensemble forecast methodologies. A drawback is that hydrologic forecaster adjustments over the observed period may be suboptimal in terms of model state changes and can be very time-consuming, making possibly critical forecast updates infeasible.

14.3 Pakistan's Needs

Adams (2019) identifies major water resources issues that must be included in a comprehensive hydrologic modelling and forecast system for Pakistan. These include:

- Reservoir simulation, including storage to meet irrigation requirements, hydropower generation, and flood control operations.
- Rainfall-runoff modelling.
- Snow accumulation and melt modelling.
- Dam break modelling.
- Modelling of Glacial Lake Outburst Floods.
- Flash flood guidance (FFG).
- One-dimensional hydraulic routing through major tributaries and irrigation channels, including flood inundation modelling with levee failures.
- Water resources Decision Support Systems for water management of irrigation allocations and reservoir releases for hydropower generation.
- Groundwater modelling for the assessment and prediction of groundwater levels and water quality changes due to withdrawals for irrigation.

The overall aim is to outline a roadmap for the development and implementation of a risk-based, real-time, operational system that, at its core, is used for the prediction of streamflow and river stage at pre-defined locations of interest. Broadly, the prediction system should include:

1. Estimation of prediction uncertainties of flow and stage at forecast point locations, as depicted in Fig. 14.4.
2. Integration with Decision Support Systems at external agencies to assess what actions to take.
3. Archiving of observed and forecast data in a relational database management system.
4. Development of real-time quality control (QC) capability.
5. Development of real-time forecast assessment tools.
6. Development of forecast verification capabilities.
7. Development of real-time system monitoring tools.

Perhaps the most critical component of the system is real-time monitoring (Item 7). System failures occur, due to a range of issues, including interruptions in the flow of input data to the modelling system, model coding bugs, data storage failures, etc., which require both automatic monitoring and notification of system problems. Human monitoring is particularly important because human expertise can best

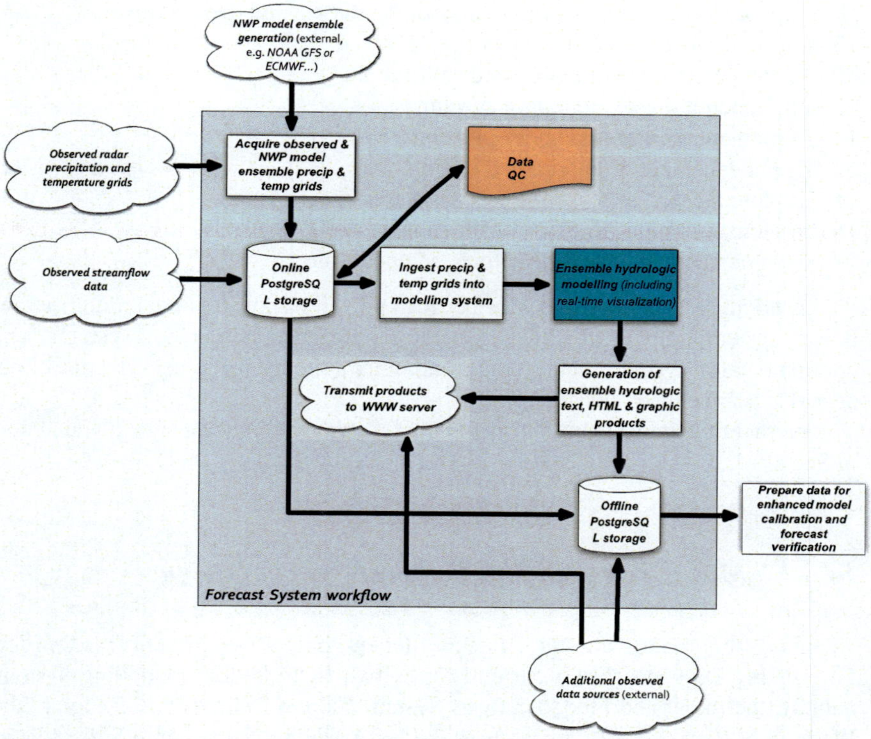

Fig. 14.4 Generalised proposed forecast system workflow

assess the reasonableness of predicted quantities that may be difficult to gauge in fully automated systems (See Sect. 14.3.9).

Generally, in operational systems, data flow into the system is continuous, or nearly so. So, system monitoring and logging of system processes should be continuous as well. Operational centres should be staffed on a daily basis and, when not, remote access for human monitoring should be available with email and possibly SMS notifications for system warnings and errors.

In detail, the forecast system should include the following components, developed and implemented sequentially to some degree:

1. Model selection and forecast system design.
2. Delineation of the modelling domain (principally from DEM data).
3. Development of Concept of Operations (CONOPS).
4. Commencement of staff training.
5. Expansion and integration of existing station observation networks.
6. Development and implementation of precipitation processing system.
7. Implementation of real-time observational data quality control.
8. Development of database architecture.

9. Implementation of real-time relational databases, and their archival.
10. Establishment of real-time data flow.
11. Model parameter estimation and calibration.
12. Implementation of modelling system.
13. Implementation of system monitoring.
14. Commencement of operational testing.
15. Implementation of real-time forecast assessment system.
16. Implementation of forecast verification system.
17. Commencement of real-time forecast operations.

An additional major need is electrical power backup, using uninterruptable power systems and local power generators, for computer, radar systems, telecommunication systems, data storage, and other systems due to periodic power outages that plague the country.

Several of the items identified as forecast system components are expanded upon in Sects. 14.3.1–14.3.10.

14.3.1 Model Selection and Forecast System Design

Adams (2019) reports that the Pakistan Meteorological Department (PMD) Flood Forecasting Division (FFD) currently uses two hydrological modelling systems, namely, the Integrated Flood Analysis System and the Deltares-based Flood Early Warning System (FEWS), which uses the NOAA/NWS Sacramento Soil Moisture Accounting (SAC-SMA) (Burnash et al. 1973; Burnash 1995) hydrologic model and the Deltares SOBEK 1-D unsteady flow, hydrodynamic model. Both modelling systems are currently used only during the monsoonal flood season. Flood forecasting at the FFD continues to rely on empirical regression relationships that translate observed upstream flows to downstream locations with a 24-h lead time. This technique is similar to the method used by Kohler (1944) in the NOAA/NWS in the 1940s. As previously identified, additional modelling capability is paramount to adequately account for the water resources complexities found in Pakistan.

14.3.2 Concept of Operations (CONOPS)

Concept of Operations (CONOPS) development is, ideally, a first step in the process of developing a complex system, such as to modernise National Meteorological and Hydrological Service operations for effectively handling the broad range of water resources forecasting and management activities. Not only does the development of CONOPS facilitate systems engineering for any needed infrastructure, it also provides a methodology with which to validate the success of that effort once the system is operational. A well-developed CONOPS helps to lower the risk of technical, political and economic failure by conceptualising the fully integrated, end-to-end systems operations as a means to guide implementation of equipment, hardware, software and training packages. The CONOPS describes the desired operation of a system by using the terminology of its users, and provides important information for the acquisition and/or development of new components. Given that a CONOPS answers who, what, when, where, why and how for a system, it should be a document accessible and useful to all stakeholders, regardless of their technical background or role within the system.

More specifically, the CONOPS documents should contain answers to the following range of questions:

- Why is the system needed?
- What does the overall system look like?
- What is the full system life cycle from deployment through disposal?
- What are the different aspects of system use (including operations, maintenance, support and disposal)?
- What are the different classes of user (e.g. operators, maintainers, supporters), and their different skills and limitations?
- What are the possible environments in which the system is used and supported?
- What are the boundaries of the system and its interfaces and relationships with other systems and their environments?
- When will the system be used, and under what circumstances?
- How and how well is the needed capability currently being met (typically by existing systems)?
- What type of scenarios exists that illustrate specific operational activities involving the use of the system?

A CONOPS is a comprehensive document that fully describes the end-to-end operational process, including the datasets and models used by PMD/FFD to inform operational decision-making. The CONOPS serves many purposes including use as an authoritative guide for the public and others to understand the operational process and as a training tool for new staff to learn the entire scope of operations. More importantly, the CONOPS serves as a focus for the organisation to fully lay out the design of the emergency management and response enterprise to meet national priorities. To this end, end-users must be included in the development of the CONOPS, to ensure that timely, understandable, and actionable forecasts are delivered to them. From a technical standpoint, the CONOPS provides a thorough explanation of data collection, the data collection network, transmission, quality control, storage for real-time data operations, and long-term archiving for model calibration, analysis, and validation. This includes the communication of clear technical standards for any equipment used and data transmission requirements. The methodologies in which the data is processed prior to their use in FFD models must also be clearly outlined. The models and operational system used need to be identified and the organisation and operational use of the forecast system fully explained. Finally, it is important to frame the CONOPS as a dynamic and evolving document that necessarily changes over time in response to changes in operational procedures, models, and the system as a whole. Such a document (see https://ral.ucar.edu/projects/assessing-the-viability-of-over-the-loop-streamflow-forecasting) should include details of roles and responsibilities of everyone involved in the end-to-end forecast process, including emergency operational procedures and other emergency scenarios that could affect operations. Those that should be contacted and their contact details should be clearly identified for emergency contact and during periods of both elevated operational readiness and during extreme events. The CONOPS should include facility design and routine maintenance procedures.

A standard CONOPS document may be structured around the following sections:

1. **Scope** – Scope provides the vision for this system, the outline of the contents of the CONOPS document, the purpose of the hydro-meteorological system, data and equipment standards needed to drive modernisation, the intended user community, and resultant benefits received from products and services and limits of the system performance.
2. **Knowledge References** – Describes the experts and methods consulted, through discussions with stakeholders, academics, and other experts. It describes studies of systems around the world, analyses mission requirements, operational needs and recommendations by vendors and product materials.
3. **Operational Description** – Describes the system from the users' perspectives, and includes a summary of each user's role and activities, clarifies order or sequence of user operations, summarises operational process procedures (SOPs) and has descriptions showing flow diagrams that indicate how organisational decision-making takes place with management structure.
4. **System Overview** – A high-level description of the mission requirements and interrelationships of key system components. It provides specific goals and objectives that are measurable and time-bound, describes interdependencies between sub-systems, and confirms that the system's capabilities will satisfy its mission.
5. **Operational and Support Environments** – Describes the infrastructure associated with each sub-system, including facilities, equipment, hardware, software, personnel, operational procedures and maintenance, and training and support requirements.
6. **Operational Scenarios** – Describes the system in action using one or more representative extreme event scenarios to reflect a range of stakeholder perspectives, a range of stress-failure scenarios, and both typical and extreme circumstances.

14.3.3 Staff Training

Training is a critical task in making the forecast system operational and must include the development of a Concept of Operations (CONOPS) and forecast system trouble-shooting guide used by staff to identify problems and procedures for correcting system related issues. The CONOPS details the workings of the system, notification procedures, staffing, and other matters related to the complete operational environment and related procedures.

14.3.4 Precipitation Processing System

Precipitation processing systems are critical to the generation of accurate hydrologic forecasts (Adams and Dymond 2019). They are needed for:

1. The integration of multi-sensor observational systems, which combine radar, rain gauge, and satellite precipitation estimates into a coherent, seamless, national precipitation field.
2. Data quality control.
3. Bias removal.

A multi-sensor precipitation observation system is essential for Pakistan to bias-correct radar precipitation estimates over non-mountainous regions of the country and for satellite-based estimates over mountainous areas and Indus Basin areas outside of Pakistan's borders.

14.3.5 Real-Time and Archive Relational Databases

Archiving of observed and forecast data in a relational database management system is necessary for the purposes of future model re-calibrations and verification of forecasts for reporting on the modelling prediction system performance. Forecast verification is useful for identifying where improvements can be made and to demonstrate forecast capabilities for other potential users of the system. The PostgreSQL open source relational database is recommended for archiving station data. Gridded (radar and temperature) data should be stored both in their native formats and the format used by the modelling system, if data reformatting is necessary (probably likely).

14.3.6 Model Parameter Estimation and Calibration

14.3.6.1 Parameter Estimation

Model parameter estimation is largely a geographic information system (GIS) exercise that typically depends on the availability on regional or national scale datasets. Generally, the process can be automated through scripting, using an open source

GIS package, such as GRASS GIS (https://grass.osgeo.org) or QGIS (https://qgis.org/en/site/). Major areas needed for parameter estimation include:

- Hydrologic modelling component.
- Snow model parameters.
- Collection of reservoir simulation information.
- Estimation of hydrologic and hydraulic model parameters, such as Manning roughness coefficients.

The uncertainty in model response to precipitation, irrigation, and other modelled system components can be accounted for through the ensemble modelling process, possibly with varying initial states (Fig. 14.5).

14.3.6.2 Calibration

Model calibration is critical to the use of hydrologic models, especially in an operational context. Duan et al. (2003, p. v) state:

> *… all models are approximations of the real world, model equations and associated parameters are idealized representations which are not directly (unambiguously) related to measurable watershed properties. Furthermore, there is a variety of errors in the model structure and uncertainties in the data used for parameter estimation, which introduce considerable inaccuracy into model behavior. These factors have made it difficult to develop reliable procedures for model parameter estimations, and to provide suitable estimates of uncertainties in the resulting model predictions.*

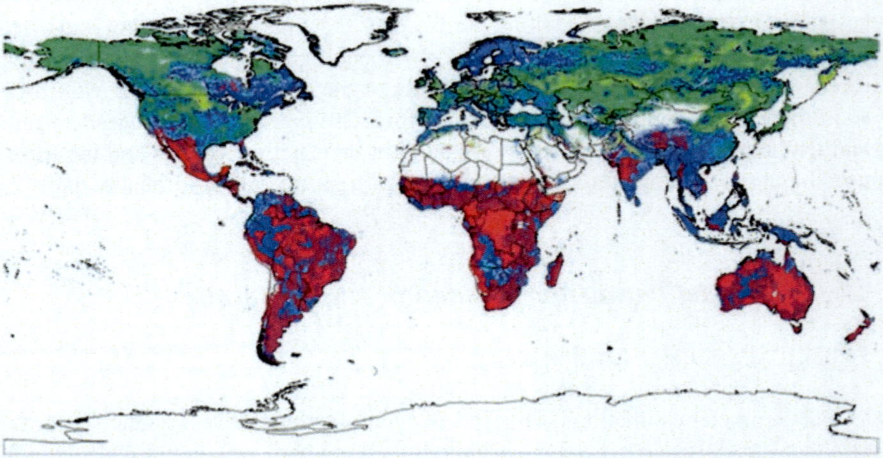

Fig. 14.5 Example NOAA/NWS RDHM SAC-SMA global LZTWM parameter estimates from soil physical properties with 1 km spatial resolution

Various authors in the hydrologic literature have addressed fundamental concerns in the calibration of hydrologic models. Many of these issues are discussed in Duan et al. (2003), who suggest the following questions be asked:

1. What constitutes the best estimates for the parameters of a watershed model?
2. What computational procedures are necessary to ensure proper model calibration and meaningful evaluation of model performance?
3. How are calibration methods developed and applied to watershed models?
4. What calibration data are needed, and how are these data obtained and analysed, in order to obtain reliable parameter values?
5. How can model parameters be estimated using a combination of expert knowledge of model physics and a priori knowledge of land surface characteristics?

A critical step in model calibration is the process to remove model input biases and re-formatting data for use in the model calibration system, as shown in Fig. 14.6, Steps 1, 2, and 4. A separate model calibration system is probably not feasible, so calibrations would use the forecast system run offline in some capacity. The data

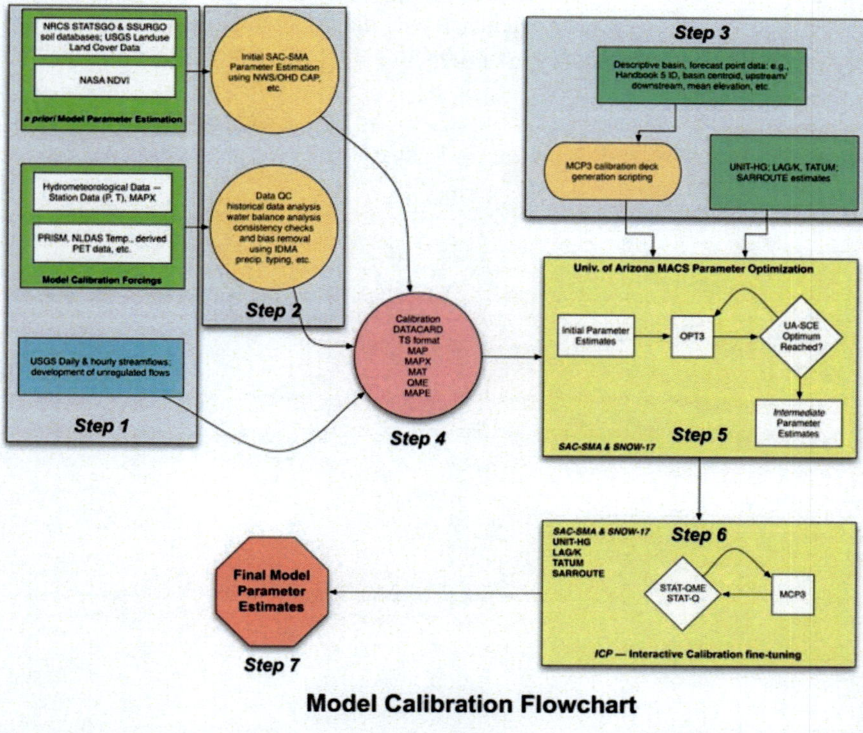

Fig. 14.6 Example of the model calibration process (from US NOAA/NWS) that could include a parameter optimisation process. Details of data acquisition depend on the availability of datasets in Pakistan

pre-processing is aimed at removing model input forcing biases and other errors, including handling of missing data, to produce continuous time.

14.3.7 Modelling System

A risk-based approach to the hydrologic prediction system should follow the ensemble methodologies proposed by HEPEX. HEPEX seeks to advance the science and practice of hydrological ensemble prediction and its use in impact and risk-based decision-making (see https://hepex.irstea.fr/about-hepex/), as illustrated by the NOAA/NWS Meteorological Model Ensemble River Forecast System (MMEFS – see https://www.weather.gov/erh/mmefs). The ensemble hydrologic forecast approach is illustrated in Fig. 14.7, which show ensemble inputs (precipitation and temperature) and modelled outputs, snow water equivalent (SWE), and stage/streamflow.

Figure 14.8 illustrates an operational NOAA/NWS MMEFS forecast, showing uncertainty bound and the individual hydrologic ensemble traces, with the ensemble median forecast. Treating the ensembles traces as independent forecasts allows exceedance probabilities to be calculated over the forecast horizon, which is 7 days in this example. Consequently, a probability of exceedance graphic, shown in

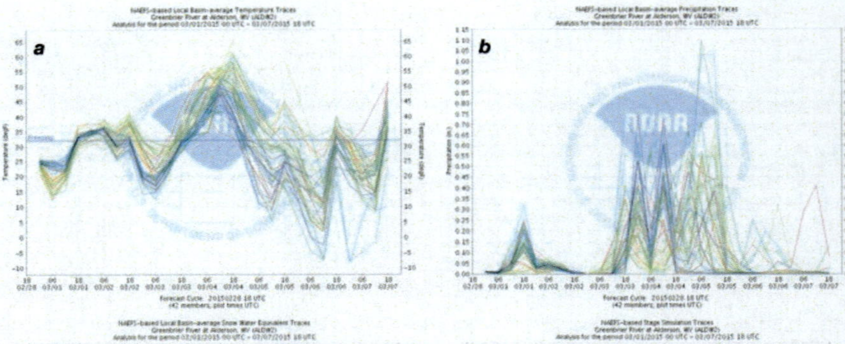

Fig. 14.7 Example ensemble inputs (precipitation (**a**) and temperature (**b**)) and modelled outputs, snow water equivalent (SWE) (**c**), and stage/streamflow (from NOAA/NWS MMEFS) (**d**), using R

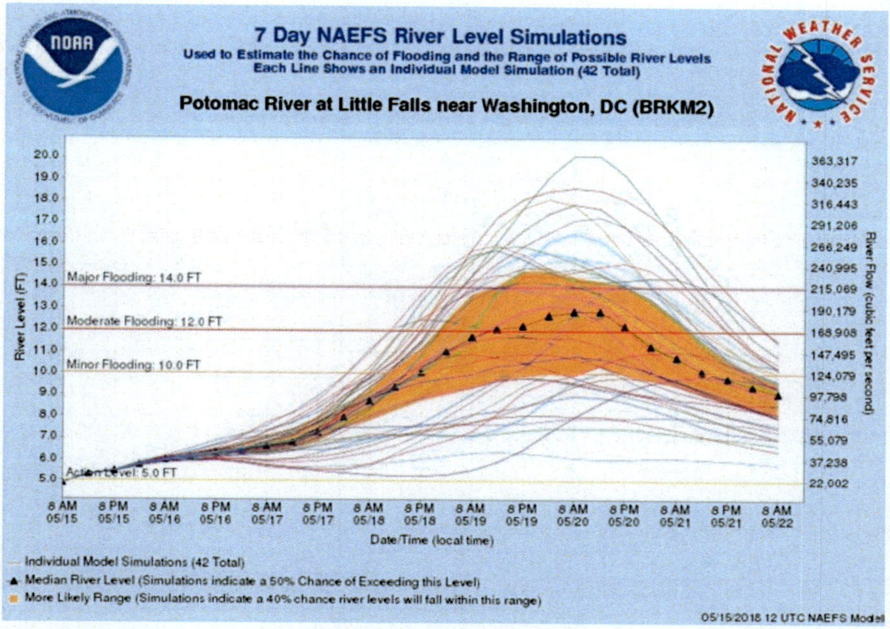

Fig. 14.8 Example ensemble output for 42 hydrologic ensemble members, showing the ensemble median (BLACK line and triangle symbols), with the GOLD shaded 40% – 80% exceedance probability range. (from NOAA/NWS MMEFS), using R

Fig. 14.9, can be generated. With this, and for any hydrologic variable, probabilities of exceedance (or non-exceedance) can be calculated. Decision makers, using a risk based approach, are then able to decide what action should be taken based on risk and potential cost and losses.

Graphical and text based products can be produced automatically, as desired, which are most beneficial to the people of Pakistan and sophisticated end-users. As a point of interest, the NOAA/NWS MMEFS workflow is very similar to that shown in Fig. 14.4.

14.3.8 System Monitoring

Real-time forecast system monitoring is necessary to identify workflow status, warnings, and errors. Two independent approaches to real-time system monitoring are suggested:

- ecFlow, described in Sect. 14.4.1 and depicted in Fig. 14.10
- FEWS, described in Sect. 14.4.2 and depicted in Fig. 14.11

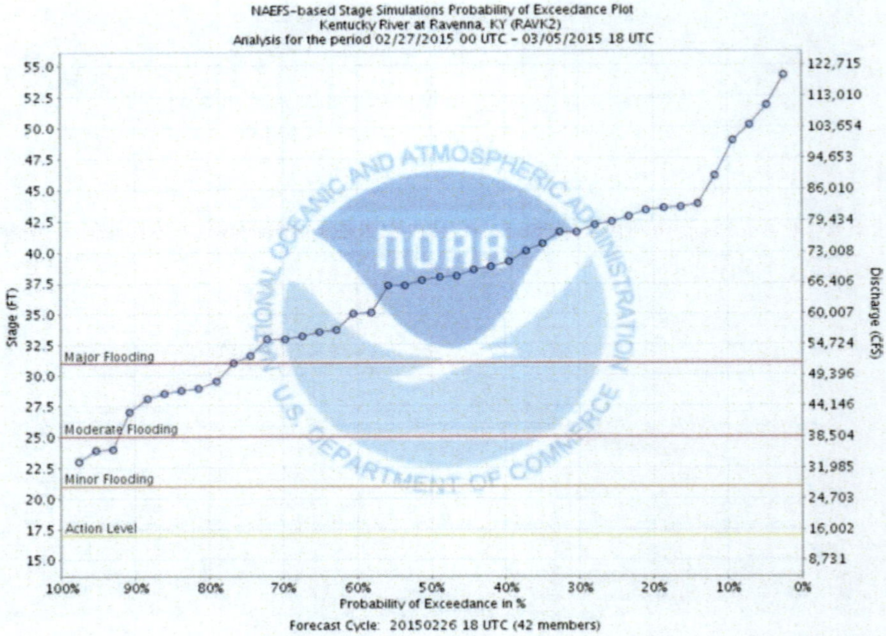

Fig. 14.9 Example probability of exceedance graphic to illustrate derivation of risk values (from NOAA/NWS MMEFS), using R

These two approaches significantly define the overall software system architecture. The advantages and disadvantages of the two divergent approaches are discussed in Sects. 14.4.1 and 14.4.2.

14.3.9 Real-Time Forecast Assessment

Real-time assessment is needed to evaluate whether or not the forecast system is producing reasonable results, which are generally identifiable from both the model inputs, largely in the form of precipitation, and predictions of modelled flow. An example of a visualisation tool used for assessing the progression of a deterministic hydrologic flow/stage forecast is shown in Fig. 14.12. With this example, forecasted precipitation – known as Quantitative Precipitation Forecast – is initially underestimated, as shown by comparisons against observed river stage values (assuming the hydrologic modelling system is well-calibrated).

Figure 14.13 shows an example of a means to visualise an ensemble forecast, though this is not intended to be a substitute for ensemble forecast verification. This type of graphic allows a forecaster to more easily visualise the ensemble forecast in near real-time, since observed data is needed for an extended period, thus it is useful

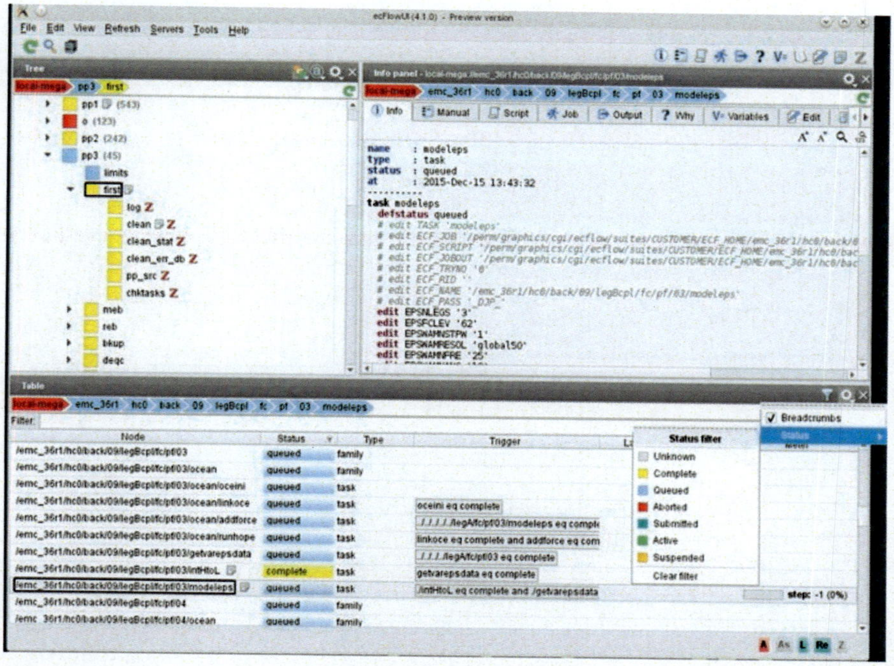

Fig. 14.10 Example ECMWF ecFlow GUI for monitoring system status

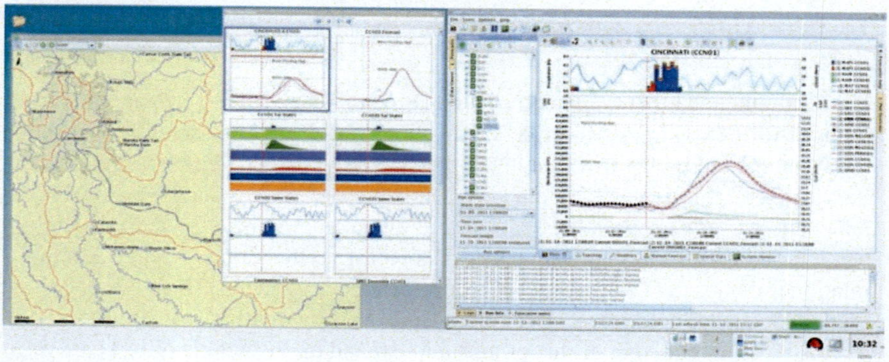

Fig. 14.11 Example of FEWS GUI for model execution, visualisation, and monitoring system status

more in hindsight. Ideally, with this example, it would be preferred that the ensemble spread contained all the observed data, especially within the CYAN shaded regions of the boxplots. Actual ensemble forecast verification requires the accumulation of years of ensemble forecasts to ensure sufficient sample sizes.

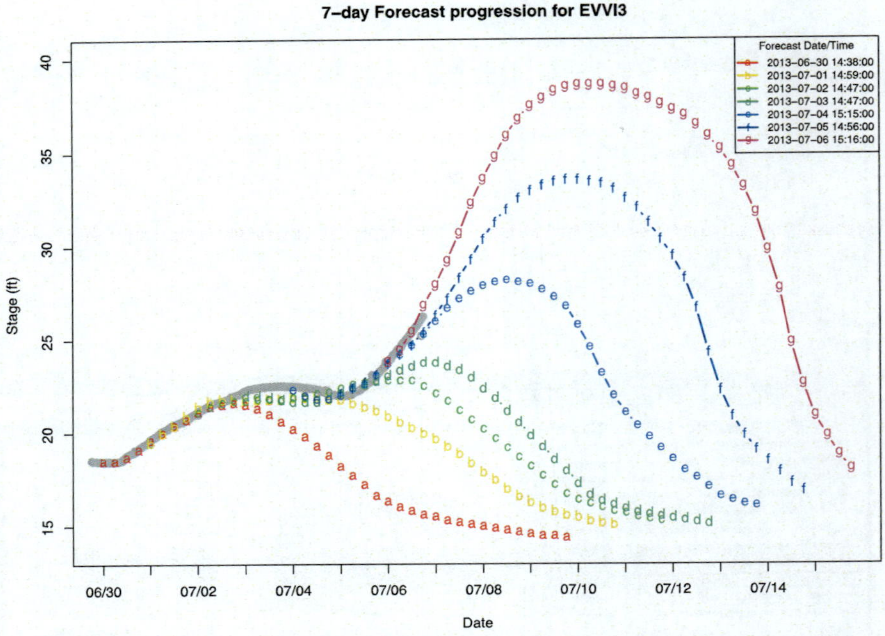

Fig. 14.12 Example real-time deterministic forecast assessment – forecast progression. Grey circles are observations and coloured lines show the progression of subsequent forecasts

14.3.10 Forecast Verification System

It is recommended that the R-contributed (R Core Team 2017) verification package be used for deterministic forecast verification. For ensemble forecast verification, the recommended software is the Java-based US NOAA/NWS Ensemble Verification Service (see http://www.nws.noaa.gov/oh/hrl/general/HEFS_doc/EVS_MANUAL.pdf). These software packages can be run either interactively or automatically in batch mode with scripting. Data used for forecast verification should be stored in a relational database (PostgreSQL is recommended). Storage of both observed and predicted values should be stored in real-time, as shown in Fig. 14.4.

It is beyond the scope of this proposal to expand on the details of forecast verification except to note that the science of hydrologic forecast verification is well-established.

14.4 Implementation Considerations

A significant advantage of both approaches are built-in capabilities for system state monitoring and reporting. The two approaches are discussed in Sects. 14.4.1 and 14.4.2.

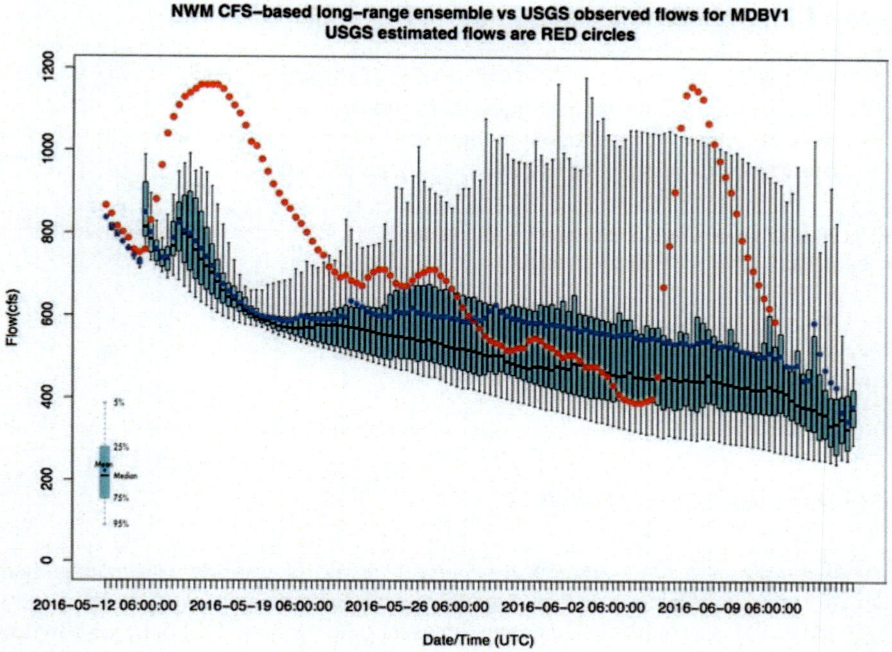

Fig. 14.13 Example real-time forecast assessment for a single, long-range (~30-day) ensemble forecast. The ensemble mean is shown as BLUE circles and the ensemble median is depicted as BLACK horizontal lines. The boxplots represent the 95%, 75%, 25%, and 5% probabilities of exceedance

14.4.1 ecFlow Approach

The European Centre for Medium-range Weather Forecasting (ECMWF) based approach uses an open source software tool called ecFlow. ecFlow (https://confluence.ecmwf.int//display/ECFLOW) is a work flow package that enables users to run a large number of programs (with dependencies on each other and on time) in a controlled environment. It provides tolerance for hardware and software failures, combined with good restart capabilities. It is used at ECMWF to run all operational suites across a range of platforms. ecFlow submits tasks (jobs) and receives acknowledgements from tasks when they change status and when they send events. It does this using using child commands embedded in the scripts. ecFlow stores the relationship between tasks and is able to submit tasks dependent on triggers. The ecFlow graphical user interface (GUI) is shown in Fig. 14.10.

14.4.1.1 ecFlow Advantages

- Relatively easy implementation as a series of workflow steps.
- Stand-alone GUI for monitoring system status.
- No requirement for additional model coding.
- Built-in notification, such as SMS messaging or email.

14.4.1.2 ecFlow Disadvantages

- No built-in analytical capability, such as QC.
- No built-in modelling/mapping visualisation.
- Small user community.

14.4.2 FEWS Approach

Delft-FEWS (https://oss.deltares.nl/web/delft-fews/) is an open data management platform initially developed as a hydrological forecasting and warning system. Essentially it is a sophisticated collection of modules designed for building a hydrological forecasting system that can be customised to meet user-specific forecast system requirements. Because of its unique characteristics concerning data importing and processing and model connections, Delft-FEWS has also been applied in a wide range of different operational settings. Examples are water quality forecasting, reservoir management, operational sewer management optimisation, and even peat fire prediction. An example of the Delft-FEWS GUI is shown in Fig. 14.11.

Delft-FEWS offers many options for the user to interact with the system. For a modern operational (forecasting) system this interaction is crucial. In water management, and other sectors, different types of models are being used to simulate real-world processes. Delft-FEWS is capable to connect too many of these models, and new connections can be made easily. It is being developed at Deltares, an independent institute for applied research in the field of water, subsurface and infrastructure. A new version of Delft-FEWS is released every 2 years containing new features and bug fixes.

14.4.2.1 FEWS Advantages

- Comprehensive data management.
- Built-in analysis and QC tools.
- Large international user community (e.g. NOAAA/NWS, UK Environment Agency, Australian Bureau of Meteorology).

14.4.2.2 FEWS Disadvantages

- Difficulty of system implementation with expert training required.
- FEWS model adapter needs to be developed for models to function within FEWS.

14.5 Conclusion and Recommendations

Various contributors in Adams and Pagano (2016) demonstrate the similarities between hydrologic forecast systems across many NHSs. The WMO (2011) identifies key components of a robust flood forecasting and warning system that are present in most national and regional forecast systems. While some details may differ in local implementations to meet specific needs, such as coping with Glacial Lake Outburst Floods or irrigation systems, general characteristics and requirements remain essentially invariant across large-scale forecast system implementations.

Open source software solutions for relational databases, geographic information systems, statistical analysis software, hydrologic and hydraulic models, and visualisation software are robust. These models and software packages are easily brought together with tools such as the ECMWF ecFlow workflow package. Moreover, with local development of the forecast system, capacity building is enhanced within Pakistan and the broader open systems community. A representative system can be found with the University Corporation for Atmospheric Research System for Hydromet Analysis, Research and Prediction (see https://ral.ucar.edu/projects/system-for-hydromet-analysis-research-and-prediction-sharp), which was designed to facilitate the demonstration and evaluation of variations in the real-time forecasting workflow for short, medium and seasonal range streamflow predictions.

Recommendations

1. Comprehensive staff training and capacity building of staff within PMD and FFD is critical, with every effort taken to retain capable and talented staff.
2. Model implementation within the forecast system must be comprehensive in its scope, including all facets of water resources forecasting and management, spanning flash flood to monthly and seasonal time horizons;
3. The implemented forecast system should focus on the use of ensemble forecasting methodologies (see Sect. 14.1.3).
4. An over-the-loop forecast approach should be used (see Sect. 14.2.2).
5. The modelling system should be structured to facilitate rapid computational turn-around; for example, 1-D versus 2-D hydrodynamic models are recommended.
6. Development of a comprehensive CONOPS is essential (see Sect. 14.3.2).
7. A robust system to monitor forecast system integrity needs to be used to handle data flow interruptions or other system failures.
8. Open source software solutions are highly recommended.

9. Electrical power backup must be installed, using uninterruptable power systems and local power generators.
10. Linux-based hardware platform is recommended due to the ease with which open source software can be installed and integrated.

References

Adams TE (1991) Graphical user interface concepts for hydrologic forecasting in the modernized weather service. In: Seventh international conference on interactive information and processing systems for meteorology, oceanography, and hydrology. American Meteorological Society Preprints, New Orleans

Adams TE (2016) Flood forecasting in the United States NOAA/National Weather Service. In: Adams TE, Pagano TC (eds) Flood forecasting: a global perspective. Elsevier, New York, pp 249–310

Adams TE (2019) Water Resources Forecasting within the Indus River Basin: a call for comprehensive modeling. In: Khan S, Adams TE (eds) Indus River Basin: water security and sustainability. Elsevier, New York

Adams TE III, Dymond R (2018) Evaluation and benchmarking of operational short-range ensemble mean and median streamflow forecasts for the Ohio River Basin. J Hydrometeorol 19(10):1689–1706

Adams TE III, Dymond RL (2019) Possible hydrologic forecasting improvements resulting from advancements in precipitation estimation and forecasting for a real-time flood forecast system in the Ohio River Valley, USA. J Hydrol 579:124138

Adams TE, Pagano TC (eds) (2016) Flood forecasting: a global perspective. Elsevier, New York

Adams T, Smith G (1993) National Weather Service interactive river forecasting using state, parameter, and data modifications. In: Proceedings of the international symposium on engineering hydrology. American Society of Civil Engineers, San Francisco

Andermann C, Bonnet S, Gloaguen R (2011) Evaluation of precipitation data sets along the Himalayan front. Geochem Geophys Geosyst 12(7):Q07023

Andrew Kruczkiewicz, Thomas Adams, and Robert F. Adler, Town Hall Meeting: Challenges, Opportunities and Advances in Global Flood Forecasting, AMS 2017 Annual Meeting, Monday 23 January 2017: 12:15 PM-1:15 PM, URL:. https://ams.confex.com/ams/97Annual/webprogram/Session41766.html

Bhatt BC, Nakamura K (2005) Characteristics of monsoon rainfall around the Himalayas revealed by TRMM precipitation radar. Mon Weather Rev 133(1):149–165

Burnash RJ (1995) The NWS River forecast system – catchment model. Water Resources Publications

Burnash RJ, Ferral RL, McGuire RA (1973) A generalized streamflow simulation system: Conceptual modeling for digital computers. Technical report, U.S. Department of Commerce National Weather Service and State of California Department of Water Resources

Clark RA, Gourley JJ, Flamig ZL et al (2014) CONUS-wide evaluation of national weather service flash flood guidance products. Weather Forecast 29(2):377–392

Demargne J, Wu L, Regonda SK, Brown JD et al (2014) The science of NOAA's operational hydrologic ensemble forecast service. Bull Am Meteorol Soc 95(1):79–98

Duan Q, Gupta HV, Sorooshian S et al (eds) (2003) Advances in calibration of watershed models, water science and application. American Geophysical Union, Washington DC

Emerton RE, Stephens EM, Pappenberger F et al (2016) Continental and global scale flood forecasting systems. WIREs Water 3(3):391–418

Giupponi C, Sgobbi A (2013) Decision support systems for water resources management in developing countries: learning from experiences in Africa. Water 5(2):798–818

Herold C, Rudari R (2013) Background paper PREPARED for the global assessment report on disaster risk reduction, improvement of the global flood model for the GAR 2013 and 2015. Technical report, U.N. Office for Disaster Risk Reduction and Global Assessment Report on Global Risk Reduction, Geneva, Switzerland

Kauffeldt A, Wetterhall F, Pappenberger F et al (2016) Technical review of large-scale hydrological models for implementation in operational flood forecasting schemes on continental level. Environ Model Softw 75:68–76

Khan SI, Adams TE (eds) (2019) Indus River Basin: water security and sustainability. Elsevier, New York

Kohler MA (1944) The use of crest stage relations in forecasting the rise and fall of the flood hydrograph. Technical Report (mimeo), U.S. Weather Bureau, September

Köppen W (1936) Das Geographisca System der Klimate. In: Köppen W, Geiger G (eds) Handbuch der Klimatologie, Vol. 1, chapter C, 1–44. W 35 Koester Ufer. Verlag von Gebrüder Borntraeger, Berlin, p 17

Kottek M, Grieser J, Beck C et al (2006) World map of the Köppen-Geiger climate classification updated. Meteorol Z 15(3):259–263

Lehner B, Liermann CR, Revenga C et al (2011) High-resolution mapping of the world's reservoirs and dams for sustainable river-flow management. Front Ecol Environ 9(9):494–502

R Core Team (2017) R: a language and environment for statistical computing. R Foundation for Statistical Computing, Vienna

Sampson CC, Smith AM, Bates PD et al (2015) A high-resolution global flood hazard model. Water Resour Res 51:7358–7381

Sampson CC, Smith AM, Bates PD et al (2016) Perspectives on open access high resolution digital elevation models to produce global flood hazard layers. Front Earth Sci 3:85

Sene K (2010) Hydrological forecasting. In: Sene K (ed) Hydrometeorology. Springer, Dordrecht

UNFCCC (2013) Decision2/CP.19 Warsaw International Mechanism for Loss and Damage Associated with Climate Change Impacts. Technical report, U.N. Office for Disaster Risk Reduction and Global Assessment Report on Global Risk Reduction, Geneva, Switzerland

United Nations (2015) Sendai framework for disaster risk reduction 2015?2030 A/- CONF.224/L.2. Technical report, U.N. Office for Disaster Risk Reduction and Global Assessment Report on Global Risk Reduction, Geneva, Switzerland

United Nations and World Bank Group (2016) Joint statement of the high level panel on water. Call to action. In: High level panel on water. Unpublished ms., New York

USGS (2006) Benefits of the USGS stream gauging program? Users and uses of US streamflow data. Technical report, United States Geological Survey (USGS), Reston, VA

Ward PJ, Brenden J, Frederiek SW et al (2013) Assessing flood risk at the global scale: model setup, results, and sensitivity. Environ Res Lett 8(4):044019

Ward PJ, Jongman B, Salamon P (2015) Usefulness and limitations of global flood risk models. Nat Clim Chang 5(8):712–715

Weerts A, Molenaar R, Schellekens J, van Dijk A (2017) Continental scale data assimilation of discharge across CONUS: first results, 12th international conference on hydroinformatics. Unpublished ms., HIC 2016

World Meteorological Organisation (2006) Guidelines on the role, operation and management of National Hydrological Services. (WMO-No. 1003). Technical report, World Meteorological Organisation, Madrid conference statement and action plan, Madrid, Spain, 19?22 March 2007. Secure and Sustainable Living: Social and Economic Benefits of Weather, Climate and Water Services

World Meteorological Organisation (2008a) Guide to hydrological practices, volume I, hydrology – from measurement to hydrological information, WMO-No. 168. Technical report, World Meteorological Organisation, Geneva, Switzerland

World Meteorological Organisation (2008b) Capacity assessment of National Meteorological and Hydrological Services in Support of disaster risk reduction: analysis of the 2006 WMO disaster risk reduction country-level survey. Technical report, World Meteorological Organisation, Geneva, Switzerland

World Meteorological Organisation (2009) Guide to hydrological practices, volume II, management of water resources and application of hydrological practices, WMO-No. 168. Technical report, World Meteorological Organisation, Geneva, Switzerland

World Meteorological Organisation (2011) Manual on flood forecasting and warning. WMO-No. 1072. Technical report, World Meteorological Organisation, Hydrology and Water Resources Programme, Geneva, Switzerland

Chapter 15
Ways Forward to Improve Water Security in Pakistan

Stephen Davies, Muhammad Arif Watto, and Erum Sattar

Abstract For a country to be water secure, it must balance water access for diverse human needs, including for health, and must manage the resource to preserve a sustainable human-nature ecosystem. Recent water policies recognise this diversity and related challenges. Despite these advances, Pakistan's history includes many failures to move related programs forward. The governance structures of the main policies, such as the National Water Policy and the Punjab Water Policy, implicitly assume that current departments or ministries are adequate to the job. With the preponderance of irrigation and agriculture representatives on councils and committees, this is unlikely to be easy to achieve. This chapter uses a socio-hydrological framework to argue that the contribution of water to sustainable development needs to be based on accurate measurement, valuation, and decision-making grounded in a system that takes account of the complex interplay of human actions and water systems – and that these will only be effective if they are embedded in strong consultative governance institutions. This framework was used to assess the way forward using two examples, groundwater management and the challenge of allocating water to the Indus Delta.

Keywords Integrated water resource management · Water policy · Policy implementation issues · Water legislation · Water reforms · Pakistan

S. Davies
International Food Policy Research Institute, Islamabad, Pakistan
e-mail: S.Davies@cgiar.org

M. A. Watto (✉)
University of Agriculture Faisalabad, Sub-campus Depalpur, Okara, Pakistan
e-mail: arif.watto@uaf.edu.pk

E. Sattar
Sustainable Water Management Program, Tufts University, Medford, MA, USA
e-mail: Erum.Sattar@tufts.edu

M. A. Watto et al. (eds.), *Water Resources of Pakistan*, World Water Resources 9,
https://doi.org/10.1007/978-3-030-65679-9_15

15.1 Water Security and Pakistan's Challenge

In Chap. 4, water security was defined to show that a water secure country must balance water access for diverse human needs, including for health, and must be managed to preserve a sustainable human-nature ecosystem. Pakistan is a country with diverse livelihoods, ecosystems and socio-economic conditions, and like many other countries, exhibits a wide range of spatial and temporal variability in its water resources and uses.

Most of the approaches to Pakistan's water security focus on agriculture, hydropower, and related structural development requirements (such as canals, canal lining and mega dams), and, as noted in Chap. 4, the system is "viewed from the lens of a bureaucracy that is disconnected from society." A broader view would focus more on adequate institutional capacity, establishing environmental water needs as a priority, and using water efficiency assessments and objectives to guide allocation and to determine the relative efficiency of different water users (Speed et al. 2013). This chapter looks at recent policy and implementation issues from these broader perspectives.

In numerous ways, Pakistan is not water secure (Young et al. 2019). The available water resources are under great pressure from a growing population, various sectoral water demands, climate change, and the continuous degradation of ecosystem services, some of which are made worse because the exact impacts on water resources cannot be seen. Inadequate planning and management further raise the challenge of making Pakistan's water secure.

Some sources of insecurity are related to physical distribution and its effects on water quality in Pakistan. Irrigation infrastructure is outdated and poorly maintained, which affects its operational performance and causes huge conveyance losses. Similarly, urban water supply infrastructure is either outdated or non-existent. Only 27% of the country's residents have access to piped water supply (Mansuri et al. 2018). Highly skewed water supplies impacted by climate change lead to "too much water, too little water." "Too little water" due to increased severity and frequency of drought hurts irrigated agriculture, which uses 96% of the diverted water resources, while "too much water" causes huge economic damage, such as the devastating floods of 2010, with an estimated US$10 billion in damage (Khan and Adams 2019).

It is estimated that Pakistan treats only 1.2% of its urban wastewater, or about 4.5 km^3 in 2010 (Thebo et al. 2017; Qureshi et al. 2010), which is more than half of the active storage capacity of the proposed Kalabagh dam. Thus 5.28 BCM of wastewater finds its way into freshwater bodies and ultimately into groundwater aquifers. While this is a small amount compared to the estimated 1250 BCM in freshwater storage in aquifers underlying the Indus River system, it is likely to create significant problems in selected areas and is very difficult to reverse.

The remainder of the paper looks at the history and adequacy of the policies developed in Pakistan to address the lack of water security and charts some ways forward to take Pakistan closer to being a water secure country.

15.2 Re-thinking the National Water Policy 2018

The year 2018 could turn out to be a pivotal year in Pakistan's water governance history, as it saw the unveiling of the country's first comprehensive National Water Policy (NWP) (Ministry of Water Resources 2018a), which may open more actionable debates over future pathways to sustainable water management in Pakistan. The NWP addresses numerous fundamental water management challenges faced by Pakistan, including sustainable groundwater use, water data scarcity, service inefficiencies in water delivery systems, stakeholder and end-user engagement, and climate change adaptation. However, effective pathways to a water-secure Pakistan need to address some critical gaps within the NWP and the water management framework of the country.

The most prominent gap in Pakistan's history of water governance and policy is the failure to move related programs forward. Regarding policy formulation, it took 70 years after independence to establish the NWP. Moreover, there is a grave danger that implementation plans and other policy-related actions fall to the wayside. An important historical example, which is reviewed later in this chapter, is the Interprovincial Water Apportionment Accord (IWAA) of 1991, which, to date, has not implemented the mandate to provide adequate environmental flows to the Indus Delta, i.e., three decades after the IWAA (Anwar and Bhatti 2018). Steps to ensure financial sustainability of the system, such as water pricing, have not generally been taken. Thus, it is essential that pathways to better water governance and management in Pakistan be built on a far stronger commitment to follow through on existing policies and post-policy implementation plans, with well-defined timelines and robust accountability mechanisms.

Post-policy implementation plans are a critical component of any successful policy framework. While such implementation timelines are present in the NWP, most focus on structural improvements. The key targets for the 2018–2030 NWP timeframe include canal lining, mega dam development, irrigation infrastructure development and improved river water monitoring. Unsurprisingly, these targets lean heavily towards infrastructure development, whereas issues pertaining to sustainable management practices, service delivery, climate adaptation and stakeholder engagement are not associated with timelines or accountable implementation plans. Pakistan's water policy and governance has historically been dictated by the bureaucracy, which tends to emphasise such structural and civil development projects (Siddiqi et al. 2018), and remains too unconnected from the varied water challenges faced by different communities and end-users (Mustafa 2002).

Thus, given the complexity of Pakistan's waters resources system, and that its uses and values are so diverse, there is an urgent need to increase focus on the broader, deep-rooted and persistent water sector challenges and issues. Some of the larger and more critical areas are legislation for, and implementation of, sustainable management of groundwater (Qureshi et al. 2008; Siddiqi and Wescoat 2013; Watto and Mugera 2016), restructuring water sector institutions to promote technology, performance and knowledge-centric management practices (Young et al. 2019), and emphasising service delivery and accountability in canal operations (Siddiqi et al. 2018). There is another key dimension of governance reform that is often missing and goes beyond the key metrics just identified. Other basins have shown huge forward movement in overcoming longstanding problems such as the potentially negative consequences of long-term drought implying a 'new normal' of a permanently reduced supply. A comparative of the Indus and Colorado River systems showed that the Indus has much to learn from the institutional reform processes undertaken with a commitment to transparency and trust-building in the Colorado's multi-state federal and institutional complex colloquially known as the Law of the River (Sattar et al. 2018). With these complex technical and institutional dimensions in mind, the remaining sections of this chapter provide selected perspectives on the requirements for a successful way forward.

15.3 The Institutional Structures and Issues in Recent Water Legislation

A logical place to start examining governance and ways forward is to review the implied governance design in the two most prominent policy initiatives, the NWP and Punjab Water Policy. In terms of topics, both are wide ranging and cover most major uses of water, including agricultural, industry, and the environment. The former also focuses on waterlogging and salinity, as well as the negatively impacted Indus Delta. The areas of domestic use, groundwater, water quality, health and water safety and issues of floods, drought and climate change are addressed to varying degrees. In each area, there are lists of objectives as well as strategies for performance. Thus, there is a general recognition that it is a complex system with multiple objectives that needs to be managed.

The national and Punjab policies are similar in their governance recommendations. For the Punjab Water Policy (PWP), a Water Council is proposed to monitor the policy's implementation progress, while another cell, the Water Policy Implementation Committee, is created to support the Council (PID 2018). The Council includes the Chief Minister and Chief Secretary of the province, senior officials in irrigation and agriculture, the secretary of local government, and other external and government officials. The Implementation Committee includes a broader group of secretaries and ministers to represent areas affected by the policy, such as environment, industry and health. Neither group meets more than several times a year, nor do they have substantial technical support teams.

The Strategic Planning and Reform Cell in the Punjab Irrigation Department is given secretariat duties for the Council, but there are only vague references to the extra capacity development envisioned, and there is no identification of the specific activities that the Cell will perform. Without this specificity, it is not likely that adequate resources will be devoted to capacity enhancement, nor is it probable that the Implementation Cell will perform the range of duties needed to support the integrated water resource management (IWRM) approach that is promoted as the broad strategy in the PWP. There is an underlying question that remains unaddressed in the overall approach of IWRM that the PWP has adopted, which raises potentially compelling challenges for its suitability to the task at hand. This could be better addressed by the new approach of socio-hydrology. Such an approach takes account of the key interconnections of hydrological conditions and human actions and may thus be a better option for Pakistan to adopt for the purposes of its water-food-energy planning. The approach brings together hydrology, nature, society and humanity. The key to advance understanding and to move policy forward in virtuous ways is to tackle a key challenge that the study of socio-hydrology offers: recognising that while human activity impacts the hydrological cycle and hydrological conditions, hydrological conditions do not necessarily always trigger changes in human systems (Garcia et al. 2016). As can be seen, this has huge implications for policy design as it points to the potential great unknown of human response to changed hydrological conditions. It is quite likely that human actions will continue in ways that will fail to adapt to changing conditions because of the difficulty in discerning the rate and directionality of change. For our present purposes, and for the purposes of the design of Pakistan's water policy at federal and provincial levels, it may be beneficial to look to the insights from socio-hydrology while designing future water policy.

The NWP has a similar structure to the PWP, with the National Water Council including the Prime Minister and key federal ministers from water resources, power, finance, and planning. Given this, key insights from the socio-hydrological approach may be beneficial here too. In the current design, there is explicit identification of a steering committee, where the members include the secretaries of the above ministries and other key representatives. The policy implementation cell is to be housed in the Ministry of Water Resources, and its capacity building is a policy target. At a fundamental level it must be understood that because interactions between humans and the environment are occurring on an unprecedented scale, such that nothing close to it has ever been seen before given our modern-era capacities to affect hydrological systems, the decisions that impact water also impact people at an immense scale (Blair and Buytaert 2016). Acknowledging these interactions and effects means that any water policy design framework must move beyond a narrow focus on limited engagement within existing institutional design towards putting in place structures that will enable a wider dialogue with the humans impacted by decisions about water. This means that the NWC and its attendant consultative structures must be widened to include a broader range of stakeholders impacted by its decisions.

The setup of these governance structures implicitly assumes that the current departments or ministries are adequate to the job, even though the PWP recognises that the current institutions need to move to an IWRM approach. With the preponderance of irrigation and agriculture representatives on the councils and committees, and the fact that the PWP is housed in the Punjab Irrigation Department (PID), substantial changes in resource management are unlikely to proceed very quickly. Indeed, the PWP even notes that the Punjab Irrigation Development Authority

(PIDA) system of devolved participatory irrigation management did not "get ownership" from PID and hence was not successful. The fact that the Asian Development Bank put together a significant program to shift the orientation of the irrigation department to an IWRM focus is just another indicator of the distance these institutions are from the broad multi-objective, multi-stakeholder approach demanded in IWRM, and which is set forth in the PWP (Asian Development Bank 2016). As we have indicated, it may be necessary to go beyond the IWRM framework towards a broader socio-hydrological framework that will make it possible to link the attainment of sustainable development goals with water-sector planning. Moving beyond a narrow policy and institutional framework becomes more necessary given the need for sustainable development as defined in the 1987 Brundtland Report, *Our Common Future*, which emphasises that the needs of the present generation be met while also recognising the needs of future generations (World Commission on Environment and Development 1987). In essence this means that the narrower sectoral focus will be inadequate to broader sustainable development goals.

It is apparent from this review that these policies, based on what is written in them, just provide initial structures for sustainable water management. In particular, any policy and institutional design is unlikely to succeed in tackling present and future challenges without a deep appreciation of historical context and the attendant contemporary challenges that Pakistan's Indus Basin faces ranging from a falling water table, low productivities, environmental degradation, polluted water supplies, and growing conflicts from inter-provincial to local levels (Briscoe 2014). One of the main insights of the Harvard Water Security Initiative and the comparative five-basin Water Federalism Project that also looked at the Indus is particularly relevant to policy. This is the finding that given all policy design and legal and institutional solutions are provisional, systems need to be created that are dynamically adaptive to changing conditions. This is important as society's values around its water resources change over time, particularly as they relate to levels of economic development. One example of such an adaptive agricultural system is one that is responsive to issues of water pollution effects on agricultural yields, food production and soil erosion, such as that witnessed over the last 40 years in Britain which has devastated the health of the nation's soils. This growing awareness means that the issues of reservoir sedimentation and ecological damage to watercourses from nutrient enrichment and pesticides have risen in importance as policy concerns. There is a clear shift to concerns about a healthy and sustainable environment which includes soils (Boardman 2013). While Pakistan's existing institutions at the provincial and national levels are not currently tasked with monitoring such concerns, it is recommended that they be assigned such additional capacities to be able to adopt more holistic irrigation and agricultural practices. We also reiterate that with the noted delays that have transpired in the Water Accord and in water pricing, it is useful to go beyond those broad policies and examine the interdependence of legislated institutional structures with important policy support activities of measurement, valuation and evidence-based decision-making often highlighted in the policies themselves. On the Water Accord itself, there are additional significant areas that need attention and reform that go to the heart of inter-provincial mistrust. This includes the three-tier method of water allocation, along with such technical issues of the prediction of water availability, flow monitoring, and increasing conveyance losses, as well as institutional issues of an absence of an overarching water regulatory framework and weak enforcement (Condon et al. 2014).

15.4 Leverage Points for Better Water Policy and Achieving Water Security

A recent Science article examining the importance of valuing water to meet sustainable water security and related development provides important perspectives for Pakistan (Garrick et al. 2017). Their analysis forms a structure to help chart our view of the way forward. First, the article suggests that measurement underpins valuation, noting that measurement extends beyond just volumes, flows, and quality, and should reflect series that help capture the potential value of water in its broadest context. Logically, this leads into valuation, which is "difficult but necessary" and without it, water use can easily become unsustainable. A broad approach to valuation, including ecosystem services, cultural values, and equity perspectives are necessary. As we've suggested, a socio-hydrological approach may be helpful, as it helps bring together human values, social organisations, markets, public policy, climate, hydrology, toxicology and ecology (Gober et al. 2017). Given that humans are an endogenous component of the water cycle, it is imperative that the effect of human actions on the water cycle are fully understood for the purposes of water-sector engagement and public policy planning (Garcia et al. 2016).

In the process of creating values, trade-offs inevitably present themselves, and so a third area needing greater emphasis is the governance structure used to make decisions. The article notes that, in fact, decision science has expanded to incorporate a broader set of values and to emphasise that such decisions cannot be made without all stakeholders included.

Therefore, the critical last input is to have robust and trusted institutions that rely on ongoing partnerships, that reliably measure the value of water in different uses, and that have a structure to make evidence-based decisions. This means that governance institutions must be designed to take advantage of the most advanced interdisciplinary frameworks available to evaluate the complex interplay of society and nature, including frameworks for social-ecological systems analysis, coupled human and nature systems research, and complex systems science (Di Baldassarre et al. 2019). This will mean that there will have to be an acknowledgement that Pakistan needs to undertake significant institutional redesign and restructuring to follow new paradigms, including, perhaps, the creation of new institutions with a fresh ethos embedded in their foundational structures. It is clear that new social science frameworks are needed for the purpose (Sanderson et al. 2017).

The puzzles with governance are major, with vested interests, incomplete information, and historical path dependency all at work. Scaling up approaches, such as managing a common pool resource, will inevitably require ongoing and sustained efforts. However, any way forward towards water security needs to include and prioritise improved institutional structures jointly with more technical and infrastructure efforts. As has been shown, the country can learn from other basins such as the Colorado's institutional and technical innovations to tackle systemic and long-standing challenges and design appropriate solutions across a wide domain (Sattar et al. 2018).

In this chapter, we contend that the first three steps – measurement, valuation, and decision-making – will only be effective if they are embedded in effective governance institutions. There is a tendency for policies and the bureaucracy to look heavily towards technology and to infrastructure development. An instructive recent publication looking at decision support systems (DSS) generally concluded that these complex data management and integration systems designed to support decisions typically had disappointing outcomes (Giupponi and Sgobbi 2013). The authors concluded that rather than overemphasising such efforts, "improving the effectiveness and applicability of integrated water resource management legislative and planning frameworks, training and capacity building, networking and cooperation, harmonisation of transnational data infrastructures and, very importantly, learning from past experiences and adopting enhanced protocols for DSS development" (p. 798) might be a more productive effort. Another recent study applied relevant lessons from the Colorado Basin to the Indus Basin and found that innovations centred around allocational flexibility, shortage-sharing, privileging demand management over hardware solutions, water banking, and water marketing may be very usefully adapted (Sattar et al. 2018). To provide some concrete examples here, the next sections lay out some of these dimensions and interplay between them for the issue of groundwater and its conjunctive use with surface water. The issue is especially important to think through and is ripe for potentially large gains to be had given the country's increasing reliance on groundwater and its significant price-differential from surface water irrigation supplies (Sattar 2019, 2020).

Groundwater Management Challenges One of Pakistan's great challenges in water resources management is to improve the performance of groundwater in the country. The unsustainability of groundwater use, as well as the dependence of agriculture and domestic uses on it in many parts of the country, is well understood now. One heroic attempt to piece together the overall balance and dynamics of groundwater abstraction in Punjab province suggests that pumping costs will rise 270% within the next 20 years (Khan et al. 2016). While any such analyses have great uncertainty, if there is truth to the analysis, the situation is dire and the likely negative impact on agriculture is a real risk, especially recognising that Punjab is often seen as the breadbasket of the economy. Even the country's NWP highlights the immense price differential between canal water and groundwater, indicating that planners are aware of the negative externalities that have been created in this realm. Moreover, entire cities such as Lahore and Quetta depend on groundwater. As the World Bank's authoritative *Running Dry* study highlighted, the dire threats to Quetta's confined aquifer should be of grave concern particularly given that the city may run out of even the deep fossil groundwater on which it relies (Briscoe et al. 2005). Accordingly, it is insufficient to spend the next decade or more deriving accurate measurements of groundwater at the expense of other management dimensions. The slow managerial response to environmental flows, and the inability to raise water prices, are cautionary tales about the speed at which these policies have made impacts historically.

A recent World Bank study makes recommendations for four groundwater policy areas: (1) institutional reforms and regulation; (2) data for integrated basin planning; (3) conjunctive management, waterlogging, salinity and depletion; and (4) groundwater quality. Within these four areas, there are 16 more detailed steps (See Table 15.1) (World Bank 2020). This proposed management strategy is close to the framework we presented above, as the four areas focus on institutions, measurement for data collection, and two areas where key decisions need to be made: conjunctive use and water quality. In this chapter, we stress the cross-category dependencies and the need to work on multiple fronts at once to improve the odds of success. These are separate parts of the whole, with the implication that, for each part to be successful, others need to be making progress as well. Holistic management of the resource is particularly critical given that groundwater accounts for nearly a third of irrigation water in what has been termed a regime of reliance without regulation (Sattar 2019).

Looking first at institutional reforms, Nabeel (2020) notes that there has been a proliferation of related institutional developments recently affecting groundwater. In the NWP, the one real new institution proposed at the national level was a Groundwater Authority, which is designed to monitor groundwater but which might

Table 15.1 The World Bank (2020) recommendations for groundwater management

Institutional reforms & regulation	Conjunctive management, waterlogging, salinity and depletion
Appoint provincial lead agencies and agree on scale of management	Develop water budgets for management areas
Regulatory reform	Colosely monitor water balance to ensure irrigation deliveries do not exceed drainage capacity
Interim management reforms	Demand management and managed aquifer recharge
Stakeholder/community engagement	Stakeholder/community engagement
Data for integrated basin planning & resource modelling	**Mapping groundwater quality**
Lead agency responsile/accountale for data	Identification of local water quality parameters and regular water quality monitoring
Structured capacity building and development of data architecture	User participation to monitor and control land use practices and effluent disposal
Consolidated surface and groundwater data (static and dynamic data)	Treatment of drinking water
Data sharing with managers and users	Data sharing with managers and users

give the federal government a bigger role in that area and could complicate the provincial role. Similarly, the Punjab Water Act promulgated at the very end of 2019 gives its implementation to a new organisation, the Punjab Water Resources Commission, "headed by the chief minister and the irrigation minister of Punjab, with members from all stakeholder departments of the provincial government." At one level, these institutional developments, coming fast and furiously, are a good thing, but it is not clear whether their features are compatible and whether they can lead to a more sustainable and responsive water management system. For example, the Punjab Water Act takes a major step to separate land and water rights. This is in contrast to the long-standing colonial legislation in the Indian Easements Act of 1892, which allowed landowners to have access to all groundwater under their property (Nabeel 2020), such that unlimited pumping within a landowner's property was permissible, and, by implication, collective action would be difficult given users do not need to take into account others' interests in sharing the resources of the aquifer. Here too a turn to a socio-hydrological approach may be beneficial as it would mean taking account social aspects, such as interests, motivations, abilities to influence, attitudes and access to and power of individuals and groups, rather than the present or proposed regime that treats society as an exogenous factor to the changes taking place (Mostert 2018).

While separating water rights from land rights is most likely positive, there are subtleties and equity issues that require an effective water management institution be established (Nabeel 2020). There is also a question about how to merge the current rights under the old law with the new separation of rights. The new water act in Punjab simply stipulates that all groundwater is a controlled resource and there is no reference to the older legislation. It is unclear whether a command-and-control regime as envisaged will be effective or desirable, especially given the command-and-control nature of the extant surface water regime of *warabandi*, which has had implementation shortfalls since its inception (Bandaragoda 1998).

One of the most important areas is the management of water as a conjunctive resource. The World Bank proposes that: (1) water budgets are developed for various management areas covering both surface and groundwater; (2) these budgets are dynamically monitored so that withdrawals do not exceed the drainage capacity in the area; (3) demand management is successfully implemented to maintain sustainability within the area; and (4) a significant community and user participation is included, as managing these dimensions can only be done with stakeholder input (World Bank 2020). For the first two objectives on measurement and data compilation, it is likely that community and user participation will be essential given the potential costs of such a system. Also, if there is any hope to have successful demand management, an in-depth understanding by participants is required, and hence their inclusion in the data and resource management assessments would be beneficial. Of course, the challenges of this should not be underestimated given the present capacities of the bulk of small and medium farmers.

If there is confusion about the ownership of groundwater, then these steps become much harder to accomplish. Designing sufficient data collection systems with farmers as partners is more complex than are the challenges associated with telemetry or automating discharge measurements between provinces, which are themselves controversial and complex endeavours. Then, in addition, creating a strong lead institution is critical to handle data collection, financial support, management capacity enhancement, stakeholder engagement, and effective demand management. In combination, these activities are quite tall orders, especially given that federal-provincial capacities and relations will play a large role in any such an endeavour, and that the installation and operation of a reliable telemetry system for surface water flows has been so protracted and largely unsuccessful. Given the estimated 1.1 million tube-wells spread across the country, the regulatory task becomes ever more complex (Sattar 2019).

These steps will only be successful if the lead institution and other attendant institutions that will need to be created are adequately financed, technically strong and sufficiently independent, while, at the same time, able to work alongside partners to make decisions that at times will be controversial. The lead agency needs to be established, be up and running, and be able to decide where to invest and where to initiate management actions. Again, it has to fit into a collaborative decision-making and implementing structure. Managing demand requires making decisions, limiting flows, and perhaps creating sanctions. With the past inability to raise *abiana* at the provincial level or handling environmental flows across provincial boundaries, taking these steps is truly a grand set of challenges.

This is not to say that moving in this direction is hopeless. An initial step of raising awareness and recognising the critical situation of groundwater in the country has succeeded. Also, the broad policy setting has improved, and elements of new institutions have been proposed. There are also other logical starting points to move this effort forward. While the lead agency for Punjab at least will need to be adequately funded and staffed at the beginning, it is possible to start with modest size management areas as pilots, and then work with selected groups to design and test approaches before scaling out. Of course, we should be very mindful of the long gap between pilot programs and larger scaling out that have occurred in other areas of irrigation reforms, such as irrigation management turnover, which continues to have a chequered history going back decades. There is always a concern for sufficient and ongoing funding and adequate capacity of regulatory agencies. There may be a case for special courts and regulatory bodies like those existing in the past in Pakistan, and in other countries. For example, in the Colorado Basin there is a specially dedicated water court which bases its decisions on defensible measurements and ownership rights. It is critical to recognise that a collaborative approach to problem-solving has been spurred in the Colorado over the past two decades that has led it towards a relational understanding of water law and policy, resulting in it becoming the most institutionally encompassed basin in North America (Sattar et al. 2018).

Uses of Water Below Kotri Barrage and the Interprovincial Water Apportionment Accord (IWAA) An important historical example referred to earlier is the IWAA of 1991 (Ministry of Water 2018b), which was signed with an understanding that post-accord studies would be conducted to ascertain needed environmental flows in the Indus Basin downstream of Kotri Barrage, and that implementation would follow (Anwar and Bhatti 2018). However, while such a study was eventually conducted in 2005, no implementation of minimum flows is enforced even today, i.e., three decades after the IWAA. The implications of this non-action in the Indus Delta have been immense (See Chap. 4) (Memon and Thapa 2011). In this section, this issue is seen through the same lens applied above, where the steps forward need to be a combination of institutional reform based on improved measurement, consequent valuation, and ultimately better decisions.

To see the challenges and the above interplay between the needed steps, it is useful to break the perspective into at least three different economic activities that are supported by this flow: agricultural, urban, including domestic and industrial, and environmental flows. In Chap. 4, the authors show that on average about 10 MAF a year of surface water is diverted into canals supporting mostly agricultural activities, while other diversions provide water to Keenjhar Lake and the Hub Dam reservoir, which then flow to Karachi for domestic and industrial purposes. The remaining water mostly flows below Kotri Barrage into the delta to support ecosystem services, protection against seawater intrusion, and various economic activities. In an era of rising sea levels, this threat becomes ever-more important to guard against, as loss of land to sea becomes a distinct possibility. Stunningly, these latter flows are estimated to be zero 64% of the time. Together these demonstrate the complexity of institutional gaps, measurement issues and an inability to value many of the flows. It is suggested that key issues here are inter-provincial mistrust and contestation, an inability to move beyond a narrow provincial lens, and appreciating environmental flows as critical for national efforts at sustainability (Habib 2007).

Simple calculations from Chap. 4 lay out the proportions going to various uses. The canal diversions from Kotri Barrage for agricultural purposes are 10 MAF on average, but with a huge variation year to year. The flows into Keenjhar Lake and the Hub Dam reservoir to divert water for domestic purposes in Karachi account for about 0.67 MAF, while the flows below Kotri, during the 35% of the year that they exist, may average perhaps 200 thousand AF, so together the latter two uses add up to about 8% of agricultural uses.

The water allocations at Kotri Barrage could in principle be shifted out of agriculture to provide for environmental, domestic and industrial purposes. While current infrastructure determines much of the possible reallocation, the value of water used in industry or for the environment may be much higher than in agriculture. Significant shifts within the current infrastructure still could be done, as nearly 92% of the flows from the barrage seem are allocated for agricultural purposes. However, the value of water for uses outside of agriculture is very poorly understood, and such knowledge would aid decision-making about these allocations. For example, a hypothetical 5% reduction in industrial productivity growth from consequences of water-constraints leads to a GDP loss of US$84 billion by 2055, equivalent to 90% of the GDP contributions of Pakistan's major crops (Young et al. 2019). Moreover, these valuations can help guide decisions about new investments and policies as they affect various sectors. Of course, it is important to point out that these complex decisions cannot be solely based on economic considerations. Environmental needs must also be given strong protections, especially when it may be expedient to sacrifice long-term environmental protection for temporary economic gains. It is precisely at such junctures that a strong protective state that is cognizant of its overall responsibilities, and is able to balance the short, medium and long-term implications of present actions, is required.

These issues are similar at the basin level, where decisions need to be made about diversions to each canal along with how much should flow below Kotri, which may be considered as one of the responsibilities of IRSA. At the heart of this is the need for basin-level management (Young et al. 2019). To adjust flows at the end of the river requires coordination across all provinces and uses, with a basin-scale planning institution being a key part. As evident from the requirements at Kotri Barrage and below, this institution will be better if it can make decisions based on valuations and a firm knowledge of the consequences of altering flows to different potential uses.

15.5 Conclusions

For a country to be water secure, it must balance water access for diverse human needs, including for health, and must manage the resource to preserve a sustainable human-nature ecosystem that is viable across longer time horizons. Pakistan is a country with diverse livelihoods, ecosystems and socio-economic conditions, and like many other countries, exhibits a wide range of spatial and temporal variability in its water resources and uses. Recent policies, particularly the NWP and PWP, recognise this diversity and related challenges, and yet do not go far enough in adopting best practice as described above. Despite advances, Pakistan's history of water governance includes many failures to move related conceptions and programs forward. It took 70 years after independence to formalise the NWP, and yet the environmental stability of the Delta has not been achieved three decades after the adoption of the IWAA, and water pricing cannot cover even one fifth of the operation and maintenance costs of provincial irrigation infrastructure.

The NWP and PWP acknowledge the importance of these management challenges and address many key areas. The governance structures of NWP and PWP implicitly assume that the current departments or ministries are adequate for the job, even though the PWP recognises that the current institutions need to move to an IWRM approach. Even this would be an advance and, as we have pointed out, policy-making needs to be based on newer paradigms such as socio-hydrological concepts. We suggest that given the preponderance of irrigation and agriculture representatives on the councils and committees, this is unlikely to be easy to achieve. However, we reiterate that nevertheless it must be the direction in which policy formulation moves.

In this chapter, a framework is used to assess the way forward using two examples, groundwater management and the challenge of allocating water to the Indus Delta. Our approach recognises three steps – measurements, valuation, and decision-making – that will only be effective if they are embedded in effective governance institutions. These are important to Pakistan's way forward towards water security, as there is a tendency for policies and the bureaucracy to look heavily towards technology and to infrastructure development. Given the relevant lessons from the Colorado Basin, we reiterate that it is only when planners recognise the need to move forward with an awareness of the relational aspects of water management, and that the interconnections of various sectoral users became central to effective policy formulation, that an institutionally robust and responsive water management regime can be developed.

The chapter revisits the recommendations of the recent World Bank (2020) study involving four groundwater policy areas, and argues that these steps will only be successful if a lead institution is adequately financed, technically strong, and sufficiently independent to make decisions that at times will be controversial. Without this, measurement and valuation exercises will not lead to effective decisions that maximise the value of this scarce resource. We also reiterate that, as in the Colorado, a possible lead institution to achieve its given objectives will also need to be collaborative and consultative in its ethos to ensure that lessons are learnt while progress is made in the long-term.

The chapter applies the same framework to the issue of environmental flows below Kotri Barrage that was required by the IWAA. Most water coming into Kotri Barrage – over 90% – was diverted into agriculture, with environmental uses neglected: 65% of the year there are no flows below Kotri for environmental purposes. A critical gap at the heart of the IWAA has still not been settled. From which party's share will environmental flows be taken such that they will be calculated on an all-Pakistan basis, or will it be taken from the share provided to Sindh province such that it will have to reduce Sindh's agricultural users to meet the needs of the delta? Presumably this is one thing that IRSA could help develop consensus around, but it has not taken up the challenge in any meaningful way to date. Until trust is built to tackle this challenge head-on, it is likely that this problem will continue to thwart the search for a just solution impeding progress towards other goals. As evident from the requirements at Kotri Barrage and below, this institution will be better if it can make decisions using valuations and knowledge of the consequences of altering flows to different uses – and again, building consensus here will be a key. One analysis showed reduced industrial productivity growth from water constraints led to a GDP loss equivalent to 90% of the GDP contributions of Pakistan's major crops. It is likely to be the case for environmental uses as well such that these valuations need to be conducted fairly and should be done keeping in mind sound principles of environmental stewardship and the long-term implications of climate change. Strong institutions, with adequate capacity, and an enlightened ethos that can work to build consensus across the country's federal constitutional structure, will be the only way to attain a water secure country.

Acknowledgement We thank Taimoor Akhtar, College of Engineering and Physical Sciences, University of Guelph, Ontario, Canada for his input in this chapter.

References

Anwar AA, Bhatti MT (2018) Pakistan's water apportionment Accord of 1991: 25 years and beyond. J Water Resour Plan Manag 144(1):1–13. https://doi.org/10.1061/(ASCE)WR.1943-5452.0000831

Asian Development Bank (2016) Islamic Republic of Pakistan: institutional transformation of the Punjab Irrigation Department to a water resources department. Capacity Development Technical Assistance Project Number 49048-001

Bandaragoda DJ (1998) Design and practice of water allocation rules: lessons from Pakistan's Punjab. IIMI Research No. 17, International Irrigation Management Institute (IIMI), Colombo, Sri Lanka. https://doi.org/10.3910/2009.002

Blair P, Buytaert W (2016) Socio-hydrological modelling: a review asking "why, what and how?". Hydrol Earth Syst Sci 20(1):443–478. https://doi.org/10.5194/hess-20-443-2016

Boardman J (2013) Soil erosion in Britain: updating the record. Agriculture 3(3):418–442. https://doi.org/10.3390/agriculture3030418

Briscoe J (2014) The Harvard Water Federalism Project – process and substance. Water Policy 16(S1):1–10. https://doi.org/10.2166/wp.2014.001

Briscoe J, Qamar U, Contijoch M et al (2005) Pakistan's water economy: running dry. World Bank, Washington, DC

Condon M, Kriens D, Lohani A, Sattar E (2014) Challenge and response in the Indus Basin. Water Policy 16(S1):58–86. https://doi.org/10.2166/wp.2014.004

Di Baldassarre G, Sivapalan M, Rusca M et al (2019) Sociohydrology: scientific challenges in addressing the sustainable development goals. Water Resour Res 55(8):6327–6355. https://doi.org/10.1029/2018WR023901

Garcia M, Portney K, Islam S (2016) A question driven socio-hydrological modeling process. Hydrol Earth Syst Sci 20(1):73–92. https://doi.org/10.5194/hess-20-73-2016

Garrick DE, Hall JW, Dobson A et al (2017) Valuing water for sustainable development. Science 358(6366):1003–1005. https://doi.org/10.1126/science.aao4942

Giupponi C, Sgobbi A (2013) Decision support systems for water resources management in developing countries: learning from experiences in Africa. Water 5(2):798–818. https://doi.org/10.3390/w5020798

Gober P, White DD, Quay R et al (2017) Socio-hydrology modelling for an uncertain future, with examples from the USA and Canada. Geol Soc Lond, Spec Publ 408(1):183–199. https://doi.org/10.1144/SP408.2

Habib Z (2007) Depletion of water resources and environmental concerns in the Indus Basin: problems and politics of water sharing and management in Pakistan. Islamabad Policy Research Institute, Islamabad

Khan SI, Adams TE (eds) (2019) Indus River Basin: water security and sustainability. Elsevier, New York

Khan HF, Yang YE, Ringler C et al (2016) Guiding groundwater policy in the Indus Basin of Pakistan using a physically based groundwater model. J Water Resour Plan Manag 143(3):05016014. https://doi.org/10.1061/(ASCE)WR.1943-5452.0000733

Mansuri GF, Sami TTD, Dung P, Pandey, AM (2018) When water becomes a hazard: a diagnostic report on the state of water, sanitation and poverty in Pakistan and its impact on child stunting. WASH Poverty Diagnostic, World Bank, Washington, DC. http://documents.worldbank.org/curated/en/649341541535842288/When-Water-Becomes-a-Hazard-A-Diagnostic-Report-on-The-State-of-Water-Supply-Sanitation-and-Poverty-in-Pakistan-and-Its-Impact-on-Child-Stunting

Memon JA, Thapa GB (2011) The Indus irrigation system, natural resources, and community occupational quality in the delta region of Pakistan. Environ Manag 47(2):173–187. https://doi.org/10.1007/s00267-010-9569-0

Ministry of Water Resources (2018a) National water policy and water charter. Ministry of Water Resources, Government of Pakistan. https://mowr.gov.pk/index.php/national-water-policy-2018/. Accessed 10 Aug 2020

Ministry of Water Resources (2018b). The water accord 1991. Ministry of Water Resources, Government of Pakistan. https://mowr.gov.pk/wp-content/uploads/2018/05/Water-Accord-1991.pdf

Mostert E (2018) An alternative approach for socio-hydrology: case study research. Hydrol Earth Syst Sci 22(1):317–329. https://doi.org/10.5194/hess-22-317-2018

Mustafa D (2002) Theory versus practice: the bureaucratic ethos of water resource management and administration in Pakistan. Contemp South Asia 11(1):39–56. https://doi.org/10.1080/0958493022000000035

Nabeel F (2020) Pakistan meets climate SDG, but that is not enough. The Third Pole, July 29. https://www.thethirdpole.net/2020/07/29/opinion-pakistan-meets-climate-sdg-but-that-is-not-enough/

PID (2018) Punjab water policy, Punjab Irrigation Department (PID), Government of Punjab, Lahore, Pakistan

Qureshi AS, McCornick PG, Qadir M, Aslam Z (2008) Managing salinity and waterlogging in the Indus Basin of Pakistan. Agric Water Manag 95(1):1–10. https://doi.org/10.1016/j.agwat.2007.09.014

Qureshi AS, McCornick PG, Sarwar A, Sharma BR (2010) Challenges and prospects of sustainable groundwater management in the Indus Basin, Pakistan. Water Resour Manag 24(8):1551–1569. https://doi.org/10.1007/s11269-009-9513-3

Sanderson MR, Bergtold JS, Stamm JLH et al (2017) Bringing the "social" into sociohydrology: conservation policy support in the Central Great Plains of Kansas, USA. Water Resour Res 53(8):6725–6743. https://doi.org/10.1002/2017WR020659

Sattar E (2019) Pakistan's water challenges: how we got here and how we should move forward. Cover Story, Newsweek Pakistan, January 26–February 9, 2019

Sattar E (2020) Pakistan's fundamental water governance challenges – and how to overcome them. In: Zahoor B, Rumi R (eds) Rethinking Pakistan: a 21st century perspective. Anthem Press, London

Sattar E, Robison J, McCool D (2018) Evolution of water institutions in the Indus River Basin: reflections from the law of the Colorado River. Univ Mich J Law Reform 51(4):715–775. https://repository.law.umich.edu/mjlr/vol51/iss4/3

Siddiqi A, Wescoat JL (2013) Energy use in large-scale irrigated agriculture in the Punjab province of Pakistan. Water Int 38(5):571–586. https://doi.org/10.1080/02508060.2013.828671

Siddiqi A, Wescoat J, Muhammad A (2018) Socio-hydrological assessment of water security in canal irrigation systems: a conjoint quantitative analysis of equity and reliability. Water Secur 4(5):44–55. https://doi.org/10.1016/j.wasec.2018.11.001

Speed R, Yuanyuan L, Zhiwei Z et al (2013) Basin water allocation planning: principles, procedures and approaches for basin allocation planning. UNESCO, Paris. https://www.adb.org/sites/default/files/publication/30247/basin-water-allocation-planning.pdf

Thebo AL, Drechsel P, Lambin E, Nelson K (2017) A global, spatially-explicit assessment of irrigated croplands influenced by urban wastewater flows. Environ Res Lett 12:074008. https://doi.org/10.1088/1748-9326/aa75d1

Watto MA, Mugera AW (2016) Groundwater depletion in the Indus Plains of Pakistan: imperatives, repercussions and management issues. Int J River Basin Manag 14(4):447–458. https://doi.org/10.1080/15715124.2016.1204154

World Bank (2020) Groundwater in Pakistan's Indus Basin: present and future prospects delivered under the South Asia Water Initiative (SAWI) Indus Basin Focus Area

World Commission on Environment and Development (1987) Our common future. Oxford University Press, Oxford

Young W, Anwar A, Bhatti T et al (2019) Pakistan: getting more from water. World Bank, Washington, DC. https://openknowledge.worldbank.org/handle/10986/31160